The Political Philosophy of Francis Bacon

Tom van Malssen

The Political Philosophy of Francis Bacon

On the Unity of Knowledge

Back cover image and frontispiece by Anima van Malssen.

Published by State University of New York Press, Albany

© 2015 State University of New York

All rights reserved

Printed in the United States of America

No part of this book may be used or reproduced in any manner whatsoever without written permission. No part of this book may be stored in a retrieval system or transmitted in any form or by any means including electronic, electrostatic, magnetic tape, mechanical, photocopying, recording, or otherwise without the prior permission in writing of the publisher.

For information, contact State University of New York Press, Albany, NY
www.sunypress.edu

Production, Jenn Bennett
Marketing, Michael Campochiaro

Library of Congress Cataloging-in-Publication Data

Malssen, Tom van, 1982–
 The political philosophy of Francis Bacon : on the unity of knowledge / Tom van Malssen.
         pages cm
 Includes bibliographical references and index.
 ISBN 978-1-4384-5417-7 (hc : alk. paper)—978-1-4384-5416-0 (pb : alk. paper)
 ISBN 978-1-4384-5418-4 (ebook)
 1. Bacon, Francis, 1561–1626.2. Political science—Philosophy. I. Title.

B1198.S8M35 2014
192—dc23   2014002781

10 9 8 7 6 5 4 3 2 1

For Anima
ἀρχή ζῴων

Contents

Acknowledgments		ix
Abbreviations		xi
Introduction		1
Chapter 1	The Art of Transmission	11
Chapter 2	The Baconian Turn	33
Chapter 3	The Trinity of Philosophy	93
Chapter 4	The Masculine Rebirth of Time	167
Chapter 5	The Perfective Son	201
Epilogue		237
Notes		243
Bibliography		311
Index		319

Acknowledgments

As the debt most easily acknowledged is the one that need not be acquitted, let me begin by expressing my gratitude to Heinrich Meier, who arranged my dialogical confrontation with Francis Bacon, knowing that encountering one's opinions in a philosopher who seems to confirm them is the best way to start questioning them. He thus expanded my horizon when I was on the verge of narrowing it, although it was only by following the wise dialogical rule of advanced contraction for the sake of suspended dilation that I was able to reverse direction and set out for open seas. Most importantly, however, he made me doubt the good of philosophy at the very time when I myself most believed in it, thereby causing me not only to experience that the good is not good for him to whom its goodness is unknown, but also to see the contrary danger of the good becoming its own evil by self-elevation. It is therefore to Heinrich Meier that I attribute any good that may come from reading this book.

Mentioning the director of the *Carl Friedrich von Siemens Stiftung* almost naturally brings me to that remarkable and unique Baconian institution itself, whose official objective to advance the sciences goes hand in hand with the unofficial objective of its incomparable staff to advance the comfort of the scientist, and without whose generous support I would not have been able to contribute to the advancement of the science of Bacon.

My gratitude also goes to Thomas L. Pangle and Lorraine Smith Pangle, co-directors of the Thomas Jefferson Center for the Study of Core Texts and Ideas at the University of Texas at Austin, who offered me a Postdoctoral Fellowship for the academic years 2011–2013, thus enabling me to make the necessary revisions and additions to an earlier version of this manuscript, which was accepted as a doctoral dissertation (summa cum laude) by Ludwig-Maximilians-University Munich on July 26, 2011.

I benefited greatly from the reports of Rémi Brague, Ralph Lerner, and Devin Stauffer, all three of whom carefully read, relentlessly criticized, and understandingly commented on drafts of the book at hand, thereby allowing me to advance in learning while preventing me from committing unnecessary blunders. I also profited from the two anonymous reviews that were submitted at SUNY Press. The assistance, moreover, by Michael Rinella and his colleagues at the aforementioned Press contributed in many ways to the present work finding its way through the publishing process as smoothly as possible.

I thank my parents, my mother and my late father, for the unconditional love and support they showed me throughout the process of giving birth to the present work. My mother deserves special praise for her infinite readiness to improve my English through reading the early proofs of my English, although it is I alone who is to be held responsible for any errors this book may still contain. I extend my thanks to some of my friends, who, always in their own way, and sometimes without knowing it, contributed to making my life more agreeable.

Finally, I thank Anima for saying yes to me in all my imperfections. I dedicate this book to her.

Abbreviations

—*The Works of Francis Bacon*. Fourteen vols., edited by James Spedding, Robert Leslie Ellis, and Douglas Denon Heath. London: Longmans & Co, 1870: *Works* (volume number and page number added; e.g., *Works* VII, 124).

—*The twoo Bookes of Francis Bacon. Of the proficience and aduancement of Learning, diuine and humane* (1605): AL. Because the first book and the Epistle Dedicatory to the second book of the *Advancement of Learning* found their way into the Latin version of 1623 without any essential changes or additions having been made, I have decided to consult the English original. I have used the critical, unmodernized edition of Michael Kiernan[1] because it is the most faithful to the original text. When I refer to the first book, I refer to the number of the book, the number of the paragraph (103 in all), and the page number of the Kiernan edition, as well as to the volume number and the page number of the *Works* edition (e.g., AL 1.62, 36, III: 299). In the case of the Epistle Dedicatory to the second book (first sixteen paragraphs) I do the same (e.g., AL 2.16, 61–2, III: 328–9). In the case of the second book itself, however, I have decided to consult the Latin version, that is, books II–IX of *De Augmentis Scientiarum*, because Bacon tells us that the Latin version of the second book contains "great and ample additions" to and "enrichment" of the second book, causing it to be "so enlarged as it may go for a new work."[2] I only refer to the second book of the *Advancement of Learning* (Kiernan, 62–193; *Works* III, 321–491) in order to illustrate where and to argue why books II–IX of *De Augmentis* deviate from it. In this case I refer to the number of the book and the page number of the Kiernan edition, as well as to the volume number and the page number of the *Works* edition (e.g., AL 2, 129, III: 411).

—*An Advertisement Touching a Holy War* (1629): AHW (*Works* VII, 17–36); I refer to the number of the paragraph (twenty-five in all), the volume number, and the page number (e.g., AHW 13, VII: 21).

—*Cogita et Visa: de Interpretatione Naturae, sive de Scientia Operativa*: CV (*Works* III, 591–620); I refer to the number of the paragraph (nineteen in all), the volume number, and the page number (e.g., CV 3, III: 593).

—*Cogitationes de Natura Rerum*: CNR (*Works* III, 15–35); I refer to the number of the cogitation (ten in all), the volume number, and the page number (e.g., CNR 4, III: 21–2).

—*A Confession of Faith*: CF (*Works* VII, 217–26); I refer to the number of the paragraph (twenty in all), the volume number, and the page number (e.g., CF 5, VII: 220).

—*De Dignitate et Augmentis Scientiarum Libri IX* (1623): DA (*Works* I, 423–840); I refer to the number of the chapter, the volume number, and the page number (e.g., DA 3.2, I: 545).

—*The Essayes or Counsels, Civill and Morall* (1625): E (*Works* VII, 371–517). I have used the critical, unmodernized edition of Michael Kiernan[3] because this is the edition that is most faithful to the original text. I refer to the number and the title of the *Essay* in the 1625 edition, to the page number of the Kiernan edition, as well as to the volume number and the page number of the *Works* edition (e.g., E II "Of Death," 9 ff., VI: 379–80).

—*Filum Labyrinthi*: FL (*Works* III, 496–504); I refer to the number of the paragraph (ten in all), the volume number, and the page number (e.g., FL 8, III: 502–3).

—*Instauratio Magna* (1620): IM (*Works* I, 119–45); I refer to the Proemium as IMProemium, to the Epistle Dedicatory as IMEpD, to the Praefatio as IMPr. (paragraph number added, six in all), and to the Distributio Operis as IMDO (paragraph number added, thirty-one in all). In all cases I add the volume number and the page number (e.g., IMDO 30, I: 145).

—*Meditationes Sacrae* (1597): MS (*Works* VII, 231–42); I refer to the number and the title of the Meditation, the volume number, and the page number (e.g., MS 10 "De Atheismo," VII: 239–40).

—*New Atlantis* (1627): NA (*Works* III, 127–68); I refer to the number of the paragraph (fifty-nine in all), the volume number, and the page number (e.g., NA 17, III: 151 ff.).

—*Novum Organum, sive Indicia Vera de Interpretatione Naturae* (1620): NO (*Works* I, 147–365); I refer to the number of the book, the number of the paragraph, the volume number, and the page number (e.g., NO 1.65, I: 175–6). To the Praefatio I refer as NO Pr.

—*De Principiis atque Originibus, Secundum Fabulas Cupidinis et Coeli*: POFCC (*Works* III, 79–118); I refer to the volume number and the page number (e.g., POFCC, III: 113).

—*Redargutio Philosophiarum*: RP (*Works* III, 557–85); I refer to the volume number and the page number (e.g., RP, III: 581).

—*De Sapientia Veterum* (1612): DSV (*Works* VI, 617–86); I refer to the number and the first part of the title of the fable, the volume number, and the page number (e.g., DSV XXXI "Sirenes," VI: 684 ff.). The Epistles Dedicatory are referred to as DSV, EpD1, and DSV, EpD2. I refer to the Praefatio as DSV Pr.

—*Temporis Partus Masculus sive Instauratio Magna Imperii Humani in Universum*: TPM (*Works* III, 527–39); I refer to the volume number and the page number (e.g., TPM, III: 538).

—*Valerius Terminus of the Interpretation of Nature; with the Annotations of Hermes Stella*: VT (*Works* III, 215–52); I refer to the volume number and the page number (e.g., VT, III: 241).

All translations from Latin are mine.

Introduction

> I am come in the name of the Father, and ye receive me not: if another shall come in his own name, him ye will receive.
>
> —John 5:43

It is common knowledge that modern science's claim to philanthropy or charity can be traced back to Francis Bacon.[1] But what is not universally known is that it was also Francis Bacon who foretold the historical potentiality of Daedalus, the man whose "wicked industry and pernicious genius" were responsible for the "infamous and infelicitous" birth of the Minotaur; the man, moreover, who "devised and constructed" the labyrinth as a security measure against the evil he himself midwifed but wanted to conceal; the man, finally, who pointed out the use of the "ingenious thread" by which the maze of the labyrinth could be unraveled, because he desired to be known "not only for evil arts." Since Daedalus thus incorporates science's potential to "serve for the harm as well as for the remedy," and its potential to "almost dissolve and undo its own virtue," Bacon's parable could already have conveyed the lesson that science can be turned to "ambiguous uses" (DSV XIX "Daedalus," VI: 659–60). It was, however, not until the parable's potential was actualized in history that modern science proved itself potentially destructive on account of its ultimately instrumental and, therefore, not truly neutral nature. And it was only the actualization of the parable that prevented the obvious truths about Bacon from becoming self-evident, and the redemptive claims of modern science from becoming too successful. For it was precisely the loss of an unqualified belief in Baconianism that enabled us to ask ourselves, for the first time in history without partiality and prejudice, how this historically singular movement came into being, and why it turned out to be so successful.

The present work grew out of the firm conviction that only through answering this question can we come closer to understanding the duality of man that underlies the ambiguity of science. The gravity of what we may thus call our guiding question is affirmed by no less a person than Bacon himself, as it is Bacon himself who emphasizes that the "greatest things are indebted to their beginnings" (DA 9.1, I: 837); the same Bacon, it is true, who warns us that the road toward these beginnings will be long, rocky, and full of obstacles, seeing that "times abound with history now" (DA 8.2, I: 769); but also the same Bacon who instills into our minds the strength that will guard and guide us on the way. For the man who took fortitude to be the "Vertue of *Adversity*" assures us that it is adversity that "doth best discover Vertue" (E V "Of Adversitie," 18–9, VI: 386).

Taking what is first for us as our point of departure, we encounter our first obstacle. For although one contemporary scholar claims that the fact that the twentieth century denied Bacon any "philosophic relevance" has led scholars in recent decades to "concentrate on other directions of research" in order to "account for the otherwise inexplicable place Baconian ideals had held in European thought,"[2] it is not at all clear how this academic trend justifies such a shift of focus without explaining what it presupposes: the distinction between philosophy and non-philosophy.[3]

In order to shed some light on this somewhat dark issue, we should focus our attention on the leading scholar who is said to have inaugurated and consummated the turn just referred to.[4] For in addition to having developed the most recent and most elaborate conception of Baconianism, he had, we may presume, a thorough insight into the reasons for announcing a turn subsequent generations of scholars have taken for granted. According to this scholar, Baconianism consists mainly in the "criticism of tradition," and a new idea of knowledge and science. Knowledge is no longer understood as "contemplation or recognition of a given reality," but as an "exploration of unknown lands," and the goal of science becomes the transformation of the "condition of human life on earth." Although science itself is "internally value-free," it is "not in reality indifferent to the values of ethics and the reaches of political and social life," and it "always has a precise practical function." More specifically, Baconian science is an "instrument constructed by man with a view to the realization of the values of fraternity and progress," values that in their turn must "be strengthened and reinforced by that same science." This is one of the reasons, our scholar goes on to explain, why the extension of man's power over nature by means of a "new logic of invention and discovery useful in the construction of works" is "never the work of a single investigator," but always "the fruit of an organized collectivity of scientists financed by the State or by public bodies." Stronger even, any attitude that would "substitute the wisdom of one man for the organized efforts of humanity is to be rebuffed," and any doctrine that "places science in the service of some one man rather than in the service of the whole human race is to be rejected." But notwithstanding this diminution of the status of the one man, an exception is to be made for one specific man. For our authority on twentieth-century Baconianism concludes that although Francis Bacon was a "great" but "isolated, disappointed philosopher," whose "restricted" views were "a reaction to contemporary culture," and who was "temperamentally and educationally unfit for the solving of scientific problems that were being raised in that dawn of modern times," he foresaw the revolutionary nature of the radical changes occurring in Europe, changes that did "not depend on philosophy." Bacon's true greatness thus consists in his having been the herald of the "new world" that was arising.[5]

Impressive claims, but the fact of the matter is that this analysis of Baconianism and Baconian science, the latter of which can and cannot account for itself, does not account for its possible inauguration as a result of its name giver having explored, investigated, recognized, and contemplated the realities of the condition of human life on earth; a possibility that gives some credit to philosophy, and that this book will show to be a probability. But even if we confine ourselves to taking a retrospective view of the announcement of the restricted views of the great and far-sighted herald of Baconianism, we cannot but conclude that the consequent turn was an afterthought of a scholar of the history of ideas.

To increase confusion for the sake of clarity, we should draw attention to the fact that recent scholarship has also proved itself unable to develop a coherent view as to Bacon's philosophic relevance; an inability that we believe to have ensued from a denial of relevance to the meaning of "philosophic" relevance, and that constitutes a serious drawback for our study of a movement commonly acknowledged to be philosophically relevant.

One scholar, according to whom Bacon's "main and permanent significance" consists in his being "a thinker about science," describes Bacon as "never a detached philosopher contemplating the human or natural world from a haven of serene seclusion," yet "invariably a philosopher in dealing with any question," and in addition to being a philosopher, lawyer, political man, and statesman, a "very conscious, highly gifted literary artist," in whom "thinker and artist" and "reason and imagination" were "inseparably fused."[6] Others dismissed these and similar oscillations as "clearly incompatible versions of Francis Bacon," and decided to look for coherence in Francis Bacon "as he survives in archives and contemporary printed texts,"[7] even though Bacon's own contemporary William Rawley, also his secretary and chaplain, tells us that Bacon's books were the "image of his brain."[8] And although there are some according to whom philosophy was "only one facet" of Bacon's life,[9] there is one scholar who considers philosophy to be a "distinctive" way of engaging problems, and who looks upon the philosopher as "someone who has a particular standing." This scholar credits Bacon with having been the "first engineer" of the transformation of the philosopher into the natural philosopher or what later came to be known as the scientist. He claims that Bacon made the first "systematic, comprehensive attempt to transform the epistemological activity of the philosopher from something essentially individual to something essentially communal," as a result of which the philosopher came to be understood no longer as an "individual seeker after the arcane mysteries of the natural world," but as a "public figure in the service of the public good."[10] But even if this transformation has been correctly described, and even if Whitehead is only partly right in saying that "[s]cience repudiates philosophy," and that it "never cared to justify its faith or to explain its meanings,"[11] our confusion as to Bacon's philosophic relevance only increases. This is due to the fact that the transformation of philosophy into science cannot account for its own success, if only because Bacon himself considered philosophy and science the same thing ("eadem re," DA 2.1, I: 495). Be that as it may, the same scholar goes on to say that "Christianity set the agenda for natural philosophy in many respects," adding that this was particularly the case in Bacon's own seventeenth century England.[12] That science and biblical religion were indeed closely intertwined can be illustrated by the well-known fact that in the immediate aftermath of his life Bacon's writings "came to attain almost scriptural authority."[13] It remains an open question, though, how science justified itself before it became oblivious to the need for its own justification.

In pursuing our quest for the roots of the success of Baconianism, it stands to reason that we take a closer look at the coming into being of Bacon's reputation. According to the Counter-Enlightenment thinker Joseph de Maistre, the Englishman's fame commenced with a crime, considering that the *Encyclopédie ou Dictionnaire Raisonné des Sciences, des Arts et des Métiers* (1751–1772) of Diderot and d'Alembert was the "greatest and most redoubtable conspiracy that had ever been formed against religion and thrones."[14] But despite or because of his moral indignation, de Maistre puts us on the right track, as Diderot tells us in his *Prospectus* (1750) that if the *Encyclopédie* becomes successful, the *encyclopédistes* will be "principally obliged" to the "extraordinary genius" of the far-sighted chancellor Bacon, "who drew up the plan of a universal dictionary of the sciences and the arts at a time when, so to speak, there were no sciences and arts." In his *Discours Préliminaire* (1751) d'Alembert adds that the *encyclopédistes* even owe the "tree" of the *Enyclopédie* to Bacon, although they decided to cut off some branches. But the panegyric on Bacon also serves another purpose. For d'Alembert points out that Bacon's untimely works for the times were performed in the service of emerging philosophy: "At times when poorly educated or badly intentioned adversaries of philosophy openly make war against philosophy, philosophy takes refuge, so to speak, in the works of some great men who, at a distance, in the shadow and in silence, and without having the dangerous ambition of tearing off the blindfold of their contemporaries, prepare the light by means of which the world is to be enlightened little by little and by degrees insensible." As the "immortal chancellor" of England is "at the head of these illustrious personalities," d'Alembert concludes, "one is tempted to regard Bacon, who was born in the deepest night, and to whom philosophy was that part of knowledge which ought to contribute to making us better or happier, yet who sensed that there was not yet philosophy, as the greatest, the most universal, and the most eloquent of philosophers."[15] Voltaire, a forerunner of the *encyclopédistes*, had called Bacon, who lived in an age when "good philosophy was even less known than the art of writing well," only a "precursor of philosophy," who "did not yet know nature," but who "knew and pointed out all the roads that lead to nature."[16] It is, however, not the best-known forerunner but the most profound critic of the *encyclopédistes* who in his *Discours sur les sciences et les arts* (1750) throws us into complete confusion by calling Bacon "perhaps the greatest of philosophers."[17] But the fact that the philosopher Jean-Jacques Rousseau never mentioned Bacon again in his works after d'Alembert had almost repeated his eulogy only proves what we had already come to suspect: the political relevance of the question of Bacon's philosophic relevance.[18]

Having arrived at the dawn of the Modern Era, we find Baconianism merging with the greater movement called Modernity. But what does "Modernity" mean? According to Hans Blumenberg, one of the most influential historians of early Modernity, it was the internal contradictions of the Christian Middle Ages regarding especially the issues of omnipotence and providence that imposed on man the "burden of self-assertion," a burden that necessitated and legitimated a self-conscious opposition to the antecedent tradition in the form of the "immanent self-assertion of reason by means of the control and transformation of reality."[19] This discontinuity between the new and the old caused "worldliness" to become the major characteristic of Modernity. Blumenberg explicitly denies the possibility of there having been a secularization of some sort at work at the beginning of Modernity, as any kind of secularization would allow theology to dismiss the claim—taken by the historian to be historically false—of philosophy having been the founding agent of Modernity as a kind of "providential resistance of the indispensable."[20] But does not Blumenberg himself argue in the service of some kind of providence by saying that, although man does not make an epoch, Modernity was the first and only epoch that understood itself as an epoch?[21] It is no less a person than Blumenberg himself who sustains this suspicion by taking as one of his major premises that there is "no fixed canon of great questions, which with consistent urgency and throughout history engage the human appetite for knowledge and motivate the claim to the interpretation of the self and the world."[22] Since Blumenberg thus unquestionedly rejects the possibility that what inaugurated Modernity could have been a "self-assertion" inaugurated by one or more active and "self-conscious" agents,[23] rather than a passive and as it were "semi-conscious reception of self-consciousness,"[24] the possibility remains actual that what was new at the time, or not coeval with human thought, was not providential, though neither accidental.

Allowing a modern author to have a say in the matter, we find Francis Bacon arguing that if true antiquity means the "old age of the world," our age is the older age, considering that it is "enlarged and heaped up with infinite experiments and observations" (NO 1.84, I: 190). On the assumption that all authors recognize that good authors make self-explanatory statements, we would not generate dismissive objections if in elaboration of this argument we took the Baconian observation that "in the mind" you can "hardly blot out the old" except by "writing in the new" (RP, III: 584) to imply that "writing in the new" was Bacon's way of "blotting out the old," especially if we refrained from stating the subsequent and far more objectionable but nowise impossible hypothesis that calling truth the "daughter of time" (NO 1.84, I: 191) was Bacon's way of blotting out the very difference between new and old. But since we should not have recourse to assumptions, we limit ourselves to adducing Bacon's mere statement that "time is the author of authors" (NO 1.84, I: 191) in justification of the conclusion that the only way for us to judge Blumenberg's claim that philosophy has not inaugurated Modernity by means of an active secularization is to investigate what "newness" could have caused Bacon to become the author of Baconianism.

In an ambitious study on the origins of Modernity, a contemporary scholar tries to look behind the veil with which, according to him, Modernity has concealed its own origins. He criticizes Blumenberg for not having recognized that Modernity's response did not so much consist in human self-assertion as in a "realization of the metaphysical and theological possibilities left by the antecedent tradition."[25] According to this scholar, it was the crisis within Christianity about the nature of God that gave rise to the Modern Age. The nominalist God, who wills what He wills, who is unconstrained by nature and reason, and who is no man's debtor, had "turned the order of nature into a chaos of individual beings." But it was on the ontological basis of nominalism that Modernity came into being as a series of attempts to solve the conflict between the Reformation and humanism with their respective ontic priorities of God and man. The aforementioned scholar claims that the common denominator of these attempts was the assertion of the "ontic priority not of man or God but of nature,"[26] but unfortunately enough, he does not adduce arguments in support of his postulate that Bacon "accepts the nominalist vision of the world" in his attempt to find a solution to its "fundamental problems" by investigating how nature works.[27] We therefore cannot but conclude that the origins of Bacon's modernity remain as yet to be unveiled.

But we should leave the beginnings of Modernity under the veil that hides them, and bring together the threads of our descent into the history of Baconianism; threads that lead us to find that there seems to be only one possibility left to coherently conceive of Baconian philanthropy in connection with the probability of Bacon's philosophic relevance and the actuality of its political implications: the possibility that Bacon was a political philosopher. After all, if it is true that the philosophic understanding of political subjects constitutes political philosophy, and that man is the paramount political subject because he is by nature a political animal,[28] Bacon may very well have loved man because he loved political subjects, that is, because he was a political philosopher.[29] On the other hand, if Bacon's philanthropy, or his "affecting of the Weale of Men" (E XIII "Of Goodnesse and Goodnesse of Nature," 38, VI: 403), was ultimately a consequence of his being a political philosopher and not an anticipatory cause of his becoming a political philosopher, Bacon did not love man because he desired man to love and admire him for his philanthropy, but because his philosophical *eros* extended to political subjects and therefore to man.[30] In this line of argument, Bacon did not love man because his consciousness of man's natural resources of love and admiration made him aware of a poverty that induced him to anticipate man's love and admiration as the fulfilment of his own natural need. Stronger even, love that is reciprocated would never have been able to fulfill this need,[31] and "if there be no fullnesse, then is the Continent greater, than the Content" (AL 1.6, 6, III: 265). This line of argument would thus lead us to the conclusion that, although Bacon self-consciously shaped the philanthropic formula *par excellence* of the "reliefe of Mans estate" as the "last or furthest end of knowledge" (AL 1.48, 31–2, III: 294), to Bacon the political philosopher this could not have been an end in itself; a conclusion that would have far-reaching implications for our understanding of the nature of philanthropic science. But irrespective of this conclusion and its implications, the only way to come closer to answering our guiding question seems to be to approach it on the assumption that Francis Bacon was a political philosopher, an assumption whose foundation therefore constitutes our principal subject of investigation.

There is a preliminary problem, though. For although Jonathan Swift tells us that in being a "Modern," Bacon has a "tendency towards his Center," and although his Aesop adds in a comical fashion that there is nothing so modern as a spider who "Spins and Spits wholly from himself, and scorns to own any Obligation or Assistance from without,"[32] the truth of Aesop's tragi-comical conclusion that the materials out of which the "modern web" is spun are "nothing but dirt" can only be determined after having located the modern center, which is the center of Bacon.[33] It is thus with a view to making navigation possible that I have to devote my first chapter to investigating the contemporary hermeneutical prejudice that in studying the surface center of the thought of almost every thinker of the past, being his own writings, we need assistance from without, because we cannot by ourselves trace the threads of the web the writer has spun on the sole basis of a philosophical concern with the writings' philosophical claim to truth. Fortunately enough, though, it is Bacon himself who assists us by providing important clues as to how he thought such a philosophic concern should be translated into hermeneutics. Put into concrete terms, the first chapter begins with a discussion of the necessity and limits of a historical contextualization of philosophic writings, to be followed by a detailed treatment of Bacon's self-explanatory statements and allusions on hermeneutics. Applying these statements and allusions to some of Bacon's most important works, the chapter concludes by demonstrating that without a previous understanding of the art of attentive reading an adequate understanding of the English author's writings is simply not possible.

Returning to our principal subject of investigation, I argue in the first part of the second chapter that a philosophical concern with truth in Bacon would necessarily have caused him to become a political philosopher, and I retrace the genealogy of Bacon's political self-understanding in order to prove that his works testify to a turn to political philosophy: a turn that results in a political understanding of philosophy and a philosophical understanding of politics, along with an understanding of the tension between philosophy and politics. Whereas the ordering principle of the first half of the chapter is Bacon's threefold political apology of philosophy as it is undertaken in the *Advancement of Learning*, the second half treats the turn to political philosophy as it finds expression in *De Sapientia Veterum*, Bacon's most important but least understood work.

In the third and central chapter, I attempt to show that the question: "What is a god?" the question, to put it differently, of philosophy's rational grounding, which is the question for the sake of which philosophy must become political, constitutes the center of Bacon's political philosophy. *De Augmentis Scientiarum*, Bacon's most theoretical work, serves as the framework for the development of the chapter's argument, whose starting point is the distinction between philosophy and inspired theology, made by Bacon at the beginning of the book on philosophy. Bacon's concept of *Philosophia Prima* is discussed, which is defined as a "universal science," but which can be taken as a guiding principle of Bacon's political philosophy as it turns out to be a measuring instrument of the power of nature and therefore of the power of God. After a detailed interpretation of Bacon's discussion of "natural theology," the natural standard of the divine, of physics and metaphysics, and of the eternity or createdness of the world, the second, more historical half of the chapter treats Bacon's understanding of prophecy, and of the incarnation, resurrection, and second coming of Christ, especially as they are discussed in the highly important but universally ignored *Confession of Faith*. A careful treatment of Bacon's interpretation of divine justice, love, and mercy then leads to a discussion of his typology of believing man, before the chapter ends with an analysis of Bacon's understanding of the relationship between morality and inspired theology.

In the fourth chapter, I investigate how the truth about political subjects reached on the basis of Bacon becoming a political philosopher may have resulted in him conceiving the *Instauratio Magna*; I try to explain, in other words, how a political action on the part of philosophy can be said to have ensued from philosophy's knowledge of politics. Taking as its guide the Prometheus fable from the *Wisdom of the Ancients*, which prepares the project as a whole, the argument moves from a discussion of the rhetoric of indignation with which Bacon recruits for the project, through the plea for justice with which he attempts to win the minds of its potential benefactors, to the project's Machiavellian foundations. The equivocality of the term *Instauration* as well as its theological history are discussed, before the pillars of the project, the Baconian version of the theological virtues, are considered more closely: especially charity, which in the form of philanthropy holds together the parts of the "Kingdom of Man."

The fifth and final chapter is devoted to what may be called Bacon's political testament: his account of the future of his project, his message to his successors, and his reassurance of his followers. Since this testament has been transmitted to us by way of two posthumously published and seemingly unfinished works, *New Atlantis* and *An Advertisement touching a Holy War*, it is the careful interpretation of the narrative of these works of ministerial poetry that constitutes the subject matter of the final chapter. The epilogue is reserved for reflection and retrospection.

CHAPTER 1

The Art of Transmission

> The thing that hath been, it is that which shall be; and that which is done is that which shall be done: and there is no new thing under the sun.
>
> —Eccl. 1:9

In his last will Bacon states that the "durable part" of his memory consists in his "works and writings."[1] But how are we to approach these works and writings in light of the vast period of time that separates us from them? It is to answer this question that we turn to one of our authorities on philosophical hermeneutics, Hans-Georg Gadamer, who teaches us that, since writing is a kind of "alienated speech," in a "special sense" hermeneutics depends on the "art of writing,"[2] and who thus invites us to take our primary hermeneutical task to consist in a revivification of the writings' speech by means of an interpretation of the art by which it was written down.

But let us take one step back. For it is Bacon himself who considered securing the "durable part" of his memory a subject matter important enough to devote a separate book of *De Augmentis Scientiarum* to what he calls the "art of transmission" (DA 6, I: 650–712), which he classifies as one of the "rational arts" (DA 5.1, I: 616) because it includes all the arts relating to "words and discourse," and because reason, whose "footsteps" are words, is as the "soul of discourse." It consists in "disclosing and enunciating what has been discovered, judged, and laid up in memory." As its "principal part" concerns the writing of books, so its "relative part" turns on the reading of books (DA 6.1, 6.4, I: 651, 654, 708). And it was a "great reader of books"[3] who observed that the "Images of mens wits and knowledges remaine in Bookes." For, as our principal writer explains himself, "so immortall and incorruptible a thing did knowledge" seem to be to the "Philosophers which were least diuine, and most immersed in the sences," and who "denied generally the immortality of the soule," that they thought that the "motions" of the "vnderstanding" might "remaine after death" (AL 1.102, 52–3, III: 318). But if this is true, and a book is as it were the mirror of its author's soul, then hermeneutics ultimately becomes the art of re-ensouling Bacon's writings by reading the thought that is transmitted by their speech.

We went a few steps too far, though. For if all human thought is ultimately determined by the historical horizon within which it takes place, the act of transmission can never or never wholly be consummated. Although according to Bacon himself reading is either "guided by teachers," or "accomplished [perficitur] by man's own industry" (DA 6.4, I: 708), our "historical conscience" teaches us that our only true teacher is the "historical consciousness" becoming aware of its historical horizon, within which it can become truly "conscious of itself." Our own "industry," it is true, enables us to transmit the author's writings to our time by transposing ourselves and our own horizon into the author's horizon, but it is the historical consciousness that relates both horizons to itself before effecting a new hermeneutical horizon that makes understanding possible. And after having effected a fusion of our horizon and the author's horizon that overcomes their respective particularity, the hermeneutical circle's prime mover comes full circle by acquiring a higher universality within particularity.[4]

But the act of making the historical consciousness transmission's principal agent presupposes that the nature of the subject matter of Bacon's writings is determined by Bacon's particular horizon. It therefore presupposes what could only be the outcome of a hermeneutic effort. Besides, it was Bacon himself who put the horizon of his writings in a broader context by emphasizing that books are "capable of perpetuall renouation" if "exempted from the wrong of time" (AL 1.102, 53, III: 318). We can therefore agree with Gadamer's hermeneutical principle of the "anticipation of completeness [Vorgriff der Vollkommenheit]"[5] only to the extent that it becomes coeval with a hermeneutical openness to the possibility that we can only or only completely understand Bacon's writings after having understood what Bacon considered his horizon to be. And we can endorse Gadamer's emphasis on the importance of "prejudices" for hermeneutics only in the sense that the principal prejudice that could prevent us from understanding the subject matter of Bacon's writings is the prejudice that a mere consciousness of history suffices to effect a historical consciousness.[6]

Context

In any event, it is Bacon himself who testifies to the illuminative power of the writings' historical context for hermeneutics by explicitly counselling readers of books to take the "times" within which they were written into consideration, because these "in many cases giue great light to true Interpretations" (AL 2, 131, III: 414). It therefore seems to be the right time to put Bacon's early intellectual upbringing in context.

Sir Nicholas Bacon, who was Lord Keeper of the Great Seal during the reign of Queen Elizabeth I (1558–1603), must have intoxicated his youngest son with politics almost prematurely, as Rawley tells us that Her Majesty often termed young Francis *the young Lord-keeper.*[7] Lady Anne Bacon, formerly Anne Cooke, was the sister of the second wife of William Cecil (Lord Burghley), who was Queen Elizabeth's Secretary of State and Lord Treasurer. She was the daughter of Sir Anthony Cooke, a former tutor of Prince Edward, who subsequently became King Edward VI. The Protestant Sir Anthony Cooke fled to Geneva in 1554 to become one of the Marian Exiles after Mary Tudor had acceded to the throne in 1553, and had restored England to Roman Catholicism. A pious and zealous Puritan and a fervent supporter of the Nonconformists, Lady Anne was widely esteemed for her learning, and she is still known for her English translation (1564) of the *Apologia Ecclesiae Anglicanae* (1562), written by John Jewel, Bishop of Salisbury. The *Apologia*, which became one of the most influential attempts to establish the Church of England in the early Elizabethan years, tried to counter accusations of heresy by emphasizing the continuity and inner harmony of Christianity. Lady Anne took a personal interest in the translation, considering it to be the performance of her religious duty.[8] As it was she who saw to the early education of Francis and his older brother Anthony, the brothers were, in the words of a contemporary biographer, given "a solid grounding in the severer sort of radical Protestantism," the importance of which in later life would turn out to be "considerable."[9]

After having spent some of their early years (probably 1566–1569) under the tuition of John Walsall, a scholar from Christ Church, Oxford, Francis and his older brother Anthony were sent to Trinity College, Cambridge in 1573, where they were left in the care of John Whitgift, Master of Trinity and later Archbishop of Canterbury. In his capacity as Archbishop of Canterbury (1584–1604) Whitgift, whom Macaulay describes as a "tyrannical priest,"[10] was to pursue a line of strict orthodoxy. His intransigent policies were directed against the Nonconformists, and they provoked a pamphlet war that came to be known as the "Marprelate Controversy"

(1588–1589). The main combatants in this war of libelers were an unnamed Puritan writer who employed the pseudonym Martin Marprelate, and the Established Church with its leading figure Richard Bancroft. Bancroft was Whitgift's chaplain and his successor as Archbishop of Canterbury (1604). In his notorious *Sermon preached at Paules Crosse* (1589), he had approvingly referred to John Jewel's apology "to the justifieng of our doctrine," which had, ever since its publication, "obtained principall commendation amongst all the apologies and confessions" set forth "by any church in christendome."[11]

In 1575, Bacon left Trinity College, and in 1576 Sir Nicholas sent him to France, where he spent almost two and a half years (1576–1579) under the tuition of Sir Amias Paulet, ambassador to the French King and a fanatical Puritan. Bacon mainly lived in Paris, the center of the French Wars of Religion. It was the immediate aftermath of St. Bartholomew's Day Massacre (1572), in which thousands of Huguenots had been slaughtered by a Roman Catholic mob. Less than ten years after his return to England, Bacon closely witnessed the attempted invasion of his home country by the Spanish Armada (1588), to which Pope Sixtus V had granted a large subsidy—the same pope who had renewed the papal bull *Regnans in Excelsis* (1570), which had been issued by Pope Pius V, and which had declared Queen Elizabeth a heretic.

These experiences and observations, in conjunction with our author's upbringing, which had already made him well versed in the effects of religion on politics, can only have increased the confidence that speaks from his first work: *An Advertisement touching the Controversies of the Church of England* (1589). For although his two most influential teachers among contemporaries must have advised him otherwise, albeit for contrary reasons, Bacon decided not to pay attention to the inner voice of Lady Anne's and John Whitgift's tuition when he directed his *Advertisement* against both the Nonconformists and the Established Church. We therefore should not be surprised to find that Lady Anne, who wanted her sons to "serve the Lord duly and reverently," would soon afterward complain to Anthony that his brother Francis was "too negligent herein."[12] Bearing witness to a moderation reminiscent of the credo "Mediocria Firma,"[13] which had been chosen by Sir Nicholas as the family motto, Bacon's pamphlet aimed at finding "correspondence in their minds which are not embarked in partiality," and which "love the whole better than a part."[14]

Many other events that left an imprint on the young Bacon's mind should be mentioned in the context of the foregoing, events ranging from the Battle of Lepanto of 1571 to the assassination of King Henry III in

1589, and including the series of religiously motivated plots and conspiracies that led up to the execution of Mary Stuart in 1587. But this rudimentary sketch of Bacon's early intellectual biography may suffice to justify the conclusion that even from a tight correlation between Bacon and the times in which he lived and wrote, we cannot infer the nature of the influence these times exerted on him, unless we put Bacon's historical horizon in the context of what Bacon himself considered his horizon to be.[15]

Let us therefore transpose ourselves once more into the context of the art of transmission. At first sight it seems strange that it is only after having arrived at the "Appendices" of his book on the art of transmission that Bacon sheds light on the art of reading his transmitted writings, especially as it is Bacon himself who remarked that it is "strange, how long some Men will lie in wait, to speake somewhat, they desire to say; and how farre about they will fetch; And how many other Matters they will beat over, to come neare it." But it is also Bacon who considered this strategy to be a "Thing of great Patience, but yet of much Use." Moreover, it is in the same context that Bacon speaks of a person he knew, who used to "put that which was most Materiall, in the *Post-script*" of his writings, "as if it had been a By-matter" (E XXII "Of Cunning," 71–2, VI: 429, 431). Having learned that a thinker is to be read in context, we are willing to be patient enough to let Bacon speak of the writings of the person he knew best.

Immediately after having distinguished between the "guidance of teachers" and "man's own industry," Bacon takes over guidance by teaching us how to use the critical part of our own industry. The first part of the "critical part" of the art of transmission concerns the "refined correction and amended edition of approved authors" (DA 6.4, I: 708). Bacon points out that the "rash diligence" of some men has done "no little damage" to these studies. For, he explains, when coming across something they do not understand, many critics "at once suppose there is a mistake in their copy." According to Bacon, this perverse habit of critics has resulted in the fact that the "most corrected copies are often the least correct," as "someone [non-nemo] prudently remarked" (DA 6.4, I: 708). Strangely enough, though, Bacon illustrates this point by misrelating and misquoting a passage from Tacitus's *Historiae*, and by misquoting and falsely criticizing an unnamed critic for having expunged a word and having replaced it by another word.[16] Since in the *Advancement of Learning* our approved author had illustrated the same point by making a similar mistake in order to illustrate precisely the danger of making such mistakes (AL 2, 131, III: 414),[17] we may reasonably assume that these obvious blunders were intentional. Besides, Bacon elsewhere emphasizes that it contributes more to practice when the "discourse attends on the example" than when the "example is subjugated to the discourse." For, he explains, when the example is used to "serve as the basis of the discourse," it is presented "with all the attendant circumstances," which may "sometimes correct and sometimes supply the discourse, as a pattern for imitation and practice" (DA 8.2, I: 769). In other words, it is only by considering the "discourse" in light of the "example" that the reader is drawn into true discourse. But although the aforementioned examples only teach us not to confuse pedantry with understanding, the fact that modern editors have "diligently" supplied them with corrective comments only proves the ongoing relevance of Bacon's warning against mistaking the "most corrected" for the "most correct."[18]

In the context of the second part of the "critical part" of the art of transmission, which concerns the "interpretation and explication of authors," Bacon explicates his intention to some extent. He points out that in labors of this kind some critics have been visited with the "bad disease of passing over many of the obscurer places, while expatiating and lingering to the point of nausea on those places that are likely enough [in satis vero]," as if the critic's object were "not so much to illustrate the author as to show off on every possible occasion his own erudition and various reading." Bacon adds that it were especially desirable, and that it is a matter pertaining "not to the appendices but to the principal part of the art of transmission," if every writer who treats "arguments of the nobler and obscurer sort" should "subjoin his own explications," so that "the text is not interrupted by digressions and explications, and the notes do not deviate from the writer's intention" (DA 6.4, I: 709). But since Bacon never subjoined notes to his arguments, and the blunder immediately preceding this sole "note" of his was "obscure" precisely because it was so obvious, his principal intention in stimulating our critical faculties seems to be to encourage us to look for "internal" explications when he treats arguments "of the nobler and obscurer sort." If this is true, then Bacon is right in counselling that "*Optimi Consiliarii mortui*; *Books* will speak plaine, when *Counsellors* Blanch" (E XX "Of Counsell," 67, VI: 426).

But the majority of modern editors and commentators neither blanched nor blushed when they vindicated their own erudition by "correcting" some of the many obscurities contained in Bacon's writings. Sometimes out of negligence, more frequently out of indifference disguising itself as diligence, or out of a scholarly conscience unaware of its paternalistic pride, they imposed their own standard of careful writing on Bacon by assuming that his blunders were due to a fault of his memory.[19] But if they had taken into consideration the most obvious internal explication, which consists in a careful comparison of the wording and the respective contexts of the original quotation and the misquotation, they would have noticed that simply leafing through Bacon's works already indicates the implausibility of their claim. For Bacon sometimes employs long Latin quotations without making any mistake,[20] and he sometimes omits part of a quotation in order to adapt it to the context in which he uses it.[21] Moreover, Bacon's commonplace book, *A Promus of Formularies and Elegancies* (1594–1596), which contains quotations and misquotations of Virgil, Horace, Erasmus, and Ovid, among others, as well as of Proverbs, Ecclesiastes, and Psalms, turns out not to be the store room of Bacon's memory that some scholars thought it was. According to one contemporary scholar, the commonplace book illustrates Bacon's "method of gathering texts from the past," a method that became "intimately related to the development" of Bacon's "philosophical and scientific methods."[22] Spedding also looks upon the commonplace book as an "illustration of Bacon's manner of working." Since many of the quotations are "slightly inaccurate," the great nineteenth-century editor of Bacon's works concludes that Bacon was "in the habit of sitting down from time to time, reviewing in memory the book he had last read, and jotting down those passages which for some reason or other he wished to fix in his mind."[23] But how to interpret the fact that Bacon never uses the vast majority of the quotations and misquotations contained in his commonplace book, whereas the vast majority of the quotations and misquotations that he does use are nowhere to be found in the commonplace book? And how to interpret the fact that, although the commonplace book contains correct quotations that are subsequently used correctly, it also contains misquotations that are subsequently used as misquotations in the context of an argument that makes one suspect that Bacon already had that or a similar argument in mind when he wrote the misquotation down in his commonplace book?[24] Finally, how to interpret the fact that the commonplace book contains correct quotations that were subsequently distorted by Bacon in the context of an argument that makes the distortion almost immediately intelligible?[25]

For those editors of Bacon's books or interpreters of his mental faculties, however, who still remain unwilling to concede there is the *Comentarius Solutus* (1608), a collection of private notes from which it appears that Bacon composed four notebooks on the basis of his readings. In these notebooks, of which the commonplace book was the least important, our author carefully differentiated between different kinds of arguments, distilled from different kinds of books, and meant for different kinds of purposes.[26] That Bacon reserved the most important arguments for his most carefully written books is already indicated by his counsel that only "some Few" books are to be "read wholly, and with Diligence and Attention," and that only in the case of "the lesse important Arguments, and the Meaner Sort of *Bookes*" "Extracts" may be made by "Others." For else, he warns, "distilled *Bookes,* are like Common distilled Waters, Flashy Things" (E L "Of Studies," 153, VI: 498).

Method

Fortunately enough, the dead counselor is alive to our need for guidance on how to read his most carefully written books, as appears from the fact that he devoted a whole chapter of the sixth book of *De Augmentis* to what he calls the "doctrine concerning the method of discourse." Since placing the method of discourse "in the service of other arts" leads to "passing over many things relating to it that are useful to know," Bacon decided to turn it into a "substantive and principal doctrine." He emphasizes, however, that it is pointless to speak of a "unique method," because "uniformity of method cannot be accommodated to the multiformity of matter." Those who employ a unique method "torture their object with the laws" of this method, and if the object "does not aptly fall into the dichotomies of method," it is "either laid aside or forced out of its nature." It is therefore method that must "accommodate itself to the subject matter that is treated," which goes a long way toward explaining why Bacon calls the method of discourse the "prudence of transmission" (DA 6.2, I: 662–3, 666). But notwithstanding method's adaptive potential, Bacon must have known that it is only the coincidental that can make discourse actual.[27]

The first difference of method consists in the difference between the "magisterial" and the "initiative" method. Whereas the initiative method "discloses and lays bare the very mysteries of the sciences" by transmitting them for examination to the "sons of science," the magisterial method "conforms the sciences to the vulgar" in order to have the vulgar use the sciences "as they are in their current state" (DA 6.2, I: 663–4).

The second difference of method, the "exoteric" and the "acroamatic" or "enigmatic" method, is "affinitive to the first as far as its intention is concerned," although "in itself [reipsa] it is almost its contrary." For although the two methods have in common that they both separate vulgar from select auditors, they are opposed as regards the manner of transmission. Whereas the former method employs a manner of transmission "more open than usual," the latter employs a manner "more concealed." The intention of the exoteric–acroamatic or exoteric–esoteric method is to "move the vulgar away [summoveantur] from the secrets of the sciences," and to "admit" only those who have either "received the interpretation of parables through the hands of teachers," or who have "wits of such sharpness and discernment as can pierce the veil" (DA 6.2, I: 664–5).

The problem underlying a mixture of commonality and contrariety could hardly have been indicated more aptly than by means of a cursory reference to the one instrument the two contrary methods have in common. For it is not only the exoteric–esoteric method but also its contrary that uses parables, as appears from Bacon's discussion of "parabolic poetry." After emphasizing that he considers poetry not with respect to "words," but with respect to "matter," Bacon points out that parabolic poetry, which is the "most eminent form" of poetry, "reduces objects of the intellect to the sense."[28] It is used "ambiguously and for contrary purposes." For parables are used either for "illustration and as a method of teaching," or for "enfoldment and as an artifice for concealment." As a method of teaching they were "much used in ancient times," because the "wits of men were hardly subtle enough to conceive the discoveries and conclusions of human reason that are now vulgar and trite, but that were at the time new and out of the common, unless they were reduced to the sense by images and examples." Bacon adds that "even now, and at all times, the vigour of parables is exceptional, because arguments cannot be made as perspicuous, and true examples cannot be made as apt as parables." Parables that serve for "enfoldment," on the other hand, are used for "such things the dignity of which deserves that they be discerned as it were through a veil," which is the case when the "concealments and mysteries of religion, politics, and philosophy are covered by fables or parables" (DA 2.13, I: 518, 520–1).[29]

To simplify matters for the sake of clarity, one could say that parables as a "magisterial" device "reduce objects of the intellect to the sense" in order to let the imagination of the vulgar tide over the time that is needed for certain "discoveries and conclusions of human reason" to become "vulgar and trite" enough to actually unite them with the sense (cf. AL 1.29, 23, III: 284–5). They are used in order to gradually unveil to the vulgar what has to be provisionally veiled from them, and they ultimately serve to bridge a provisional gap between the initiates and the vulgar to the largest extent nature allows. As spokesman of the initiates Bacon confirms this point in *Novum Organum* by saying that the initiates are pleased that the sciences are used as they are in their current state, because what the initiates bring about "cannot be wholly reduced to vulgar apprehension, except by means of effects and works" (NO Pr. 4, I: 153; NO 1.128, I: 220). As works require time in order to be effected, parables make the imagination of the vulgar already perceive as sensible what by means of works does not become perceptible to the senses until after a certain period of time. Parables as an artifice for concealment, on the other hand, permanently veil from the vulgar what they want the select to unveil by reuniting with their own intellect the objects that the author's intellect reduced to the sense. They ultimately serve to widen the perennial gap between the select and the vulgar as far as prudence counsels, although the actual width of this gap is inversely proportional to the degree to which the select sensed it from the beginning. One could therefore say that the gap separating the select from the vulgar is ultimately identical to the difference in consciousness of those idols that can "never wholly be torn out" (IMDO 14–5, I: 139; DA 5.4, I: 643–6; NO 1.38, 41 ff., 45–60, I: 163–72). But the fact that the difference between the initiates and the vulgar is almost as bridgeable as the difference between the vulgar and the select is unbridgeable already indicates the natural tendency of commonality to obliviate the consciousness of contrariety.

We are running ahead of things, though. Because the difference between the vulgar, the select, and the initiates made his subject matter as it were multiform, Bacon had to paint his veils with multiple colors in order to prevent them from being lifted by the wrong persons or in the wrong manner. And because our author reserved the treatment of the perennial problems of politics, philosophy, and religion for his books, and books are available to every man who can read, this implied that he had to take long-term measures in order to secure that the reading man who is vulgar would or would ultimately be "moved away from" the very book the reading man with a "piercing" wit similar to his own was to be wholly drawn into. In other words, Bacon had to speak more than once by speaking once, that is, he had to conceal from the vulgar what he wanted to say to his equals in a speech that is and is not directed to both.

Although Bacon underlines this most important point by discussing the difference between the exoteric and the esoteric method as a difference within one method,[30] in an earlier version of the argument he made the same point in a less concealed way by describing the exoteric–esoteric method as "publishing in a manner whereby it shall not be to the capacity nor taste of all, but shall as it were single and adopt his reader" (VT, III: 248). The exoteric veil ultimately moves away vulgar readers by seeming to be in conformity with their opinions and prejudices, whereas only wits of such sharpness and discernment as can truly pierce the exoteric veil can unveil the esoteric center. The fact that Bacon discusses or in his expressions even conforms to the opinions and prejudices of the vulgar therefore does not justify the conclusion that Bacon's horizon was absorbed by the horizon of the vulgar. After all, it was Bacon himself who emphasized that "Bookes (such as are worthy of the name of Bookes) ought to haue no Patrons, but Truth and Reason" (AL 1.24, 20, III: 281), seeing that the patronage of truth and reason consists in its refusal to submit to any patron. But at the same time it is only by understanding Bacon's reasons for conforming in expression that we can understand his reasons for diverging in thought. In other words, we can only lift Bacon's veils by piercing them after discerning his intention in drawing them and painting them with certain colors. We must think through the opinions and prejudices of the vulgar as well as Bacon's reasons for diverging from them without neglecting them. But since opining oneself above the vulgar is a prejudice of the vulgar, we should guard ourselves against mistaking the opinion of knowledge for the knowledge of opinion, although the road to knowledge necessarily goes through opinion. Since the esoteric center is therefore also always on the exoteric surface, we can only reach the center by continuously descending from the surface and ascending to the surface.[31] It is only by considering all the parts separately and jointly that we can unveil the whole in its unity and trace back Bacon's movement of thought. And it is only in this case that the relative part of the art of transmission is transhistorical to the extent that hermeneutics turns into the philosophic activity of tracing the threads of the web Bacon's mind has spun.

As one cannot prudently approve of the art of concealment except by concealing that one approves of it, we have to go back to one of Bacon's early and unpublished writings in order to read that exoteric–esoteric writing is "not to be laid aside, both for the avoiding of abuse in the excluded, and the strengthening of affection in the admitted" (VT, III: 248). But modern scholars tend to disapprove of exoteric–esoteric writing, as appears from the fact that they solve the problem underlying its ongoing necessity by disallowing its relevance or by disregarding it altogether.[32] It was, however, well into the modern era that John Toland observed that the exoteric–esoteric method is "as much now in use as ever; tho the distinction is not so openly and professedly approv'd, as among the Antients."[33]

Application

At the end of the *Advancement of Learning* and *De Augmentis,* after saying that he has propounded his opinions "naked and unarmed," Bacon draws our attention to the method of reading that meets the requirements of his method of writing by saying that "in anything which is well set down" he is "in good hope that, if on the basis of the first reading there emerges a scruple or an objection, the second reading will of itself make an answer" (AL 2, 192, III: 491; DA 9.1, I: 831). In other words, Bacon hopes that the second reading will "unclothe and disarm" what will still be "clothed and armed" at the time of the first reading.

We can best illustrate this procedure and the decisive importance of exoteric–esoteric writing for the possibility of understanding Bacon's thought by looking somewhat more closely into the most hazardous argument in all of Bacon's works. In *De Augmentis* Bacon observes that "in the reception and adoption of philosophic truth the same thing happens as in divine truth; *Veni in nomine Patris, nec recipitis me; si quis venerit in nomine suo, eum recipietis.*" However, so he adds, if we consider that this "heavenly aphorism" was applied "primarily to the Antichrist," the "greatest deceiver of all ages," we may discern well that the "coming in one's own name, without regard of paternity, is a bad sign of truth, although it is oftentimes joined with the fortune of *Eum recipietis*" (DA 3.4, I: 549). Our first reading of this argument makes it seem as if Bacon warns against receiving "antichristian" men, who come in their own names, and are to be suspected of untruth because they do not "regard paternity." By explicitly calling the Antichrist a "deceiver," Bacon draws our attention to the Heavenly Book that explicitly calls the Antichrist a deceiver (2 John 7). But when we follow Bacon's lead, we find that we have been deceived by the man who led us into following his lead. For it is exactly the fact that the Antichrist does *not* come in his own name, but goes out from the children of Christ, that makes him the "greatest deceiver of all ages" (1 John 2: 18–22). Earlier in the same paragraph, Bacon argued that where his own conceptions and notions are "novel" and "deviate from the ones received," he would "retain the ancient words with the utmost religious care" (DA 3.4, I: 548). We therefore suspect Bacon of arguing that in order to be received as a father without being suspected of untruth, he "regards paternity" by "going out from" the Father without "being in" the Father. We find this suspicion confirmed when we have a closer look at Bacon's version of the "heavenly" aphorism that constitutes the foundation of his argument. For Bacon subtly misquotes the words of Christ by omitting a most important word, since what Christ actually said was that He came in the name of *His* Father ("Patris *mei*," John 5: 43). By thus misquoting the words of God's Son, Bacon argues, we conclude, that the Father in whose name he comes is not *his* Father, although his paternity is dependent on the paternity of the Father.[34] But had we not paid attention to exoteric–esoteric writing, and had we not taken Bacon's misquotation and its context into consideration, we would not have known that Bacon deceived us into believing that he did not want to precipitate the coming of him whom he said to be detaining (cf. 2 Thess. 2: 6–7).

Since Bacon observed that the ancients "prudently and judiciously" used exoteric–esoteric writing in their books (DA 6.2, I: 664–5), it stands to reason that we look more closely into Bacon's book on ancient wisdom, especially since it is in this book that Bacon transmits his teaching in the form of parables (DSV EpD1, VI: 619).

In the final fable of *De Sapientia Veterum,* Bacon says that "we find that the *Wisdom of the Ancients* is like grapes ill-trodden: something is squeezed out, but the best parts are passed over and left behind" (DSV XXXI "Sirenes," VI: 685). As the words "Sapientia Veterum" stand out in different sorts of typeface in the early editions of *De Sapientia Veterum,*[35] and Bacon hardly ever uses the noun "veteres" to depict the ancients, we may reasonably assume that Bacon referred to his own book.

Bacon's remark calls to mind the description of exoteric–esoteric writing given by the philosopher Maimonides, who himself was a master of the art of writing: "[A] saying uttered with a view to two meanings is like an apple of gold overlaid with silver filigree-work having very small holes . . . the external meaning ought to be as beautiful as silver, while its internal meaning ought to be more beautiful than the external one, the former being in comparison to the latter as gold is to silver. Its external meaning also ought to contain in it something that indicates to someone considering it what is to be found in its internal meaning, as happens in the case of an apple of gold overlaid with silver filigree-work having very small holes. When looked at from a distance or with imperfect attention, it is deemed to be an apple of silver; but when a keen-sighted observer looks at it with full attention, its interior becomes clear to him and he knows that it is of gold."[36]

At the end of the Epistle Dedicatory to the Earl of Salisbury Bacon points out that, although to the "vulgar apprehension" his suggestions will be "vulgar," he hopes that the "deeper intellect" will not "abandon" them, but will rather be "carried along" (DSV EpD1, VI: 619–20). Bacon is articulately silent on the question whether an intellect does not reveal its true depth by its degree of readiness to abandon his suggestions after having been "carried" to the conviction that they are not vulgar, although he puts us on the way toward the answer by prefacing his book with some directions for the use of his winepress.[37] He starts by saying that "usurping a licence almost similar to that assumed by the poets themselves would be done most sanely by sprinkling over more arduous contemplations, being either one's own meditations or the readings of others, with some pleasure." Bacon wants us to know that it is not unknown to him "how freely the versatile material of which fables are made can be drawn any way it is commanded," and how "with a little convenience and discourse of wit unintended meanings can be attributed to fables." The thought also came to Bacon's mind that "the thing was abused by men of no experience in matters, nor any learning beyond a few commonplaces, who only wanted to acquire the reverence of antiquity for inventions and doctrines of their own," and who therefore "applied the meaning of parables to some generalities and vulgarities, without attaining their true force, genuine property, and deeper traces"; consequently, the "distinction and virtue of the thing were almost [fere] ruined" (DSV Pr., VI: 625, 628).

It needs no argument that it is Natale Conti's *Mythologiae sive Explicationis Fabularum* (1551), a work that went through nineteen editions between 1551 and 1627, that Bacon deprecatingly alludes to here. But the fact that Bacon diminishes the importance of the *Mythologiae* for his interpretation of the ancient fables provides us with the opportunity to elevate the work's importance for lending credibility to our interpretation of Bacon's fables. For although we believe that Bacon knew how to thread his way through the woods of ancient mythology by himself, we acknowledge the encyclopedic value of Conti's *Mythologiae*. And seeing that the credibility of our interpretation of Bacon's fables will increase to the extent to which Bacon is believed to have found his way to the ancient sources, we concede that Bacon took Conti's *Mythologiae* as his point of departure.[38]

But let us return to our discussion of Bacon's directions on how to read his book of fables. After saying about fabulists in general what he wanted to say about himself in particular, Bacon says that in some of the fables of the ancient poets, either in the "texture of the fable" or in the "propriety of names used to indicate the persons and actors" in them, he finds "such an important and evident similitude and conjunction with the thing signified, that no one can persistently deny that this meaning was designed and thought through from the beginning, and was purposely shadowed out" (DSV Pr., VI: 626). But since the examples by means of which Bacon illustrates his finding are almost all derived from his own fables, and are nowhere to be found in the ancient sources, we cannot but conclude that Bacon acted "licentiously" when he purposely "shadowed out" his "more arduous" contemplations by making them seem fabulous and therefore pleasurable.[39]

Bacon goes on to emphasize that "no one needs to be disturbed if sometimes some history is found underneath a fable, or if times are confounded, or if part of one fable is transferred to another, and a new allegory is introduced." For, he explains, this is due to the fact that it is "inevitable for such things to occur in stories invented by men who lived in different ages and had different ends, some of which were more ancient, some more recent, and some of which related to natural things, and some to civil things" (DSV Pr., VI: 626). Bacon remarks that it is another "sign of no small value of a concealed and covered meaning if a fable is so absurd and stupid upon the face of the narrative taken by itself that it shows and as it were cries out from afar that there is a parable underneath." For a fable that "sounds probable may be thought to have been composed merely for pleasure, as a similitude of history." A story that "could never have entered a man's mind either to think of or to narrate," on the other hand, which Bacon considers to be the case when "things are so monstrous and beyond all normal ways of thinking that no mortal could ever dream them," requires "some other use to be sought" (DSV Pr., VI: 627).

The usefulness of these observations for interpreting Bacon's fables can best be illustrated by the following example, an example that will receive substantial treatment in the next chapter. The narrative of the second fable presupposes the narrative of the "monstrous" and "at first hearing very foolish" thirtieth and penultimate fable. For whereas the second fable relates that, in response to Jupiter having begotten Pallas by himself, jealous Juno posed a threat to the tyrant's rule, the thirtieth fable relates the birth of the goddess of wisdom from the head of Jupiter (DSV II "Typhon," VI: 630–1; DSV XXX "Metis," VI: 683). But since the thirtieth fable is silent on Juno's jealous response to Jupiter's self-begetting act, we suspect that Juno's anger had been appeased in the meantime. It is in the sixteenth and central fable that our suspicion is proved correct. For it is in this fable that we find Jupiter seducing Juno by turning himself into a "wretched cuckoo." The sixteenth fable is the only fable in which the central character is not mentioned by name either in the title or the subtitle of the fable. It is also the only fable whose narrative is nowhere, in whatever form, to be found in the ancient sources, which raises the likelihood that there is "some history" to be "found underneath."[40] Ironically enough, the sixteenth fable is also the center of five fables beginning with "narrant poëtae," and of thirteen fables beginning with a conjugation of 'narro.' Although Bacon says that Juno represents persons of a "proud and malignant disposition" (DSV XVI "Procus Junonis," VI: 654), he did not and did not need to say that the father of gods and mortals represents the Heavenly Father. But by not mentioning the name of Jupiter in the title and the subtitle of the fable on Juno's suitor, Bacon suggests that the "proud and malignant" are seduced by an unnamed suitor, who turned himself into the shape of the representative of the Heavenly Father in order for the goddess of wisdom to be born from the head of Jupiter. It was, however, only by making this grave metamorphosis seem "absurd" that Bacon could draw our attention to its parabolic meaning.

Although to the extent that the problems of religion, philosophy, and politics are perennial, exoteric–esoteric writing is a possibility coeval with philosophy itself, the "most obvious and crudest reason," as the philosopher who retrieved the art of exoteric–esoteric writing from oblivion calls it,[41] already induced Bacon to employ exoteric–esoteric writing as an answer to the danger of religious persecution.

But despite being only the crudest reason, persecution is commonly believed to be the only legitimate reason for authors to avail themselves of exoteric–esoteric writing. To set aside the more timeless moral appeal that goes out from visible and therefore self-forgetful subversiveness, a present-day explanation for this reductionist view may be found in the tendency of liberal societies to venerate dissidence elsewhere in order to have a good conscience when suppressing dissident voices at home.

Confining ourselves, however, to the religious persecution that is commonly acknowledged to have taken place in the times of Bacon, we find that Bacon's friend Tobie Matthew, a Roman Catholic convert who had spent six months in prison on account of his conversion, repeatedly and at Bacon's explicit request "censured" his works and warned him when he fell asleep.[42] The influential Bishop Lancelot Andrewes, a former chaplain of John Whitgift who had performed a leading function in the coming to fruition of the "Authorized Version" of the Bible (1611), functioned as Bacon's "inquisitor,"[43] although in the case of *De Augmentis Scientiarum*, Bacon himself played the part of "*Index Expurgatorius*," as he writes in a letter to the Protestant King James.[44] But although the official manuscript decree *Donec corrigatur* of April 3, 1669 caused *De Augmentis* to be put on the *Index Librorum Prohibitorum*,[45] the overall reception of Bacon's works raises the question whether Bacon's precautionary measures may not have been almost too successful, either in the sense that they have left an opening only to their most acute object, or in the sense that they only seem to have left an opening to their most acute object.

Reserving the answers to this question for the places where they belong, we continue our discussion of the matter of self-censorship, which can best be illustrated by the example of Bacon's apologetic treatment of the principal matter subject to censorship. In the fifth paragraph of the first book of the *Advancement of Learning*, Bacon distorts three biblical quotations on the temptation or sin of human knowledge. In the sixth paragraph, immediately before saying: "let those places bee rightly understoode, and they doe indeede excellently sette foorth the true bounds and limitations, whereby humane knowledge is confined and circumscribed" (AL 1.5 and 6, 6–7, III: 264 and 266; cf. Eccl. 12: 12 and 1: 18; Col. 2: 8), he distorts the third of these quotations for the second time. But whereas the first time Bacon had only added the adjective "*vaine*" to the philosophy Paul had warned the believer not to let himself be spoiled by, the second time he also replaced "*spoyled*" by "*seduced*," which he must have felt confident enough to do in light of the fact that he had already obscured philosophy's seductive power by making philosophy as such seem vain. With regard to the third limitation of human knowledge, "that we doe not presume by the contemplation of Nature, to attaine to the misteries of God," Bacon explicitly says that this point "deserueth to be a little stood vpon, and not to be lightly passed ouer." The contemplation of "Gods creatures and works produceth (hauing regard to the works and creatures themselues) knowledge, but hauing regard to God, no perfect knowledg, but wonder, which is broken knowledge." But if we follow Bacon's admonition and "stand upon" this point, we detect that we almost "passed over" Bacon's earlier description of wonder as the "seede of knowledge" (AL 1.6, 7–8, III: 266–7).

In *A Tritical Essay upon the Faculties of the Mind* (1707), Jonathan Swift, in a comical fashion, provides the discerning reader of his *Essay* with counsels on how to read Bacon's *Essays*. Swift introduces his parody on Bacon by saying that he is "offended" with writers of essays "for running into stale Topicks and thread-bare Quotations, and not handling their Subject fully and closely." His own *Essay*, however, abounds with inexact and superfluous quotations, unintelligible digressions, and sentences like "BUT to return to our Discourse."[46] Divesting Swift's *Essay* of its comedy, we find that wittingly or unwittingly Swift argues that one can only fully disclose the subject matter of Bacon's *Essays* by paying close attention to how inexact quotations, apparent disorder, seeming superfluities, and meaningful digressions constitute a coherent discourse if taken in conjunction.[47]

As the first *Essay* (E I "Of Truth," 7 ff., VI: 377 ff.) sets the stage for the argument and the action of the *Essays* as a whole, a movement of speech and deed that will occupy us in chapters to come, it provides us with a good example of this order within seeming chaos.

The prospect of "Giddinesse," which goes with the perception of those "Philosophers" and "discoursing wits" who are practicing "Free-will in Thinking," induces a number of men to "fix a Beleefe" and to submit themselves to "Bondage," as Bacon makes us distil from his "giddying" amalgam of words. It is, however, "not onely the Difficultie, and Labour, which Men take in finding out of *Truth*; Nor againe, that when it is found, it imposeth upon Mens Thoughts; that doth bring *Lies* in favour: But a naturall, though corrupt Love, of the *Lie* it selfe."[48] Lucian, who had already "stood upon" this matter, and whose argument Bacon refers us to by "coming to a stand" at it, had concluded that men love lies for the sake of their substance.[49] It is therefore not the "*Lie*, that passeth through the Minde" that "doth the hurt," but the "*Lie* that sinketh in, and setleth in it," because in "mens depraved Judgements and Affections" its object supplies a natural need and is therefore "corruptly loved."

Passing in deed from "Theologicall, and Philosophicall *Truth*" to the "*Truth* of civill Businesse" he will soon pass to in speech, Bacon observes that although truth is a "Naked, and Open day light" that unclothes and discloses "the Masques, and Mummeries, and Triumphs of the world," it does not show them "halfe so Stately, and daintily, as Candlelights." For "Truth may perhaps come to the price of a Pearle, that sheweth best by day: But it will not rise, to the price of a Diamond, or Carbuncle, that sheweth best in varied lights." Bacon does not doubt that "if there were taken out of Mens mindes, Vaine Opinions, Flattering Hopes, False valuations" and "Imaginations," it would "leave the Mindes, of a Number of Men, poore, shrunken Things; full of Melancholy, and Indisposition, and unpleasing to themselves." The wine of poetry can prevent the minds of these men from becoming "giddy" with disillusionment at the "naked" sight of having been "unmasked as mummery" what was believed to be true, albeit that it can only do so by filling the imagination with the "shadow of a *Lie*."[50] But man's natural love of the lie causes a "Mixture of Falshood" to be like "Allay in Coyne of Gold and Silver," as it makes the "embased" metal that his mind is made of "worke the better." And although the "winding, and crooked courses" of untruth are the "Goings of the Serpent," the "dove" cannot remain innocent unless "serpentine prudence" teaches him about the "nature of evil," as Bacon emphasizes elsewhere (DA 7.2, I: 729–30).

There is, however, "no Vice, that doth so cover a Man with Shame, as to be found false, and perfidious." For the word of the lie is "such a Disgrace, and such an Odious Charge," that, as Bacon makes Montaigne say: "*If it be well weighed, To say that a man lieth, is as much to say, as that he is brave towards God, and a Coward towards men.*[51] For a Lie faces God, and shrinkes from Man." But if it is evil to speak evil of evil, Bacon had good reason to "shrink back" from being "found false and perfidious" by men ready to persecute him on the charge of being evil. Although he knew that by seducing men with poetic lies he would eventually have to "face

God," who is immune to lies, he was "brave toward God" by preferring to be a "coward toward men."

But to return to our *Essay*: Bacon invites the sober reader not to become giddy with the wine of his poetry,[52] and to "stay for an Answer" to the question his Pilate jested about after having received an answer from Truth Himself: What is truth?[53]

Being untrue without being "found false and perfidious," in other words, concealing one's heterodoxy by a veil of orthodoxy, is the subject matter of the sixth *Essay* (E VI "Of Simulation and Dissimulation," 20 ff., VI: 387 ff.). "*Dissimulation* is but a faint kind of Policy, or Wisedome; For it asketh a strong Wit, and a strong Heart, to know, when to tell Truth, and to doe it. Therfore it is the weaker Sort of Politicks, that are the great Dissemblers." But is it not precisely because one cannot prudently counsel on the wisdom of dissimulation except by inducing a wakefulness toward its policies that Bacon dissembles the fact that the wisdom of dissimulation primarily consists in having the "strength of wit and heart" to make one's policies depend on the knowledge of the occasion that teaches when and when not to tell the truth?

There are three degrees of "this Hiding, and Vailing of a Mans Selfe." The first is "*Closenesse, Reservation*, and *Secrecy*; when a Man leaveth himselfe without Observation, or without Hold to be taken, what he is." The second is "*Dissimulation*, in the *Negative*, when a man lets fall Signes and Arguments, that he is not, that he is." The third is "*Simulation*, in the Affirmative; when a Man industriously and expressly, faigns, and pretends to be, that he is not." Bacon makes it clear that these three degrees naturally follow from each other; although throughout the *Essay* he dissembles this fact by making the attributes of dissimulation seem mutually exclusive.[54] Because "*Nakednesse* is uncomely," it "addeth no small Reverence, to Mens Manners, and Actions, if they be not altogether Open." Dissimulation follows upon secrecy "by a necessity," because others "will so beset" the secret man "with Questions, and draw him on, and pick it out of him, that without an absurd Silence, he must shew an Inclination, one way; Or if he doe not, they will gather as much by his Silence, as by his Speech." The "affirmative" side of this picture is simulation and "false Profession," which Bacon does not consider "culpable" if the matter is "great and rare" enough.

Bacon concludes his *Essay* on dissimulating untruth with the principal counsel that the "best Composition, and Temperature is, to have *Opennesse* in Fame and Opinion; *Secrecy* in Habit; *Dissimulation* in seasonable use; And a Power to faigne, if there be no Remedy." We, for our part, conclude our discussion of Bacon's *Essays* on simulating truth with the relative part of their counsel, which is directed to those able to measure the "state of the weather" in which nakedness "covers" a man with such "shame" that "secrecy" dissembled as "openness" is the only shelter from a storm severe enough to exculpate even a "false profession," and which teaches that the "signs and arguments" by means of which the dissimulator dissimulates also state "in the affirmative" that the dissimulator is not in truth what he is to those toward whom he dissimulates.

In accordance with the teaching of the sixth *Essay*, Bacon elsewhere reveals the dissimulation of Augustus Caesar by referring to a letter of Cicero to Atticus in order to illustrate how little of a dissimulator the Roman Emperor seems to have been (DA 8.2, I: 782–3). On his entry into public affairs, "when he was still the favourite of the senate," Augustus used the following form of oath in his harangues to the people: "As I hope to attain the honours of my parent," which, Bacon interprets, were in fact "nothing less than the tyranny." At the same time, "in order to lessen the envy his hope would arouse," Augustus extended his hand toward a statue of Julius Caesar, which had been erected in the place. Men "laughed, applauded, and wondered at this," and said to each other: "What is this, what kind of adolescent is this?" Yet they thought that "a man who had spoken his feelings so candidly and ingenuously could not be suspected of any maliciousness."

In his letter to Atticus Cicero briefly mentions the *contio* delivered by Octavian on his first visit to Rome, only pointing out that Octavian made the oath: "Ita sibi parentis honores consequi liceat," and that at the same time he extended his right hand towards the statue of Julius Caesar. Cicero adds the following exclamation in Greek: "Μηδὲ σωθείην ὑπόγε τοιούτου!" [May they not be saved by somebody like that!]," which implies that he realized that Augustus may have been dissimulating his true aim of becoming the same kind of "savior" as his father.[55]

One could say that Augustus dissembled his boldness by being so bold that almost nobody suspected him of having been deliberately bold. Exoteric–esoteric writing is thus a form of Augustean dissimulation to the extent that the boldness of the exoteric surface both dissembles to the many and discloses to a few the boldness of the esoteric center. Leaving aside for now the question whether the true boldness at the center is necessarily in line with the boldness visible on the surface, we consider it certainly in line with the speech of Augustus that envious responses to the naked boldness of the center are anticipated by the embedded boldness of the surface. And since "*Envy* is ever joyned, with the Comparing of a Mans Selfe" (E IX "Of Envy," 29, VI: 394), a comparison with the person whose jealousy would be excited when seeing himself compared with the bold author should provide us with a first key to unveiling Bacon's reasons for dissembling by means of boldness the true boldness at the center of his writings. Considering that "Kings, are not *envied,* but by Kings" (E IX "Of Envy," 29, VI: 394), we find this key in the ninth and final book of *De Augmentis,* as it is in this book that Bacon compares himself with the King of Kings. For he observes that "although it is more worthy to believe than to know as we know now, it will be otherwise in the state of man glorified, for then faith shall cease and we shall know even as we are known" (DA 9.1, I: 830; cf. 1 Cor. 13: 12–3). Since the "Times, when the Stroke, or Percussion of an *Envious Eye*[56] doth most hurt, are, when the *Party envied* is beheld in Glory, or Triumph" (E IX "Of Envy," 27, VI: 393), almost nobody believed Bacon to have been so bold as to dissemble his true boldness in a book on sacred theology. But boldness "prevailith with wise men, at weake times" (E XII "Of Boldness," 37, VI: 402).

We conclude this chapter with a subject most important, but hardly credible. The supposedly unfinished *New Atlantis* contains twenty-two clearly identifiable biblical quotations or paraphrases. The word "God" is used twenty-two times in the fable. The Septuagint version of the Hebrew Bible contains twenty-two books. The Hebrew alphabet has twenty-two characters. "*Prosperity* is the Blessing of the Old Testament" (E V "Of Adversitie," 18, VI: 386), and the drama of *New Atlantis* takes place on a prosperous island with the Hebrew name "Bensalem." Since too many incredible coincidences make coincidence incredible, we cannot help noticing that Bacon employed numbers in indication or confirmation of his arguments; arguments, it needs emphasizing, which can ultimately do without the numbers that lead toward or shed light upon them.[57]

Augustine, whose *De Civitate Dei* also consists of twenty-two books, emphasizes that the meaning of numbers is "not to be contemned," because numbers and combinations of numbers are used in many places in Sacred Scripture "to convey secrets by means of similitudes." He himself discovered that the number seventeen signifies "law aided by grace." The number ten signifies the Law, and the number seven the Holy Spirit, "since God sanctified the seventh day, on which He rested from all His works." According to Augustine it is with good reason that the seventeenth Psalm (Vulgate) is the only Psalm that is completely given in the Book of Kings (2 Samuel), seeing that it signifies the "kingdom in which we shall have no enemy" (cf. Ps. 18 with 2 Sam. 22: 2–51). The Psalm is entitled: "A Psalm of David, in the day that the Lord delivered him from the hand of all his enemies, and from the hand of Saul." David is the "type of Him who, according to the flesh, was born of his seed, and who still endures the malice of enemies in His church, that is, in His body," which explains why Augustine continues by saying that "the words that fell from heaven upon the ear of the persecutor whom Jesus slew by His voice, and whom He transformed into a part of His body," were these: "Saul, Saul, why persecutest thou me?" (Acts 9: 4). But the body of Christ shall be "finally delivered from its enemies" when its last enemy, death, shall be destroyed. Augustine takes the number of 153 fishes (John 21: 11) to pertain to that time, arguing that if one assumes the number seventeen to be the side of an arithmetical triangle, formed by placing above each other seventeen rows of units, increasing in number from one to seventeen, the sum of these units is 153, the number that signifies "all the sharers in the grace of the Spirit," by which grace "harmony is established with the Law of God."[58]

But the number seventeen also has another tradition: The Greek alphabet has seventeen consonants, and philosophers have used the number seventeen to signify nature because nature is mute, as are consonants. That philosophers sometimes resort to incredible means in order to reveal their incredulity appears from the use Bacon makes of the number seventeen in the context of his seventeenth *Essay*, which is entitled "Of Superstition" (E XVII "Of Superstition," 55, VI: 416). As it is in the seventeenth *Essay* that Bacon observes the "Similitude of *Superstition* to *Religion*," it need not surprise us that the nouns "religion," "superstition," and "faith" are used in seventeen *Essays* altogether, the center of which is the seventeenth *Essay*. And as it is also in the seventeenth *Essay* that Bacon mentions "taking an Aime at divine Matters by Human" as one of the causes of superstition, we need neither be surprised overmuch to find that the word "God" is also used in seventeen *Essays*. But since the spirit of God was revealed to present itself to man (cf. Rom. 8: 9), the following should come as somewhat of a surprise to us. For by using the pronoun "I" in thirty-three *Essays*, the center of which is seventeen, while leaving it absent in *Essay* XVII, which is the only *Essay* in which the noun "Unbeleefe" is present, Bacon reveals that he was absent when God revealed His presence.

CHAPTER 2

The Baconian Turn

He commands me to go to very many cities of mortals.

—Homer, *Odyssey*, XXIII. 267–8

In a letter (1605) to Sir Thomas Bodley, founder of the famous Bodleian Library in Oxford, Bacon confesses that he thinks "no man may more truly say with the Psalm *Multum incola fuit anima mea*" than himself. He explains himself in the following way: "[S]ince I was of any understanding, my mind hath in effect been absent from that I have done; and in absence many errors which I do willingly acknowledge; and amongst the rest this great one that led the rest; that knowing myself by inward calling to be fitter to hold a book than to play a part, I have led my life in civil causes; for which I was not very fit by nature, and more unfit by the preoccupation of my mind."[1]

Strangely enough, though, in the years following his confession of the homelessness of his mind, Bacon continued to lead his life in "civil causes," and judging from the fact that he became Solicitor-General in 1607, Attorney-General in 1613, Lord Keeper of the Great Seal in 1617, and Lord Chancellor in 1618, he was even quite successful in politics. Some scholars explained Bacon's deliberate error by referring to accidents in his biography, such as his bad financial situation from the death of his father (1579) onward, thus mistaking for an effecting cause what in the highest case would be looked on as an accident at most, albeit an inconvenient one.[2] Other scholars succumbed to the pressure of popularized psychology, diagnosing Bacon as the product of what Sir Nicholas's political biography and his own experiences of court life had destined him to become—a diagnosis that denies Bacon the potential necessary to overcome what in the vast majority of cases would determine the further course of a man's life, even if its current course provided him with the incentives necessary for making the actualization of his natural potential possible.[3] And finally, there is a category of scholars that indulged in the fashion of segmenting Bacon's life and activities, a fashion that runs parallel to the ongoing segmentation and specialization of political society, but that roots in the dismissal of the quest for a unifying whole.

Preparation

It is, however, Bacon himself who in his well-known letter (1592) to Lord Burghley emphasizes that "the contemplative planet carrieth" him "away wholly."[4] And it is Bacon himself who throughout his mature life repeatedly avowed his affective preference for the planet of Saturn over the planet named after Saturn's son Jupiter, who thrust his father down into Tartarus and took possession of his kingdom (cf. DSV XII "Coelum," VI: 649).[5]

In the aforementioned letter to Burghley, Bacon makes the frequently quoted but less frequently interpreted confession: "I confess that I have as vast contemplative ends, as I have moderate civil ends: for I have taken all knowledge to be my province. . . . This, whether it be curiosity, or vain glory, or nature, or (if one take it favourably) *philanthropia*, is so fixed in my mind as it cannot be removed."[6]

But considering his unawareness of their efficient cause, the young Bacon must have been unable to harmonize his vast contemplative with his moderate civil ends, if indeed he was aware of their true vastness. In the early (1603) and unpublished *Valerius Terminus*, we therefore find Bacon searchingly and undecidedly but not surprisingly pointing out that since there is "no composition of estate or society, nor order or quality of persons, which have not some point of contrariety towards true knowledge," it is "hard to say, whether mixture of contemplation with an active life, or retiring wholly to contemplations, do disable and hinder the mind more" (VT, III: 252).

But stalemates sometimes carry the seed of their solution in the problem that causes their solution to remain a problem. For not only had Bacon come to realize that indulging in contemplations while neglecting the political order would hinder or even disable his mind, considering that it would constantly have to reckon with the possibility of contrariety actualizing itself as hostility. He had also gained the insight that simply mixing vast contemplations with an active life would be tantamount to contracting his contemplative horizon, seeing that the political order does not recognize boundaries that transgress its own boundaries by recognizing them as boundaries. This is due to the fact that "civil matters rest upon authority, consent, fame, and opinion," and not on truth and demonstration, as Bacon explains elsewhere (CV 8, III: 597; NO 1.90, I: 197–8).

By the time the *Advancement of Learning* was published (1605), Bacon must have found a way out of the impasse of contemplations failing to harmonize with the political order, as he decidedly points out that he foresees that "if contemplation and action may be more neerely and straightly conioyned and vnited together, than they haue beene," this conjunction, which is "like vnto that of the two highest Planets, *Saturne* the Planet of rest and contemplation; and *Iupiter* the Planet of ciuile societie and action," will "dignifie and exalt knowledge" (AL 1.48, 32, III: 294).

We notice in passing that Bacon passes over the question to what extent the "conjunctive two" of contemplation and action may or may also be what a philosopher once called a "disjunctive two," that is, two separate heads bound together in one unifying head. Bacon's silence is all the more deafening in light of the fact that breaking it would cause his own active or political life to come to sight as a series of political actions on the part of his contemplative life, should knowledge of ignorance turn out to be what makes the two disjunctive.

But Bacon's forbearing from discussing the unifying power of knowledge may not be wholly unconnected with the fact that it is in the same book that he says that the question touching the "preferement of the Contemplatiue or actiue life" must be decided against Aristotle. For considering that Aristotle's presentation of his preference for the contemplative life may very well have been the result of a political decision, it cannot be ruled out in advance that it was primarily Aristotle's decision to present the contemplative life as preferable that Bacon decided against.[7]

Although we are not yet in a position to judge this possibility, its plausibility is already indicated by the fact that Bacon arrived at his decision as a consequence of his observation, "being set downe and strongly planted," that "there was neuer any phylosophy, Religion or other discipline, which did so playnly and highly exalt the good which is *Communicative* and depresse the good which is priuate and particuler as the Holy faith."[8] He hastens to add, however, that with respect to the "pleasure and dignitie of a mans selfe," the contemplative life has the "preemynence" (AL 2, 136–7, III: 421). Earlier in the same section he had moreover observed that there is "fourmed in euery thing a double Nature of Good; the one, as euery thing is, a Totall or substantiue in it selfe; the other, as it is a parte or Member of a greater

Bodye" (AL 2, 136, III: 420). But the problem inherent in a twofoldness of good has largely been neglected or forgotten, as is indicated by the fact that reception history has reduced Bacon's thought to his exaltation of the communicative good.[9]

We are running ahead of things though. Nowhere did Bacon more beautifully describe his contemplative or philosophic nature than in the unpublished fragment *De Interpretatione Naturae Proemium*, probably written around 1603. Taking a retrospective view of his life up to this point, Bacon relates that he "discovered [deprehendi]" that above all other things he was "made for the contemplation of truth." Having a mind "versatile enough to acknowledge the similitude of things, which is the highest thing," a mind "fixed and steady enough to observe and distinguish the subtle differences between things," and being gifted with a "desire to seek, patience to doubt, pleasure in meditating, slowness in asserting, a facility to return to himself, a solicitude to dispose and to comprehend," and, moreover, being a man who "neither affects what is new, nor admires what is old, and who hates every kind of imposture," he judged that his "nature had a kind of familiarity and relationship with truth."

"But [attamen]" his birth and education had made him "familiar with civil things," and, young as he was, "opinions" would sometimes make him "waver." Besides, his fatherland had "some special claims" on him, that "did not equally pertain to the other parts of the world."

Having finished the description of his early development as a movement that distanced him from the inward tendency or necessity of his nature, Bacon points out that he hoped that if he obtained "any place of honour in the commonwealth," he would be aided by a larger amount of "genius and industry" to achieve his objective of "kindling a light in nature." He therefore applied himself to acquiring the "civil arts" as well as to commending himself "modestly and in all sincerity [salva ingenuitate]" to such friends as were influential. Moreover, as "these things, of whatever kind they are, do not penetrate beyond the condition and cultivation of this mortal life," he "truly [vero] hoped" that if he came to hold civil office, he might "procure something good for the health of men's souls," the "condition of religion" at that time being "not very prosperous."

"But [Sed]" his "studiousness was taken for ambition," as we learn from Bacon's description of the birth of his political self-consciousness. He therefore "wholly [totum] betook" himself to his work *De Interpretatione Naturae*.

With regard to the "injuries of time," Bacon states that he is "almost safe from them," and the "injuries of men" do "not concern" him. For, he explains, if someone accuses him of seeking to be "wise overmuch," he "simply" responds that "modesty has its place in civil things," but that "truth is at home in contemplations."[10]

This politically ignorant and therefore fundamentally inadequate apology of wisdom must have been the most profound reason behind Bacon's decision not to publish the *Proemium*. And it was with good reason that Bacon allowed his political self-consciousness to mature, since wisdom that dissociates itself from moderation cannot remain truly courageous—in deed, that is, because moderation is not a virtue of thought.[11] In other words, wisdom must cause its madness to be looked upon as health in order to prevent the health of the political order from degenerating into madness. Sometime between 1603 and 1605, Francis Bacon became a political philosopher.

Manifestation

The *Advancement of Learning* is the first work published by the political philosopher Francis Bacon, since it is the first of Bacon's works that in an elaborate manner employs exoteric–esoteric writing as a device for harmonizing civil moderation and contemplative courage. While Bacon induces the attentive reader to think through his account of having become a political philosopher, his surface treatment of the problem of the conflict between learning and the political order makes the problem appear to be nonexistent.

In the Epistle Dedicatory to King James, Bacon displays his awareness of the sovereign under whom he writes. In the first sentence of the book he points out to his king that there were "vnder the Lawe" both "dayly Sacrifices, and free will Offerings; The one proceeding vpon ordinarie obseruance; The other vppon a deuout cheerefulnesse: In like manner," he continues, "there belongeth to Kings from their Seruants, both Tribute of dutie," referring to the "businesse" of the king's "Crowne and State," and "presents of affection," referring to the king's "indiuiduall person" (AL 1.1, 3, III: 261). But although the Law that God revealed to Moses on Mount Sinai distinguished between daily offerings (Num. 28: 2) and free-will offerings (Lev. 22: 18), both kinds of offerings were made in the public context of God's interaction with His chosen people. God's Son, on the other hand, distinguishes between what belongs to Caesar and what belongs to God (Matt. 22: 21, Mark 12: 17, Luke 20: 25). Bacon could not have demonstrated the maturity of his political self-consciousness more effectively than by blending the public and the private in the context of discussing God's Law in the very first sentence of his book on human learning.[12]

Our author continues his typology of the sovereign under whose Law he writes by problematizing the attributes of his king: "[A]s the Scripture sayth of the wisest King: *That his heart was as the sands of the Sea,* which though it be one of the largest bodies, yet it consisteth of the smallest & finest portions: So hath God giuen your Majestie a composition of vnderstanding admirable, being able to compasse & comprehend the greatest matters, & neuerthelesse to touch and apprehend the least; whereas it should seem an impossibility in Nature, for the same Instrument to make it selfe fit for great and small workes" (AL 1.2, 4, III: 262). But it is in fact an "impossibility in nature" that the greatest King, who employs the "great" works of nature as an "instrument" in the service of the "greatest" matter of having lesser kings behold Him, and who gave King James a *heart* for the purpose of "compassing and comprehending" this greatest matter, at the same time gave him an *understanding* "fit for" the "small" works of nature.[13] Bacon therefore assures his king that he does not behold him "with the inquisitiue eye of presumption, to discouer" the latter's heart, "which the Scripture telleth" him "is inscrutable" (AL 1.2, 3, III: 261; cf. Prov. 25: 3).

With regard to the king's "gift of speech," Bacon calls to mind the eloquence of Augustus Caesar, as reported by Tacitus, thereby implicitly referring to the contrast between Augustus Caesar and Nero, the latter of whom was the "first emperor to be in need of another person's eloquence."[14] But because Bacon was in need of the protection of the Christian king under whose laws he wrote, he could only draw our attention to the potential "Neronian" character of the king's borrowed speech by dissimulating it in an "Augustean" manner.

Bacon goes on to observe that it is a "positiue and measured truth" that there "hath not beene since Christs time any King or temporall Monarch which hath ben so learned in all literature & erudition, diuine & humane" (AL 1.2, 4, III: 263).[15] But "measuring" this truth by its historical clause is largely left to us, although Bacon makes the truth of Christian kingship as such dependent on a threefold structural clause. For he states that his "Maiestie standeth inuested of that triplicitie, which in great veneration was ascribed to the ancient *Hermes*; the power and fortune of a King; the knowledge and illumination of a Priest; and the learning and vniuersalitie of a Philosopher" (AL 1.2, 5, III: 263). But since we do not believe it a coincidence that Bacon goes on to observe that human learning has received "discredites and disgraces" from "ignorance . . . disguised" in a threefold way (AL 1.4, 5, III: 264), we believe his ensuing, threefold apology of learning to be at the same time an attempt to reveal the "ignorance" underlying the ascription of the attributes of Hermes Trismegistus to the representative of the Trinitarian God.

Politicians

The discredits that learning has received from the "seueritie and arrogancie of Politiques" correspond with the first attribute of the "modern Hermes," the "power and fortune of a King." Bacon presents Cato and Virgil as spokesmen of Roman political virtue and as potential enemies of the learning of the Greeks, before wrongly introducing the Greek Anytus as the person to have accused Socrates of the same thing Cato and Virgil apparently accused the Greeks of (AL 1.7, 9–10, III: 268–9).[16]

A few paragraphs down Bacon returns to his examples of Cato, Virgil, and the accusers of Socrates. He points out that Cato was punished for his "blasphemie" against learning, for at an advanced age he was "taken with an extreame desire to goe to Schoole againe, and to learne the Greeke tongue, to the end to peruse the Greeke Authors." But Plutarch, who was Bacon's source, tells us that Cato thought the most honorable old age was old age busied in public affairs.[17]

With regard to Virgil, Bacon remarks that the Romans never ascended to the height of empire until the time they had ascended to the height of other arts. He illustrates this point by saying that in the time of the first two Caesars, which "had the art of gouernement in greatest perfection," there lived the best poet, the best historiographer, the best antiquary, and the best, or second best orator "that to the memorie of man are knowne." But it can hardly have been due to a temporary memory defect that Bacon fails to mention that the only man to have two natures was born in the time of the first two Caesars. We are therefore tempted to believe that our author did not consider it prudent to mention him in the context of an argument on the art of empire, if only because his empire was not of this world.

Be that as it may, Bacon continues his political apology of learning by referring to the accusation of the best philosopher, Socrates, who did not live in the time of the first two Caesars. He emphasizes that "the time must be remembred, when [Socrates] was prosecuted," which was under the Thirty Tyrants (404–403 BCE), "the most base, bloudy, and enuious persons that haue gouerned, which reuolution of State was no sooner ouer, but *Socrates*, whom they had made a person criminall, was made a person heroycall, and his memorie accumulate with honors diuine and humane" (AL 1.15, 14, III: 273–4). However, Socrates was actually prosecuted (399

BCE) by the same demos that would later embellish his memory with human and divine honors.[18]

This observation confirms what we had already been led to suspect: that Bacon's examples all allude to the general tension between political and philosophical virtue, a tension that is all the more affirmed by seeming to be denied after having been affirmed. But our author also transmits a more particular message, which can only be unveiled if we consider the implications of his blunder pointing to the ultimate similitude between tyrant and demos. For certain "Politiques" discredit learning on account of its ability to threaten the "fortune and power" of the greatest King by intimating that He was only able to secure His reign by being so merciful as to elevate the "memory" of the person whom He had allowed to descend to the status of a "criminal."[19] It therefore seems to be principally because "leisure and privateness" are conducive to human thought that it is in the context of his apology to the "Politiques" that Bacon defends learning against the conceit that it "should dispose men to leasure and priuatenesse." Bacon justifies the business of the learned by saying that "onely learned men loue businesse as an action according to nature," provided that it concerns business that can "hold or detaine their minde" (AL 1.11, 12, III: 272). But this can only be business that concerns learning.

Theologians

Although the "knowledge and illumination of a Priest" is the central attribute of the "modern Hermes," Bacon discusses the corresponding "disgrace" that learning receives from "the zeale and iealousie of Diuines" before discussing the other "discredits," thereby alluding to the fact that the divines represent the paramount disgrace of human learning.

Bacon hears the divines say that knowledge is "to be accepted with great limitation and caution," and that "thaspiring to ouermuch knowledge, was the originall temptation and sinne, whereupon ensued the fal of Man" (AL 1.5, 5–6, III: 264). But the Englishman discredits this "opinion" as erroneous and ignorant, for "it may well appeare" that these men "doe not obserue or consider" that "it was not the pure knowledg of nature and vniuersality, a knowledge by the light whereof man did giue names vnto other creatures in Paradise, as they were brought before him, according vnto their proprieties, which gaue occasion to the fall; but [that] it was the proud knowledge of good and euill, with an intent in man to giue law vnto himselfe, and to depend no more vpon Gods commaundements, which was the fourme of the temptation" (AL 1.6, 6, III: 264–5). In the context of his treatment of the "attributes and acts of God" in the central section of the first book, Bacon puts it more penetratingly by saying that the moral knowledge of good and evil induced the Fall, "wherein the supposition was, that Gods commaundements or prohibitions were not the origins of good and euill, but that they had other beginnings which man aspired to know, to the end, to make a totall defection from God, and to depend wholy vpon himselfe" (AL 1.55, 34, III: 296–7). Can a philosopher demonstrate his knowledge of good and evil more cogently than by pronouncing its supposition?

According to the second biblical account of man's beginnings, God's commandment not to eat of the tree of the knowledge of good and evil is the origin of good and evil (cf. Gen. 2: 16–7). In other words, the biblical account of the genealogy of evil traces the origin of morality back to God's prohibition on eating of the tree of the knowledge of good and evil.[20] But the possibility of morality presupposes man's freedom to act morally or immorally, or, to state it differently, man's freedom to act piously or to sin.[21] Therefore, man's freely chosen obedience to God's commandment, being his free choice of the principle of obedience, became coeval with good, and his freely chosen transgression of God's commandment, being his free choice of the principle of disobedience, became coeval with evil. Man freely chose the principle of disobedience by eating of the tree of the knowledge of good and evil. This implies that the knowledge of good and evil that is coeval with the principle of disobedience is diametrically opposed to the knowledge of good and evil that is coeval with the principle of obedience. The major theologians agreed that God's prohibition was essentially an exercise in obedience for man, and that the transgression of God's commandment was an act of disobedience, rooted in the pride of an unbelief in the obedience to God as the source of man's good, and therefore ultimately rooted in an unbelief in God.[22] Since philosophy is coeval with the discovery of nature, it is coeval with the discovery of the distinction between what is natural and what is not natural.[23] Philosophy considers the right way of life to be the life of unreserved and unassisted inquiry into what is good by nature, albeit that pronouncing this consideration already tends toward making the philosophic life good by convention. But is not the further conventionalization of the philosophic life the most profound way of bringing to life its natural difference from the philosophic lifestyle? And is not the inflationary use of the notion "way of life" an additional reason for employing it in order to prevent the philosophic life from losing its vitality?

We are drifting away from our subject. The life of obedience is the most powerful non-natural, because supranatural, answer to the question as to the right way of life. But since the life of obedience is good by the authority or Law of the originator of good and evil, philosophy cannot recognize it as good by nature.[24] Philosophy must therefore suspend its judgment on the beginnings of good and evil, which implies that in this respect philosophy is necessarily disobedient. But as life itself cannot be suspended, the judgment on the question as to the right way of life cannot be suspended. Consequently, the even more radical conclusion must be drawn that the anticipatory decision in favor of the philosophic life is a decision in favor of the principle of disobedience, and therefore necessarily a decision against God's commandment and prohibition as the origin of good and evil.[25] For this reason, from the moral point of view of the pious man who lives the life of obedience, the philosophic life is a life coeval with the principle of disobedience or with sin. But philosophy cannot avoid being sinful or disobedient in the most impious sense possible, because it cannot avoid inquiring into the nature of the being who reprobates the philosophic life as a life of sin, and who demands of the philosopher that he live a life of pious obedience. The quest for the essence of God, *quid sit deus*,[26] is coeval with philosophy itself, as the quest for God's non-natural, because supranatural, "nature" necessarily falls within the scope of philosophy's inquiry into what distinguishes the natural from the non-natural, or, to put it differently, of philosophy's inquiry into the origin of the first things. But the Bible reprobates the question as to the nature of God as an impious question, as it reprobates the life of inquiry as a life of sin.[27] Taking this most powerful reprobation of the philosophic life and therefore of the philosopher's activity radically seriously, philosophy must conduct an inquiry into the life of inquiry by reflecting on its own legitimacy and necessity. If the philosophic life is a life of sin, philosophy has no legitimacy. If philosophy is not the one thing needful (cf. Luke 10: 41–2), it is not necessary for the good life, no more than the philosophic life is good for whom philosophy is not the one thing needful.[28] In more general terms, philosophy must turn to political philosophy in order to effectively defend and rationally justify itself against conflicting claims and opposing answers to the question as to the right way of life. For it is only by reflecting on itself "politically," and elevating to the status of political things of philosophic interest all things constituting and defining the sphere in which its self-reflective undertaking takes place, that philosophy can inaugurate the turn back to itself.[29]

These points, "being set downe and strongly planted," put Bacon's apologetic distinction between the "proud knowledge of good and euill" and the "pure knowledg of nature and vniuersality" in the proper light. For since paradisiac knowledge of nature was only "pure" because it was not yet "proud," it was not yet true knowledge, which is always proud because it is never pure.[30]

But let us look somewhat more closely into the nature of Edenic man's "purity." Although man had been created in the image of God, he was to be morally responsible for the neediness that was to ensue from his downfall. God therefore formed him of the dust of the ground and placed him in a garden providing him with all things needful (cf. Gen. 1: 26 ff. with 2: 7 and 2: 19–20).[31] That Bacon understood God's reason for being generous is already indicated by his observation that "[a]fter the Creation was finished, it is sette downe vnto vs, that man was placed in the Garden to worke therein, which worke so appointed to him, *could be no other* than worke of contemplation, that is, when the end of worke is but for exercise and experiment, not for necessitie" (AL 1.55, 34, III: 296, emphasis mine).[32] But that our author read the Bible judiciously is not proved until we read his account of "the first euent or occurrence after the fall of Man."

For Bacon points out that "wee see . . . an Image of the two Estates, the Contemplatiue state, and the actiue state, figured in the two persons of *Abell* and *Cain*, and in the two simplest and most primitiue Trades of life: that of the Shepheard (who by reason of his leasure, rest in a place, and liuing in view of heauen, is a liuely Image of a contemplatiue life) and that of the husbandman; where we see againe, the fauour and election of God went to the Shepheard, and not to the tiller of the ground" (AL 1.56, 34, III: 297). But man had lost his "contemplative" privilege by becoming disobedient. Nevertheless, God's favor went to the man embodying the "contemplatiue life." Although this paradox can partly be explained by the fact that Bacon "apologized" for the contemplative life by making it seem inactive, it can only be solved if Abel was favored on account of the inactivity of his contemplations. For whereas the "tiller of the ground" depends wholly on himself, the "shepherd" counts his sheep like Adam named the "cattle, the fowl of the air, and the beasts of the field," which had been given to him for naming. The biblical account of God's justice in favoring Abel is in conformity with Bacon's interpretation,[33] as it argues that God did not regard the offer of Cain because it was by faith that the offer of Abel was "more excellent" (cf. Gen. 4: 3–6 with Hebr. 11: 4).[34] This was the second time that God disfavored a man depending wholly on himself. But it was the first time that He had a "favorite."

Philosophers

The third sort of "discredite, or diminution of credite" that learning has received corresponds with the third attribute of the "modern Hermes," the "learning and vniuersalitie of a Philosopher." But whereas the first two parts of Bacon's threefold apology of human learning concerned a defence of learning against its two strongest adversaries, the third and final part consists in a defense of the learned against themselves, as it admonishes the learned to correct their own errors and imperfections. This change of perspective underscores the commonality between kingship and priesthood as it almost silently removes philosophy from the trinity of attributes ascribed to the representative of the Trinitarian God. The consequences of this removal for the remaining attributes were already foretold when Bacon assumed ignorance to be the underlying cause of the one's threefoldness. For whereas philosophy is knowledge of ignorance, ignorance is unknown to itself.

Bacon's central remark on the "fortune" of learned men concerns the "priuatenesse or obscurenesse (as it may be in vulgar estimation accounted)" of "life of contemplatiue men." Bacon warns that "learned men forgotten in States, and not liuing in the eyes of men, are like the Images of *Cassius* and *Brutus* in the funerall of *Iunia*; of which not being represented, as many others were *Tacitus* sayth, *Eo ipso prefulgebant, quod non visebantur"* (AL 1.17, 16, III: 276). But since what Tacitus actually said was: "praefulgebant . . . eo ipso quod *effigieseorum* non visebantur [they shone forth by virtue of the very fact that *their images* were not to be seen],"[35] we suspect that what Bacon actually suggests is that if learned men have themselves represented by a vulgar "image" of themselves, they can "shine forth" in the very sense that although they are not seen, they do seem to be seen, and they themselves see everything. In any case, it is the fortune of our author's own image that makes us wonder whether he was sufficiently aware of the fact that vulgar images have a tendency to "ossify" if they do not invite "maintenance" by nonvulgar hands.

A few paragraphs down, Bacon criticizes the "manners" of the learned by pointing out that "because the times they read of, are commonly better than the times they liue in; and the duties taught, better than the duties practised," the learned "contend sometimes too farre, to bring things to perfection; and to reduce the corruption of manners, to honestie of precepts, or examples of too great height" (AL 1.20, 17, III: 277–8). But who embodies "exemplary" manners in times unreceptive to exemplary manners? Philosophers who have written on laws, have made "imaginary Lawes for imaginary common-wealths, & their discourses are as the Stars, which giue little light because they are so high" (AL 2, 180, III: 475). Although the precepts of these philosophers are excellent, they are far from useful to a commonwealth in which the rule of philosophy has become increasingly imaginary. What is more, the philosopher who said that "*Then should people and estates be happie, when either Kings were Philosophers, or Philosophers Kings,*" might be "thought partiall to his owne profession" (AL 1.68, 39, III: 302).

He was. But philosophers in modern times could not afford the kind of "partiality" to philosophy that had caused one of their ancient predecessors to shed light on the problems of law and the coincidence of philosophy and political power by making his discourses "as high as the stars,"[36] seeing that they lived in a real commonwealth ruled by a real Law that appealed to a real and very powerful imagination. In fact, one may even raise the question whether the partiality of the classical political philosophers was not a supportive cause of the seeming impartiality of their modern counterparts, if only in light of the imaginative appeal of "imaginary laws."

Be that as it may, since lawyers write on "receiued Law, & not what ought to be Law," their discourses on law shed little light on the problem of the origin of law. Therefore, the "wisedome of a Lawmaker is one & of a Lawyer . . . another." However, the exemplary wisdom of a lawmaker "consisteth not onely in a platforme of Iustice," but also in the "application thereof; taking into consideration by what meanes Lawes may be made certaine, and what are the causes & remedies of the doubtfulnesse and *incertaintie* of Law, by what meanes Lawes may be made apt and easie to be executed, and what are the impediments, and remedies in the *execution* of lawes" and the like. Seeing that the book of their times counsels the learned to follow the example of wise lawmakers by writing on laws that are useful, certain, and effective, Bacon advises them to go to the effectual truth of the matter by letting themselves be guided by "civil men." For civil men best understand "what constitutes human society, the welfare of the people, natural equity, the customs of nations, and the different forms of commonwealths." They can therefore "determine laws by the principles and precepts of both natural equity and policy" (AL 2, 180, III: 475; DA 8.3, I: 803).

It seems to be for this reason that Bacon emphasizes that "we are much beholden to the civil man Machiavelli and other writers of this kind, who openly and without dissimulation reveal [proferunt] what men do, and not what men ought to do,"[37] although Bacon's dissimulation of openness in revealing Machiavelli's doing invites us to wonder whether Machiavelli may also have dissimulated when he "openly revealed" what he was doing. But although this question bears heavily on the question whether a lawmaker can be a philosopher, answering it would require thoughts unthought in the course of writing this book. We therefore leave it at saying that a lawmaker can only be a philosopher if it is not through lawmaking that he performs the "practice of dying and being dead."

Bacon goes on to explain the importance of revealing what men do by saying that "if one does not have a deep and thorough knowledge of the nature of evil itself, virtue lies unprotected and undefended." We are inclined to believe that it is due to the fact that virtue is treated in the context of duty and not of benevolence that Bacon does not discuss the case in which a lawmaker could add "depth and thoroughness" to the knowledge of the nature of evil by making a wholly new law in order to observe what men do when they ought to do what the wholly new law commands. But however that may be, Bacon emphasizes that "treating with integrity and sincerity the argument on the nature of evil is one of the best fortifications for virtue and honesty that can be planted." Since a "good and honest man cannot correct and amend those who are evil and wicked without first exploring all the hiding-places and depths of evil," it is for their own good that good men who are learned become acquainted with the virtue of "serpentine prudence," which answers to the danger of being like sheep among wolves that are prepared to tear to pieces the "dove" of learning (DA 7.2, I: 729–30; cf. Matt. 10: 14 ff.).

Immediately after warning the learned not to neglect the times they live in, Bacon criticizes them for the fault, "much of this kind," of having "esteemed the preseruation, good, and honor of their Countreys or Maisters before their owne fortunes or safeties." Learning, more specifically, learning in conjunction with the times, "endueth mens mindes with a true sense of the frailtie of their persons, the casualtie of their fortunes, and the dignitie of their soule and vocation; so that it is impossible for them to esteeme that any greatnesse of their owne fortune can bee, a true and worthy end of their being and ordainment." For this reason, Bacon counsels the learned to be "desirous to giue their account to God, and . . . their Maisters vnder God (as Kinges and the States that they serue) in these words; *Ecce tibi lucrifeci*, and not *Ecce mihi lucrifeci*"; words that allude to the story of the servant who had been entrusted with five talents, and who gained another five by trading with the five talents he had been entrusted with. Because he had been faithful over few things, he was made ruler over many things, and was allowed "entry into the joy of his lord." The servant who had been entrusted with one talent, on the other hand, hid it in the earth because he knew his lord was a hard man, who "reaped where he had not sown, and who gathered where he had not strawed." His lord blamed him for not having put the money to the exchangers because this had caused the servant to be unable to return the money to his lord with usury when the latter came back. The servant was cast into "outer darkness," and his talent was given to the servant with ten talents. For, as the moral of the story goes, "unto everyone who has shall be given, and he shall have abundance. But from him who has not shall be taken away even that which he has" (Matt. 25: 14–30)—which brings us back to Bacon, whose allusion to the parable of the talents seems to serve the purpose of warning the learned to appear to be "faithfully" applying their "talent" for gaining for their "lord," because knowledge of their lord's nature must make them aware of the danger of having their talent and everything they have taken from them, should it turn out that their talent has never been employed for giving their lord account for it.[38] But as the man whom God had but given a "poor talent, or half talent" (AHW EpD, VII: 13) could not have given this warning prudently except by dissimulating it, he censures those who "referre all thinges to themselues," referring to them as the "corrupter sort of meere Politiques" (AL 1.21, 18, III: 278–9).

In his *Essay* on how to prudently refer all things to oneself, Bacon counsels the learned to "Divide with reason betweene *Selfe-love*, and *Society*." He warns that "*Wisedome for a Mans selfe*, is in many Branches thereof, a depraved Thing," because those who are "*Sui Amantes, sine Rivali*, are many times unfortunate. And whereas they have all their time sacrificed to *Themselves*, they become in the end *themselves* Sacrifices to the Inconstancy of Fortune; whose Wings they thought, by their *Self-Wisedome* to have Pinnioned" (E XXIII "Of Wisedome for a Mans selfe," 73 ff., VI: 431 ff.). In other words, the learned can only be "wise for themselves" if they do not "divide" between "self-love" and "society" to the extent of neglecting their "rivals in fortune." We therefore need not wonder that at the end of the section on the third attribute of the modern Hermes Bacon emphasizes that the "morigeration and application of learned men to men in fortune" is not only not to be taxed or condemned, but is even to be recommended. For "stooping to points of necessitie and conuenience cannot bee disallowed: for though they may haue some outward basenesse, yet in a Iudgment truely made, they are to bee accounted submissions to the occasion, and not to the person" (AL 1.25, 20–1, III: 281–2).

But the learned often misjudge the necessity of becoming "outwardly base," and therefore fail to apply themselves to particular persons. Bacon observes that this want of "exact application" arises because "the largenesse of their minde can hardly confine it selfe to dwell in the exquisite obseruation or examination of the nature and customes of one person: for it is a speech for a Louer, & not for a wise man: *Satis magnum alter alteri Theatrum sumus.*"[39] Therefore, "he that cannot contract the sight of his minde, aswell as disperse and dilate it, wanteth a great facultie" (AL 1.22, 18–9, III: 276–7). It needs emphasizing, though, that it is ultimately for the sake of the learned man's love that Bacon counsels him to be so wise as to apply himself to the "observation" and "examination" of the nature of a "person" he does not love.[40]

At the beginning of the second book of the *Advancement of Learning*, the learned Bacon applies this counsel to his own person by stating that he is "not ignorant, what it is" that he does "now mooue and attempt, nor insensible" of his own "weakenes, to susteine" his "purpose," but that he hopes that if his "extreame loue of learning" carries him "too farre," he "may obtaine the excuse of affection; for that *It is not granted to man to loue, and to bee wise*" (AL 2.16, 61, III: 328). It is from Plutarch that we learn that these words were once used as an excuse by the Spartan king Agesilaus. The Greek biographer relates that the philosopher Hieronymus observed that Agesilaus "sometimes paid more attention to the necessity of his affairs than to his friend." Hieronymus illustrated this observation with the anecdote that upon the sudden and disorderly removal of his camp Agesilaus left his sick friend behind him, who called loudly after him and implored his help, whereupon Agesilaus turned his back and said that "it is hard both to love and to be wise."[41] Because Bacon is not "ignorant" of what it is that he attempts to set in motion, and because he is not "insensible" of his own "weakness," he knows that he must be so wise as to submit to the necessity of becoming "outwardly base" by obtaining the "excuse of affection" for having been "carried further" by his "love of learning" than is considered wise by those whom he left behind and who do not love his love of learning.

It is in acting on this knowledge that at the end of the first book of the *Advancement of Learning* the political philosopher Francis Bacon "contracts the sight" of his large mind by being so wise as not to let the exposition of his love of learning carry him further than is prudent in view of "popular judgment," whose ineradicability he could hardly have demonstrated more effectively than by making a popular judgment out of the judgment of popular judgment. For he emphasizes that he does "not pretend," and that he knows "it will be impossible" to "reuerse the iudgement . . . of those which preferre Custome and Habite before all excellencie; or of a number of the like popular Iudgements. For these things must continue, as they haue beene" (AL 1.103, 53–4, III: 319).[42]

Education

De Sapientia Veterum is Bacon's political-philosophical work second in sequence but first in importance. This is not so much because parabolic poetry constitutes a modern philosopher's answer to an ancient quarrel, or because a modern quarrel appears as less of a quarrel in light of a modern's appeal to ancient wisdom; neither is it primarily because the sense reveals itself as the measure of man by being made the measure of things (cf. NO 1.41, I: 163–4), or because by exposing philosophy to what it is in danger of becoming parables prevent it from not becoming what it cannot help being[43]; it is first and foremost because *De Sapientia Veterum* is the first and only one of Bacon's theologico-political treatises in which philosophy consummates the political turn and turns back to itself.

That the philosopher comes home in a state of homelessness is visualized by the fact that he finds himself at open seas at the end of the book; that his new home for the rest of his life is not a place of inactivity is pictured by the fact that the book ends with philosophic activity; and that philosophic activity is inseparable from the good of the philosopher is revealed by the philosopher's confession at the end of the book that the meditations he "sprinkled with pleasure" constituted his actual pleasure (cf. DSV Pr., VI: 625 with DSV XXXI "Sirenes," VI: 686).

De Sapientia Veterum articulates the different moments of philosophy's political journey and welds them into a whole; a whole whose internal dynamic is safeguarded by the disorderly treatment of its parts. The experience of reading this exemplary work of political philosophy can best be compared with that of a seafarer on his way home, who is tossed to and fro by the waves, and carried about by the winds, before being thrown onto unknown, unexplored, and undesired islands. At first sight, these islands seem hardly connected with the book's mainland, which is the philosopher's political education. But as it is only at the end of the book, after having returned from his excursions to natural philosophy, theology, and active politics, that the philosopher gives a demonstration of self-knowledge, the reader is forced to draw the conclusion that it was the philosopher's quest for self-knowledge that ordered the parts and set the whole in motion at the beginning. And it is at the very latest after having drawn this conclusion at the end that the reader is thrown back to the beginning, where he can set out for a second sailing.

Since we are in sympathy with Bacon's preference for the order of apparent disorder, we will not treat the separate fables in their sequential order, apart from those fables whose internal order appears from their sequential treatment. For since the reasonableness of philosophy's political defence presupposes its rational justification, it is only after having treated the fables touching philosophy's rational justification in the third and central chapter that we can reasonably treat the fables concerning philosophy's political defense in the fourth chapter. And since philosophy's rational justification presupposes philosophy's turn to political philosophy, now is the time and place to focus our attention on Bacon's fabulous account of this turn.

As a general procedure, we will give a literal and uncommented summary of the narrative and teaching of Bacon's fables before interpreting Bacon's teaching. Since we believe Bacon's true teaching arises from a careful comparison of the teaching and narratives of his fables, this procedure has the advantage that it enables the careful reader to judge the truth of our interpretation of Bacon's teaching by interpreting Bacon's teaching.

Action and Contemplation

De Sapientia Veterum is dedicated to the University of Cambridge. The work proper is preceded by two Epistles Dedicatory, which are bound together by the fact that the University of Cambridge figures in both of them. But although the two epistles are an image of the "conjunctive two" of action and contemplation if they are read in outer conjunction, if read in inner conjunction they visualize action and contemplation as a "disjunctive two."

The first Epistle Dedicatory is addressed to Sir Robert Cecil, Earl of Salisbury, Lord High Treasurer of England, and Chancellor of the University of Cambridge. Bacon points out that things dedicated to the University of Cambridge accrue to the Earl, and that the Earl is personally entitled to everything that proceeds from Bacon. However, it "remains to be seen" if these things, which Bacon owes the Earl, are also "worthy" of the Earl. With regard to Bacon's "genius," the thing that is "worth least," Bacon says that the Earl's propensity toward him will "not let it get in the way." For the rest, there will be nothing that will "disgrace [dedecori]" the Earl. Although he does not want to take up an unnecessary amount of the Earl's time, as the Earl has "so many great things" on his shoulders,[44] Bacon uses much time and space in this letter to a political man to elevate the worth of philosophy. Having stated that philosophy is the "matter" of the work, Bacon calls philosophy the "second grace [decus] of life and the human soul," before emphasizing that he estimates philosophy "of all things, after religion," as the thing "most grave, and most worthy of human nature." He adds that politics "emanates from the same fountain" as philosophy, and is a "great part" of it. Bacon tells the Earl that it was his aim to discuss the "difficulties of life and the arcana of science," passing over "things obvious, obsolete, and commonplace." Although, because the work is dedicated to the Earl, he strives to add "some dignity" to it, Bacon is in danger of "transgressing the bounds of modesty," as the work is undertaken by himself. But he asks the Earl to accept the work as a "pledge of affection, observance, and devotion," and to grant it the protection of his name (DSV EpD1, VI: 619).

In the second Epistle Dedicatory, addressed to the University of Cambridge, Bacon resumes the thread of the argument of the first Epistle by elevating philosophy once more. Since without philosophy he would "not be pleased to live," he holds in great honor the University, from whom these "protections and solaces of life" have flown to him. For this reason, Bacon professes to "owe" both himself and what belongs to him to the university. But Bacon goes on to problematize philosophy by pointing out that among the "infinite number of footprints" that have gone forth from the university one "rarely" sees any footprints pointing back to her.

This metaphor reminds us of Aesop's fable of the lion's cave. An aging lion could no longer obtain his food by force and therefore resorted to trickery. He retired to a cave and pretended to be ill, but he ate every animal that came to his cave to visit him. Eventually, a fox found out the lion's trickery. He visited the lion, but remained at a safe distance outside the cave. The lion invited him in, but the fox replied that he would have come in, had he not seen a lot of footprints pointing to and disappearing into the lion's cave but none coming out again.[45]

Returning to the second Epistle, we find Bacon saying that he is of the opinion that when contemplations are "transferred into active life," they "acquire some new grace [decoris] and vigour." Having an "abundance of matter" at their disposal, they "root perhaps more deeply, or in any case grow taller and more foliaceous." But Bacon judges that the university herself does not know "how widely" her own studies extend, and to "how many things" they pertain (DSV EpD2, VI: 621).

We may divide the observations arising from a conjunctive reading of the two Epistles Dedicatory into two parts that correspond with the outer division of the two epistles. In the first Epistle, Bacon introduces philosophy into the sphere of the political. By elevating philosophy in a letter to a political man, and by addressing his epistle to the Earl of Salisbury, Bacon alludes to the fact that a political man like the Earl of Salisbury impinges on philosophy, for which reason philosophy must elevate the nature of the political man to the status of a political subject of philosophic interest. As a reader, the Earl is not "worthy" of the "genius" of Bacon, which genius is reflected in the book that is dedicated to the Earl. Stronger even, the nobleman is a potential danger to Bacon, insofar as he might feel "disgraced" by the fact that in the book that he protects with his name Bacon transgresses the "bounds of modesty." Yet at the same time it is by protecting Bacon's book with his name that the Earl protects Bacon against other potential dangers as the Earl's protection allows Bacon a great latitude to indulge in his madness, as long as he is modest enough not to disgrace the dignity of the Earl. The art of politics will therefore begin to emanate from the "fountain" of philosophy having turned to political philosophy as soon as philosophy realizes that, because the "difficulties of life and the arcana of science" are of necessity interrelated, not even the political things that are seemingly "obvious, obsolete, and commonplace" can be passed over without further inquiry.

In the second epistle, Bacon introduces the political into the sphere of philosophy. By highlighting philosophy's vulnerability in a letter to philosophy's protective institution, and by addressing his epistle to the University of Cambridge, Bacon alludes to the fact that, for the sake of protecting philosophy, a protective institution like the university must become political by making its relationship with political life reciprocal. More generally, philosophy must turn to political philosophy in order not to be torn to pieces in the "lion's cave" of the political, which is an undesirable terminal for philosophy not understood as an instrument of politics. But more substantially, it is only by turning active, that is, by turning to political philosophy, that contemplations can "root" deeply enough for their soil to be judged philosophically fertile.

Submission

The subtitle of the first fable is "Parrhesia," meaning outspokenness or frankness of speech. Bacon could hardly have given a more acute subtitle to the first fable of a book that must employ political speech as a consequence of its philosophic author having turned political. The fable itself begins with the relationship between the human and the divine. For Cassandra was loved by Apollo. By means of "various artifices she tried to elude his desires, while keeping his hopes alive until she had extorted the gift of divination from him." As soon as she had received it, she "openly rejected his suit." Although Apollo could not demand back what he had "blindly" bestowed on Cassandra, he "burned with revenge," and added to his gift the penalty that although Cassandra "would always predict truth, nobody would believe her." In other words, although Cassandra's "prophecies were true, nobody had faith in them." This proved to be the case "ever after, even with regard to the destruction of her country, about which she had given many warnings, but with regard to which she had not been able to get anybody to listen to her or believe her."

Bacon points out that the fable seems to have been devised with a view to treating the subject of "unseasonable and useless liberty in giving counsel and admonition." Those who are of a "froward and rough disposition" are "unwilling to submit to Apollo, the God of Harmony." Submitting to Apollo would mean "learning thoroughly to and how to observe time and measure of things, sharp and deep noises in discourse, differences between the more experienced and the more vulgar ear," and "learning to and how to observe times when to speak and when to be silent." Although the counsels of those who do not submit to Apollo are "sound and good," they "hardly procure anything with their advice and impetuosity," and they are "not effective in the management of things either." They rather "hasten the destruction of those upon whom they press themselves," and it is "not until after a calamity and its consequences that they are celebrated as prophets and men of far foresight."

Bacon illustrates the foregoing by the example of Marcus Cato, who "foresaw the ruin of his country as if from a watch tower, and foretold it as if through an oracle. Yet all the while he procured nothing, but instead harmed his country and accelerated the disasters that were to strike it." According to Bacon this was prudently observed and elegantly described by Cicero, who pointed out that Cato spoke "as if he were in the Republic of Plato and not in the faeces of Romulus"[46] (DSV I "Cassandra," VI: 629–30). In the *Advancement of Learning,* Bacon had already used the exact same quotation of the Roman statesman in the context of his criticism of those learned men who were unable to adapt their manners to the "faeces" of the times in which the republic founded in speech by the Greek philosopher had become even more imaginary than in the times of the city founded in deed by Romulus. It was also in the context of his criticism of the unpolitical nature of the learned of modern times that the modern Bacon made the ancient Cicero criticize the ancient Plato himself for having been unwilling to transgress the protective boundaries of his own imaginary republic, a republic of which he had been made an inhabitant before being criticized for inhabiting it. Plato "refused to beare place or office" because he found that "his owne heart, could not agree with the corrupt manners of his Country," and "*a mans Countrey was to be used as his Parents were, that is, with humble perswasions, and not with contestations*" (AL 1.20, 17, III: 277–8).[47] In actual fact, however, it was Plato's convicted teacher Socrates who told Crito that one must either persuade the fatherland of what is just by nature, or do whatever it bids, and suffer whatever it orders one to suffer.[48] Yet the Athenian gadfly was unable or unwilling to convince or persuade the jury that he was a just man. He was therefore justly convicted by the laws of his country. As Plato could neither agree with what was just by law nor with the conventional understanding of justice as such, he refused to become political by bearing place or office. Instead, he decided to write on imaginary laws for imaginary republics.[49]

But before exposing the more acute parts of a modern philosopher's answer to the ongoing demand for philosophical politics, we should draw attention to the fact that it was not solely for narrowly political reasons that Bacon left the task of reconstructing his teaching for his reader to perform. For by making the emergence of his true teaching dependent not only on the reader's willingness to do the somewhat mechanical work of comparing and contrasting the explicit teaching with the ancient and modern narratives, but also on his readiness to engage in the nowise mechanical activity of translating the thus emerging attributes and actions of the mythological characters into those of human or divine types, Bacon ensured that only readers driven by the desire to understand would be drawn into the movement of thought that found its expression in his book of fables.

Turning to the problem of outspokenness in early modern times, we find Bacon transmitting the lesson that outspokenness is "unseasonable" inasmuch as it does not adapt itself to the "faeces" of the times by submitting its speech to the God who is fabulously represented by Apollo because He is believed to harmonize truth and faith like the God of Harmony.[50] Unseasonable outspokenness is harmful and "useless" if useful predictions or prophecies are not believed in a timely manner, or if being silent or not being unseasonably outspoken is seasonable and therefore useful and harmless. Unseasonable outspokenness is even dangerous to the speaker who employs outspoken speech, inasmuch as by being unseasonably outspoken he risks the fate of the ancient Cassandra, who, as Bacon omits to tell, tried in vain to foretell the painlessness of her own death.[51] The modern and instaured Cassandra, who represents the "mouth" of the political philosopher in modern times, can therefore only seasonably tell or foretell truth or useful truth if she appeases the "vengeful" nature of God's representative's jealousy by "marrying" him in order to demonstrate that she has "submitted" to him.[52]

Although this conclusion will become more plausible as we proceed to the second fable, which treats the subject of marriage, and stimulates us to look back upon the bond between Apollo and Cassandra as a bond of marriage before we look ahead to the "marriage" whose bond it will prove to strengthen, we are aware of the great likelihood that our return to the mythological level will raise the eyebrows even of those readers who have a taste for classical mythology. But we are also aware of the paradoxical fact that some of the deeper layers of Bacon's philosophical teaching will remain hidden behind mythological veils if we do not descend to mythology in deed after having ascended from it in speech. We have already indicated that a significant part of the argument of the book as a whole only comes to sight if the fables are read in outer as well as inner conjunction.[53] But although this fact alone would justify a certain use of mythological speech on our part, it should be added that Bacon himself also had pedagogical reasons for employing the at first sight somewhat roundabout procedure of leaving the mythological level in order to return to it. For by thus stimulating the potentially philosophic among his readers to imitate in philosophical thought the education of the philosopher that is performed in mythological deed, our parabolic poet not only induced him to compare, combine, and confront the different traits of the philosophic persona, thereby enabling him to ask the question what a philosopher is in a way that should prevent him from falling into doctrinarism; he also seduced him into treating the quest for self-knowledge by engaging in it.

Adaptation

Juno was angry with Jupiter for having begotten Pallas by himself, "without her help." She "invoked all the gods and goddesses," because she wanted to bring forth something for which she would not need the help of Jupiter. When they were "weary of her violence and importunity," they assented. Juno thereupon shook the earth, causing an earthquake, during which the earth gave birth to Typhon, an "enormous and horrific monster." He was given to a serpent that was to "nurse him as his fosterfather." As soon as the monster had grown up, he waged war against Jupiter. Typhon subjected Jupiter to his power, carried him on his shoulders, and took him to a "remote and obscure region." He cut out the sinews of Jupiter's hands and feet, and "left him behind helpless and mutilated." Mercury then stole the sinews from Typhon and gave them back to Jupiter, whereupon Jupiter struck the monster with a thunderbolt.

Bacon points out that this fable was devised with a view to treating the subject of the "various fortunes of kings, and the rebellions that occur in monarchies from time to time." Like Jupiter and Juno, kings and kingdoms are "rightly regarded as being tied by the bond of marriage." It sometimes happens that a king, "depraved by the long habit of ruling," turns into a tyrant, "taking everything into his own hands," and "administering everything by his own arbitrary and absolute authority." Hereupon the aggrieved people themselves "create and elevate a head," which thereupon "attempts to stir the people." The result of this reciprocal process is a "kind of swelling of matters," signified by the infancy of Typhon. This course of events is fostered by the "innate depravity and malignant nature of the mob," which "to the king is like a most dangerous serpent." The situation eventually erupts into overt rebellion, represented by the "frightful image of the mature Typhon with his hundred heads," which stands for the "infinite evils that are inflicted both on kings and on peoples." Sometimes these rebellions grow so strong that the king is forced to abandon his kingdom, and to "betake himself to some obscure and remote province." Yet if he "bears his fortune prudently," he will "recover his sinews through the virtue and industry of Mercury." In other words, by means of "affability, prudent edicts, and benign discourses" he will "reconcile the minds and wills of his subjects," and thus "reinvigorate his authority." Having learned "prudence and caution," he will have become "unwilling to rely almost wholly [plerunque] upon the wheel of fortune" (DSV II "Typhon," VI: 630–1).

Resuming the thread of our reconstruction of Bacon's educative teaching, we find the second fable shifting perspectives in anticipation of the definitive shift of perspective which will take place in the central fable. It is, however, only by anticipating this shift that we can begin to understand the teaching of the second fable, a fable that belongs to what one might call the performed as distinguished from the more discursive fables or sections of fables. Therefore, we suggest that we transmit the fable's teaching by narrating its performance, and that we consider the political education of the fable's protagonist from the perspective of the philosopher in anticipation of it becoming part of the political education of the philosopher.

Because Jupiter the tyrant, a "depraved" version of God the King, was unacquainted with the serpentine prudence necessary to be a fortunate ruler, Typhon, who by virtue of having been nursed by a serpent had been given the opportunity to explore "all the hiding-places and depths" of the evil that is visualized by the "rebellion" of the "mob,"[54] took him to an "obscure and remote region" where he had to bear his bad fortune "prudently." After Jupiter had reflected on his situation, the "virtue and industry" of Mercury taught him to restore matrimonial harmony by modeling his marriage on the submissive marriage between Cassandra and Apollo, that is, on the marriage in speech between the philosopher and God. To this end Mercury, the messenger god of prudent speech, interpretation, and thievery, and one of the philosopher's main mythological accomplices in conspiracy, passed on to Jupiter the art of divine or political speech, which, as Bacon tells us by means of the ancient sources, he himself had either received from the God of Harmony or had stolen from him.[55]

Having demonstrated the interwovenness of the first two fables by the appearance of Apollo in both, we continue our narration with the new or born-again ruler of gods and mortals, departing from his educational prison, standing at the dawn of a long term of ruling, ready to counter the variety of his fortune by relying "almost wholly" upon the virtue of the art of speech. After returning to his kingdom, the anticipatory representative of the philosopher instaures the bond of his marriage, and invigorates his authority by appointing Apollo's wife his prophet, whose vocation consists in telling useful truths meant to prevent the depraved and malignant Juno, who as the king's wife is a depraved image of God's favorite and therefore a potential image of the exemplary believer, from rebelling again.

Bacon's implicit and explicit use of the notion of marriage in the first and second fables stimulates us to compare the horizontal, vertical, and diagonal symmetries between the marriage of Cassandra and Apollo, and that of Jupiter and Juno, a comparison that causes us to wonder whether it is not their mutual dependence that causes God and the believer who is a potential prophet to be ultimately similar. It may not be without significance to mention in this respect that when we substitute the philosopher for the two mythological figures most closely representing him, we find him once opposed to God, and once opposed to the potential believer. Neither do we believe it wholly insignificant for a better understanding of the dialectic of faith that in a way the two opposites become one when in the central fable the philosopher becomes two by imitating the One. The significance of the previous considerations themselves, on the other hand, is proved as early as in the ninth fable, as it is in this fable that the begetter of Typhon represents the "nature of the vulgar" (DSV IX "Soror Gigantum," VI: 645). For although Typhon himself stands for the "evils" inflicted on both king and people, he owes his existence to the woman who officially represents the kingdom as a whole before she falls apart into the dialectic between "people" and "head," a dialectic that can only be understood coherently if sheep and shepherd are traced back to a common root, seeing that the king himself lost his roots when he became an anticipatory image of the philosopher.

But we should turn to the part of the second fable that anticipatingly teaches that Jupiter could have prevented Juno's jealous anger as such, instead of leaving its consequences in the arbitrary hands of lesser gods. For although it is not until the thirtieth fable that this part of Bacon's teaching is made explicit, the ancient sources already tell us that Jupiter had swallowed the Oceanid Metis, who was pregnant with her first child, because she was fated to bear a daughter who would be almost as wise and strong as her father, and a son who would replace him as the ruler of gods and mortals. The daughter Athena was thereupon born from Jupiter's head, but the threatening son was never conceived.[56] Had Jupiter been able to prevent the fated danger to his rule in the first place, Typhon would never have been born, although the fact that he had to be born in order to teach the king about the nature of evil illustrates how much philosophy owes to its enemies. But there is still time left for Jupiter to learn, as we will learn by the time we arrive at the second fable's mirror image, which teaches the philosopher that his fortune depends on learning the art of "midwifery."

Introspection

The fourth fable is the only fable in Bacon's book on ancient wisdom that has a Greek subtitle: "Philautia [φιλαυτία]," meaning "self-love." Narcissus was "of wonderful beauty," but he was "enormously proud, and intolerably fastidious." He was "pleased with himself, despised others, and led a solitary life with a few companions to whom he was everything." Wherever he went, a nymph named Echo followed him. One day he came to a fountain and beheld his own image in the water. He contemplated himself and then "fell into such a rapturous and effusive self-admiration that he could not be drawn away from his image and simulacrum, but remained rooted to the spot impassively."

Bacon points out that the fable seems to represent "dispositions and fortunes of persons who perish from effusively loving themselves either on account of their beauty, or on account of a gift with which nature, unaided by industry, has adorned and distinguished them." He argues that such a state of mind commonly goes hand in hand with an "indisposition to appear in public or to engage in civil matters," for in a life of that kind these persons would "necessarily be exposed to neglect and vilifications," which would "cause their minds to become dejected and disturbed."[57] Therefore, they commonly live "solitary, private, and shadowed lives," with "small circles of chosen companions, who as devoted admirers and like an echo assent to everything they say, until they become depraved and inflated by such habits, and besotted with self-admiration fall into inactivity and idleness, and grow utterly numb and destitute of vigour and alacrity" (DSV IV "Narcissus," VI: 632–3).

Although the person of Narcissus will turn out to be the "skeleton" of the emerging political philosopher, the fact that the fourth fable politicizes the philosopher's self-love by criticizing narcissistic self-love already makes it part of the fables constituting Bacon's political teaching.

Bacon's typology of narcissistic self-love induces us to observe that a self-love that consists in admiring one's own image of self-sufficiency, knowing that this "solitary" self-admiration is being observed and admired by one's "chosen" companions, and despising those who do not "echoingly" assent to this image of self-love, does not reflect self-knowledge. Although the narcissist proudly believes that he is self-sufficient, his self-love is constituted by the very fact that he is being observed and admired by others. And it is the very same pride that constitutes the foundation of his self-love that prevents the narcissist from reflecting on the foundation of his self-love.

Passing from the criticism of narcissistic self-love to the politicization of the self-love of the philosopher, we find Bacon's fable transmitting two lessons. First and most acutely, by contemplating himself while neglecting the "public" and "civil matters," the philosopher does not gain the political knowledge necessary to effectively defend himself against "neglects and vilifications" that would "disturb and deject" minds "depraved and inflated" by narcissistic self-love. It is, in other words, only by "industriously" appearing in public and engaging in civil matters that the "naturally gifted" can gain the "vigour and alacrity" necessary in order not to fall into the "inactivity and idleness" that especially "Greek" self-love was in danger of falling into, seeing that it was a Greek philosopher who presented the contemplative life as a life of inactivity and idleness.[58] Second and most profoundly, it is only by leaving the "shadow" of unpolitical assent and entering the light of political exposure that the naturally gifted can prevent themselves from perishing without having fulfilled their natural potential. For a self-admiration that wholly builds on a series of unexposed contemplations lacks the fertile soil necessary for the "flower" of self-knowledge to strike root.

Attention

Mercury, whom we have gotten to know as the messenger and stealer of political or divine speech, returns in the sixth fable, this time representing the "divine word" as the possible father of Pan, who himself is "one of the oldest gods." Since Sacred Scripture "established beyond controversy [extra controversiam ponunt]" that Pan, that is, "the universality of things or nature," is the offspring of the divine word, Bacon argues that the sacred vision of the genealogy of Pan, officially represented by Mercury, "has been held by all the more divine philosophers" (DSV VI "Pan," VI: 635–6).

We briefly interrupt our summary of the narrative and teaching of the sixth fable in order to draw attention to the importance of paying close attention to the correlation between the action of the book and the attributes of its characters as they develop or change. For, to illustrate this point by the example in hand, if we had not noticed that Bacon subtly connects the first, second, and sixth fables by means of a character change of Mercury, we would not have known that it is in the sixth fable that the god of prudent speech, who as a compound god performs the function of messenger of the divine word as the father of the world, delivers a message from the God of Harmony on the origin of one of the oldest gods.

But before turning to his message, let us first draw the God of Harmony himself into the narrative, so that he can leave the scene for the rest of the book. Bacon relates that Pan engaged in a music contest with Apollo, and was declared victor by the judgment of Midas, for which judgment Midas had to wear the ears of an ass, albeit "in secret and not to be seen." This music contest and its consequences present a "salutary doctrine," which is "fit to curb and reduce to sobriety the unconstraint and elevatedness of human reason and human judgment." For, Bacon explains, "it seems [videtur] that there are two kinds of harmony and music," one of "divine providence" or "wisdom,"[59] and one of "human reason." To the "human judgment and the ears of mortals" the "government of the world and of things and the more secret divine judgments sound somewhat harsh and disharmonious," which ignorance "deserves to be distinguished with the ears of an ass." These ears are to be worn "secretly and not openly," though, for "the matter is not observed or noted as a deformity by the vulgar" (DSV VI "Pan," VI: 636, 640).

We now possess the "ears" to hear the God of Harmony say that it is not to be controverted that the world has been created and is still being governed by God's power, wisdom, and providence. To the ears of a non-divine philosopher, however, it sounds "disharmonious" that a wise God should pass inscrutable judgments that are judged as "harsh" by the ignorance of human reason. Bacon therefore teaches him to wear the ears of an ass in order not to hear these harsh and disharmonious sounds.[60] But as the asinine ears of the vulgar, to which the voice of human reason sounds audaciously profane, do not hear the disharmony between divine wisdom and human reason, the visible ears of an ass that are worn by the philosopher strike the vulgar eye as a deformity. It is for this reason that the newly born divine philosopher must wear his gift from the God of Harmony in secret.

In short and in anticipation of some of the next stages of his political education, we see the divine or political philosopher looming up as a man who wears the ears of an ass when he silently listens to the voice of human reason while soberly playing the music of Apollonian harmony that proclaims divine power and wisdom. He thus prevents the fate of Midas since he avoids having to pass the latter's explicit judgment. For contrary to Midas, who did not hide his ears until after having been punished by

Apollo for his judgment,[61] the non-divine philosopher avoids divine anger over human judgment by becoming a divine philosopher.

In harmony with the fact that the ancient sources tell us that Apollo originally received the gift of divination from Pan,[62] Bacon tells us that in addition to being himself a god, Pan is also the messenger of gods, "next to [proximus]" or "after [alter]"[63] Mercury. He emphasizes that this is an allegory "clearly divine," seeing that "after [proxime post]" the word of God the image of the world itself is the "proclaimer of divine power and wisdom" (DSV VI "Pan," VI: 635, 639).

But although we may say that in a way Mercury has just returned the art of divination, supplemented with the message from the God of Harmony, to its original owner, the world is not its own image. It seems to be primarily for this reason that the narrative goes on to say that Pan had "no amours, or at least very few." For "the world enjoys both itself, and in itself all things that are," as Bacon explains after having remarked that this "may seem strange among a crowd of gods so excessively amorous," which causes us to remark that a god being amorous is already excessive. He who is in love "wants to have the enjoyment of something," and "where there is abundance, desire can have no place." Consequently, the world can have "no amours, nor, being content with itself, any longing, *unless* it be a longing for discourse." Pan therefore only had a passion for Echo, who was the "only one to be chosen as the world's wife." For "true philosophy echoes most faithfully the voice of the world itself, and is written from the world's own dictation. It is nothing else than the simulacrum and reflection of the world, and it only repeats and echoes, but adds nothing of its own" (DSV VI "Pan," VI: 636, 640, emphasis mine).

By drawing Echo into the sixth fable, Bacon weaves the as yet static fourth fable and therewith the emerging political philosopher into the web of the poetic action of his book. He thus stimulates us to consider that Echo being to Pan in the sixth fable what Narcissus was to Echo in the fourth fable causes Narcissus to come to sight as the embodiment of Echo: as the maturing political philosopher who is beginning to learn to behold his reflection in the world of political things when engaging in the reflective activity that is identical to echoing nature's "natural voice" in thought, before he echoes its "political voice" in deed.[64]

Dissimulation

At the beginning of the eighth fable we learn that tradition has it that Endymion was a shepherd who was loved by the moon. Their intercourse was "of a novel and singular kind." Whenever Endymion lay reposing in a "natural cave," the moon descended from heaven, kissed him as he slept, and then returned to heaven. Endymion's "idleness and slumber were not detrimental to his fortunes," however, for the moon "ordained that his sheep should become fat, and that his flocks should extend, so that no shepherd had fatter sheep or flocks more numerous."

According to Bacon, the fable seems to pertain to the "dispositions and characters of princes." Princes who are "full of thoughts, and prone to suspicions, do not easily allow into more intimate intercourse men who are perspicacious and curious, who have minds that are always vigilant, and who never sleep." They rather choose "men of a quiet and complying disposition, who endure what pleases their princes without inquiring further, and who behave [praebant] as persons ignorant and unobserving, and as if asleep, displaying [praestant] simple obedience rather than cunning observation." With men of this kind princes "descend from their majesty, like the moon from the highest orbit, and lay aside their mask, the continuous wearing of which becomes a burden, and engage in intercourse familiarly and freely, which they think they can do safely."

Bacon illustrates these points by means of an ancient and a modern example. Tiberius Caesar favored only those who in truth knew his character, but who "pertinaciously and as it were stupidly dissembled their knowledge." The same thing can be seen in the modern Louis XI (1423–1483), a "cautious and cunning king."

Our author goes on to observe that as Endymion had a cave as a place of repose, those who enjoy "this kind of favour [gratia]" on the part of princes commonly have "some pleasant places of retirement," which, "at leisure and with relaxation of mind, and without the burden of their fortunes," they can invite their princes to. Favorites of this kind "commonly do well as far as their own affairs are concerned." They may not be "raised to honours," but since their being favorites "springs from true affection and not from considerations of utility, they are enriched with the prince's generosity" (DSV VIII "Endymion," VI: 643–4).

Continuing the drama of the philosopher's political education, Bacon teaches the still immature divine philosopher the art of dissimulation. We may divide the substance of this teaching into two parts. First of all, the philosopher can only undisturbedly indulge in his contemplations on the "universality of things" if he gains the protective nominal favor of his "prince," who is represented by the moon because it is from the dialectic between His superlunary essence and His sublunary presence that God appears as an exemplary prince. He must be a better dissembler than Lucius Aelius

Seianus, though, the ancient Tiberius's favorite, who was overthrown by his successor Naevius Sutorius Macro because Tiberius had found out that he had become too powerful.[65]

However, and this brings us to the second point, the only way for the divine philosopher in modern times to discourse on nature is on the presupposition of the "prince" of nature descending from His "majesty" and laying aside the "mask" that obfuscates His nature. For contrary to his ancient counterpart, who lived in a "natural cave," the modern political philosopher finds himself in a cave that is at least partly historical.[66] The way to the natural cave of political life therefore goes through "intercourse" with the author of history. And since this intercourse can only take place on common ground, it can only take place in the cave of God's newly born favorite, one of the very few "pleasant places of retirement."

But modern princes, that is, God's spokesmen in modern times, tend to be very suspicious, because they are as "full of thoughts" on how to be cautious enough to avoid any further inquiry into their pleasures as Bacon is full of thoughts on the nature of these pleasures. The intercourse between Endymion and the moon already being of a "novel and singular kind," at least in the version of Bacon, the dissimulating philosopher in modern times finds himself in even more singular circumstances because he must gain the favour of princes resembling the modern Louis XI, who became so suspicious and reclusive toward the end of his life that he admitted nobody near him as a favorite.[67] He can only make them descend from "burdensome" appearances by behaving like an Abel-like shepherd, displaying "simple obedience" by cunningly playing the music of Apollonian harmony. Being guided by considerations of "utility," the divine philosopher will not be loved by his princes the way the true Abel-like shepherd is loved by the "prince" who is honored by the obedience His favourites display for Him, and who loves them by holding out to them the prospect of being rewarded with His "generosity." But he will do well as far as his "own affairs" are concerned.

Determination

"Human curiosity that pries into secrets," as well as the "knowledge of secrets desired and laid hold of by an unhealthy appetite," are "restrained" by the ancients by means of two examples, or so Bacon tells us in the tenth fable. Actaeon, having "unawares and by chance" seen Diana naked, was turned into a stag, and torn to pieces by his own dogs. Pentheus, having climbed a tree because he wanted to see Bacchus's secret sacrifices, was struck with madness. He thought everything was double. He saw two suns and two cities of Thebes, and therefore "continuously and without rest kept going backwards and forwards."

Bacon points out that the first fable seems to pertain to the "secrets of princes," and the second to the "secrets of divinity." Whoever becomes cognizant of a secret "without the admission and against the will of a prince," will "certainly incur the prince's hatred." He will live the life of a stag, "full of fears and suspicions." The calamity of Pentheus is of a different kind, though. For the punishment imposed on those who, "with rash audacity and forgetting their mortality, aspire to the divine mysteries by the heights of nature and the depths of philosophy, as if by climbing a tree," is "perpetual inconstancy, and a vacillating and perplexed judgment." As the "light of nature" is one thing and the "light of divinity" another, they are "as men seeing two suns." And as "the actions of life and the determinations of the will depend upon the intellect," these men are "undecided in will no less than in opinion, and they cannot be wholly consistent with themselves." They see two Thebeses in a similar way, for Thebes represents the "ends of our actions," as it was the home and shelter of Pentheus. Therefore, "not knowing which way to turn, but being uncertain and fluctuating as to the sum of things, they are turned round among particulars, according to the sudden impulse of the mind" (DSV X "Actaeon et Pentheus," VI: 645–6).

The time spent in the cave of Endymion has made us curious to know if we were right in suspecting that the "secrets of princes" and the "secrets of divinity" are closely intertwined. In the fable on human curiosity Bacon confirms our suspicion by using the goddess Diana as the example of a prince, and the alleged god Dionysus, who allegedly came to Thebes to present his divine credentials to Pentheus, as the example of a divinity.[68] But the fates of the victims of unhealthy curiosity prove to be interrelated as well, seeing that the ancient sources also allege that Cadmus warned Pentheus that if he did not recognize Dionysus as a god, he might suffer the same fate as Actaeon, who was torn to pieces for having recognized a naked goddess.[69]

Having come to understand that Bacon's educative teaching is to be construed dialectically, we go on to observe that the philosopher is to avoid the fate of Actaeon by behaving like Endymion, who, by having been caught by the moon's love seemingly unawares, was able to engage in intercourse with his unmasked prince, thereby disclosing his most intimate secrets without incurring his hatred.[70] In other words, it is by actively prying into divine secrets while dissembling "virginal" passivity that the instaured Actaeon, that is, the philosopher having become aware of the political nature of his curiosity, is to avoid the consequences of an unawareness of the divine laws of chastity, which stand for the theological foundation of morality. Once he has gained Diana's love by becoming her favourite, the goddess who represents morality's divine sanction will reward her favorite's obedience by willingly showing herself naked in anticipation of the consummation of her love. Thus, the political philosopher is able to satisfy his curiosity without giving offense for not deserving divine love's materialization.

The calamity of Bacon's Pentheus is of a different kind, however, because the way in which he is punished is different from the way in which Actaeon, and the way in which his own ancient counterpart was punished. The ancient Pentheus had tried to put an end to the practices that were performed in the name of a god in whom he did not believe, and he was torn to pieces by his mother Agave, who in her Bacchic frenzy took him for a wild beast.[71] But instead of seeing himself sentenced to violent death for having given in to his impious desire to inquire into the divinity of a god, Bacon's Pentheus is punished with an inconsistent life. For although the "intellect" teaches that two mutually exclusive sources of light cannot be mutually illuminative, Pentheus founded the determinations of his "will" on the "opinion" that it is by the "heights of nature" and the "depths of philosophy" that one can aspire to the "divine mysteries" that emanate from the "light of divinity." And because one cannot navigate by "two suns" at the same time, Pentheus was unable to act, except on the basis of the "sudden impulses of his mind," undecided as he was as to his fundamental approach of the sum of things.

Pentheus, to summarize the educative drift of Bacon's poetic depiction of the truth that knowledge may grow out of suffering, is punished with a "vacillating and perplexed judgment" for failing to perform the action that serves as the foundation of all subsequent actions, namely the action ensuing from the anticipatory decision on the right way of life; in other words, he is punished for having tried to suspend his judgment on the one question that does not allow of suspension of judgment. The undecided philosopher must therefore learn to decide on the basis of the intellect to either wholly rely on the light of divinity, or to develop a healthy and consistent human curiosity that translates itself into a slow but "audacious" inquiry into the secrets of divinity, to the extent made possible by the "height of nature" and the "depth of philosophy." For it is only after having decided what road to walk on that he can decide if it is the road home that he walks on or if home is the road that he walks on.

Philosophy

The fable of Orpheus is "commonly known," but has "not yet in all parts been faithfully [fidum] interpreted." It seems to refer to an "image of universal philosophy," for the person Orpheus, a "man admirable and clearly divine," who, being a "master of all harmony," "conquered and pulled along all things by sweet measures," may by an "easy transition" pass for a description of philosophy.

His love of his wife, who had been snatched from him by "premature death," made Orpheus decide to descend to the infernal regions to "entreat the shades." During his descent to and stay in the infernal regions he "relied on his lyre." His hopes were not disappointed, for the shades were "so pleased and charmed by the sweetness of his singing and playing" that they allowed him to abduct his wife on the condition that she was to follow behind him, and that he was not to look back until they had reached the region of light. Yet, "impatient with love and anxiety," Orpheus did not refrain from looking back, and when husband and wife had almost reached the point of safety, the treaty was broken, and Orpheus's wife relapsed into the infernal regions. From that time onward, Orpheus, "in sorrow [moestus] and hating women [mulierum osor]," betook himself to "solitary places." By the "same [eadem] sweetness of his song and his lyre" he drew to him all kinds of wild beasts in such a manner that they "laid aside their natures, forgot all their anger and ferocity, and stood around him gently and mildly, as in a theatre." The power of Orpheus's music was so great that it moved the woods and stones, which changed places and positioned themselves decently and orderly around him. For some time this went on "happily and in great admiration," until eventually certain Thracian women, "stimulated and aroused by Bacchus," blew "such a hoarse and hideous blast upon a horn" that the sound of Orpheus's music could no longer be heard. Eventually, "the virtue that had been the bond of the Orphic order and society was broken, the disturbance began again, the beasts returned to their natures, and the woods and stones stayed no longer in place." Orpheus himself was torn to pieces by the Thracian women in their frenzy. "In grief and indignation at Orpheus's death, the river Helicon, sacred to the Muses, buried its waters under the earth, to reappear elsewhere."

According to Bacon, the fable seems to have the following meaning. The singing of Orpheus is "twofold [Duplex]," the one to "appease the shades," and the other to "draw together the wild beasts and the woods." The former refers to "natural philosophy," and the latter to "moral and civil philosophy." For, Bacon explains, "the noblest work of natural philosophy is the very restitution and instauration of corruptible things." If this "can be achieved at all, it cannot be achieved otherwise than by the due and exquisite attempering of nature, as if by the harmony and careful modulation of a lyre." But as it is "of all things the most arduous," it "commonly fails to have effect, probably from no cause more than from curious and unseasonable sedulousness and impatience." Therefore philosophy, "finding that she is almost [fere] unequal to this great thing, in a sorrowful mood, as she deserves [merito moesta], turns [vertit] to human things," and "by persuasion and eloquence instils into men's minds love of virtue, equity, and peace." She also teaches people "to assemble, to accept the yoke of laws, to submit to authority, to forget their untamed affections, and to listen to and comply with precepts and discipline." And this care for civil things is "properly and orderly represented as subsequent to the sedulous trial to restore the dead body to life, and the final frustration of this attempt." For it is true that the "clear exposition of the inevitable necessity of death sets men's minds upon seeking immortality by merit and fame." But notwithstanding the fact that the "works of wisdom are the most excellent of human things," their "periods come to an end." After kingdoms and commonwealths have "flourished for some time, there arise perturbations, seditions, and wars." Amid the uproars the "laws are put to silence first, and men return to the depraved conditions of their nature." And "if such frenzies continue, it is not long before learning and philosophy are torn to pieces in such a way that no traces of them can be found but a few fragments, scattered like planks from a shipwreck." Then "a period of barbarism is imminent, and the waters of Helicon sink under the ground until, according to the destined [debita] vicissitude of things, they break out and emanate again, perhaps not in the same place" (DSV XI "Orpheus," VI: 646 ff.).

Although the fable of Orpheus is "commonly known," it has not yet been interpreted in the era of faith, which is the era of Bacon. A commonly known ancient source relates that Eurydice, the wife of Orpheus, died a premature death by trodding on a poisonous snake as she was fleeing from Aristaios, who had tried to rape her.[72] The "faithful" interpreter of the part of the fable treating the beginnings of the evils that were to befall Orpheus will therefore look upon Eurydice as the heathen image of Adam's wife, who caused man's evil state of mortality by allowing the snake to seduce her into the evil of eating of the tree of the knowledge of good and evil (Gen. 3: 22). Assuming that Bacon took the starting point of faith as the starting point of his fable, we assume that the "restitution and instauration of corruptible things" consists first and foremost in an attempt at restoring human nature, which, as it stands, is in a corrupted state and has been so since man freely but "impatiently" and therefore "prematurely" chose the principle of disobedience by transgressing God's commandment.[73] This implies that Orpheus must become patient enough to turn his attention to uncorrupting corrupted human nature before he can "seasonably" indulge in the love of his wife, that is, in the love of corruptible things or natural bodies. The modern Orpheus must, in other words, invert the sequence of the real or apparent affective priorities of his ancient predecessor. In short and in general terms, if the natural philosopher wants to remain a genuine philosopher, he must consummate the turn to political philosophy.

But we almost pass by what Orpheus passed by. For although Orpheus was a man "clearly divine," he died a violent death by the hands of women intoxicated by a man who had claimed to be divine. It is therefore by using an epithet that one "easily passes by" that Bacon draws our attention to the fact that Orpheus is too easily transited to the times in which another divine man had become the conqueror of man's corrupted nature by the sweet measures of His grace. As a consequence of Christ's spiritual conquest, Orpheus must first learn how to become a "master of harmony" before he can reconquer human nature by the "sweet measures" of divine or political philosophy. To put it differently, it is not until the philosopher learns how "duly and exquisitely" to "attemper" nature graced with grace that he can start his attempts at restituting and instauring the grace of nature's grace. His lyre will guide him on his educative journey, which is coeval with the journey of his lyre from divine law to divine philosophy. For the modern Orpheus, that is, Bacon's political philosopher, cannot start listening to the disobedient voice of human reason until after having "carefully modulated" the lyre that his ancient predecessor received from Apollo,[74] who himself had played a divinized version of the song of divination that he had learned from its author Pan, as we learn when looking back upon the first fable from the perspective of the sixth. It is on this lyre, which we believe Bacon's Mercury to have stolen from the God of Harmony along with the art of divination and divine speech, which had been taught to Jupiter before it was delivered to Pan, who had it echoed to Narcissus, that the divine philosopher and master of harmony will be playing the music that preaches the harmony between truth and faith or between human reason and divine wisdom.

But although it is only in the thirty-first and final fable that Orpheus will actually be playing, it is in the fable named after him that he learns the art of playing, as is already indicated by the fact that the word "harmony" occurs for the third and final time in the eleventh fable, after having been used only in the sixth and first fables, that is, in the fables on nature and law. And since it is in the fable on philosophy that the philosopher is united with the instrument that symbolizes his political education, it is in the context of discussing the fable in which philosophy turns to political philosophy that we should draw a provisional sketch of the divine philosopher as he emerges in the form of a compound of Bacon's philosophic divinities. Anticipating the completion of his education, we see Orpheus, who wears the invisible ears of an ass and speaks with the submissive mouth of Cassandra and the faithful voice of Echo, coming to sight as Narcissus having become reflective, Endymion having become favorable, Actaeon having been chastened, and Pentheus having decided.

But in order to diminish the risk of confusing completion with completeness, we should confine ourselves to interpreting Bacon's most aporetic fable, whose apories are natural to the extent that they can only be solved temporarily. It is therefore especially the fable on philosophy that is to be read both in conjunction and in disjunction with the fable that will propose a political solution to its political apory, if only because it is precisely from their temporary conjunction that philosophy and politics emerge as naturally disjunctive.

Orpheus was unable to abduct his wife from the infernal regions because he had been so impatient as to rely so much on his unmodulated lyre that he was compelled to allow the infernal divinities to attach a "condition" to allowing him to abduct his wife. But apart from allowing the modern Orpheus to learn how to make the hostage takers of his wife release their hostage without conditions, in other words, to liberate natural philosophy from theology without paying the price of superstition that is represented by the "shades," Bacon also seems to have interposed the negotiations over a dead woman in the fable on philosophy in order to draw our attention to the fable treating the negotiableness of death as such. For in the fable on treaties Bacon had pointed out that it has been "interposed in several fables" that "there was one oath by which the gods bound themselves when they meant to leave no room for penitence." This oath did "not invoke as a witness any majesty of heaven, nor any divine attribute," but Styx, the river flowing in the infernal regions. Only this form of sacrament was held to be "firm and inviolable," for the gods "most feared the penalty for breaking it," as it consisted in the breaker "not being allowed access to the banquet of the gods for a certain number of years."

According to Bacon, the fable of Styx seems to have been devised with a view to treating the subject of "treaties of princes." The "solemnity and sanctity [religion] of the oath by which a treaty is confirmed is not much to build on," because "princes easily find various and specious pretexts wherewith to secure and veil their cupidity and their less than sincere faith, there being no arbiter to judge the reason of the thing." For this reason, "only one true and proper pledge of faith is to be adopted," and it is "no celestial divinity": it is necessity, the "great deity of the powerful." A treaty confirmed by necessity is a treaty "ratified, sanctioned, and confirmed by the oath of Styx." Consequently, it is a treaty "subjected to the fear of being interdicted and suspended from the banquet of the gods," for by the name of Styx the ancients signified the "rights and prerogatives of empire, affluence, and felicity" (DSV V "Styx," VI: 633–4).

We call to mind what Bacon must have called to mind when he wrote the fable on treaties, namely that after man's first parents had broken the "treaty" that had been imposed on them, the Majesty of Heaven in His mercy sent His Son to hold out to the "penitent" the prospect of being allowed access to the "banquet" of the elect one day. But the heathen Orpheus had no reason for being penitent, because his wife had died a premature death. Nonetheless, the princes of the infernal regions, who seem to represent the pagan soothsayers, and who thus seemed to have the power to decide on the possible resurgence of Eurydice, in their mercy concluded a treaty with Orpheus, the reason of which was inscrutable to him. He therefore guiltlessly broke the treaty, as a consequence of which his wife suffered eternal death.

The unwillingness or inability of the infernal divinities to resurrect Orpheus's wife leads us to the conclusion that Bacon "interposed" the lesson of necessity in the fable on philosophy in order to teach the modern Orpheus not to rely on death's negotiators, who concluded a treaty with his predecessor, possibly knowing that the latter could not but break it. They may have done this in order to veil their "cupidity," which consisted in their wanting to keep for themselves the woman who had been destined by necessity to die prematurely. But as there are no witnesses of the hearts of gods, the modern Orpheus can only avoid "bad faith" by solely relying on the "deity of the powerful," that is, by solely listening to the voice of "necessity" when he appeases the cupidity of the infernal or celestial divinities by singing of their "rights and prerogatives."[75]

We hear the voice of necessity telling us to let ourselves be carried along by the flow of the river of "abhorrence [Στυξ]" in order to find out what other lesson it could have carried along for Orpheus. The philosopher's love of his wife induced him to look for a kind of Edenic immortality that denies the abhorrent but "inevitable" and irrevocable "necessity of death,"[76] which is an immortality incompatible with the love of the woman called philosophy. Impatient with mortality, the philosopher tried to uncorrupt the corrupted human body by attempting to restore it to its corruptible state. But philosophy is "unequal" to this supernatural thing, because the natural state of the natural body whose uncorruption constitutes the "noblest work of natural philosophy" has always been a corruptible state at the mercy of the necessity of irrevocable corruption by the hands of death. Trying to achieve the wrong goal, the impatient philosopher could not but achieve disappointment, for which reason he deserves a "sorrowful" mood. Moreover, he necessarily becomes a "misogynist," because a philosopher who misunderstands the boundaries of a life that is devoted to his love cannot but come to hate his beloved for failing to reward his devotion with a removal of the boundaries that he misunderstood.[77]

Orpheus bewailed the loss of what he never truly possessed by playing the "same" song he played to please and charm the shades. Being preoccupied with his sorrow in a manner reminiscent of the immature Narcissus's unselfconscious self-admiration, he failed to notice the political nature of his song because he failed to notice its political effect on beasts, woods, stones, and Thracian women (cf. AL 1.67, 39, III: 302). In other words, he turned to political things unselfconsciously and passively. Bacon subtly draws our attention to this failure by teaching the unpolitical Orpheus, who played one and the same song, that his singing must become "twofold."

But the man who had become a misogynist because he had misunderstood the nature of the woman he loved, was torn to pieces for misunderstanding the nature of the women he hated. In other words, the hater of women who had come to hate the woman he loved because he had tried to secure his beloved by playing the wrong song to her, was torn to pieces by the women he hated because by failing to play a song that would have secured their love but would not have aroused their jealousy he failed to secure his safety. The ancient sources tell us that Orpheus was torn to pieces primarily because the Thracian women took offence at his neglecting or hating them.[78] The jealousy of these zealotic women thus primarily seems to have been generated by the fact that the sorrowful philosopher neglected them because he was absorbed by his hatred of another woman, whom he used to love, and whom he had come to hate for not reciprocating his love. But because the self-absorbed Orpheus did not notice the Thracian women, he did not notice the fact that they took offense at not being noticed. He therefore could not prevent them from drowning his music and tearing him to pieces in their "frenzy" as a way of avenging their being neglected.

By means of the political fate of the unpolitical Orpheus, Bacon seems to teach that the philosopher who is so impatient for his love that he indulges in the principle of disobedience before having understood what made corruptible men become obedient, has not yet consummated the turn to political philosophy. This turn demands, to put it in the broadest possible terms, that the philosopher become a patient but active inquirer into the causes of the corruptibleness of corruptible things. For, to narrow the causes down to their consequences, the modern Orpheus can only instaure the possibility of philosophy and effectively defend it against zealous barbarians if he follows the "traces" leading to the "planks" that carry the remains of "shipwrecked" philosophy, and secures those few "scattered fragments" of the ancient Orpheus's song that teach him how not to play the lyre of the God of Harmony.

Although Orpheus's song was so powerful that it tamed corruptible beasts by temporarily stupefying their corruptible natures, the bard could not prevent the "bond" of his unpolitical order and society from being broken by corrupted women who were singing in the language that appealed to the beasts' corruptive potential. Because Orpheus had turned to political things unwittingly and unselfconsciously, he did not speak the language of the province of political things in which he necessarily found himself.

That the modern Orpheus has good reason for learning to speak a political language is indicated by the fate of his predecessor. The reasons themselves for learning the language of the political are also indicated, but they are not explicated. Taking Bacon's indications as our starting point, we discern a twofold reason for philosophy having to learn this language. Most acutely, Orpheus will not be able to uncorrupt men's corrupted corruptible natures by making them freely submit to the "yoke of laws" until he learns to speak the same language that the snake spoke when, by expounding the reason of God's first Law to the first woman, he seduced her into freely breaking it. But because the snake's speech did not prevent men in later times from breaking human laws for the purpose of regaining access to the tree of life that had wakened first man's desire for immortality, the lawgiver Orpheus must appeal to man's uncorrupted desire for immortality by playing a song on his lyre that eloquently persuades man into "seeking immortality by merit and fame." In short, the disobedient philosopher must make formerly obedient men obedient to his laws by appealing in his song to man's natural propensity toward an obedience that anticipates immortality, the knowledge of which propensity he himself gained by becoming disobedient. He can thus prevent the political order from ending like the unpolitical order of the ancient Orpheus, although the aporetic ending of the fable on philosophy turning political already indicates that he cannot prevent it from ending as such. The ending of the philosopher's political order comes to sight as all the more definitive in light of the fact that the ending of the political philosopher's book will indicate that it can be postponed.

More profoundly, however, Orpheus will not overcome his sorrow and reconquer his love until he regains his devotion by understanding the boundaries of a life of devotion. Because the circumscription of these boundaries can only be achieved elenctically, Orpheus must radically expose his life of love to men's opinions on love and on a life devoted to the love of a woman. In other words, Orpheus must elevate to the status of political things of philosophical interest his life of love, the nature of his love, and the opinions of men leading lives that are not devoted to the love of a woman. This implies that he must learn to speak the language that all men, philosophers and non-philosophers alike, speak when they speak about political things, which presupposes that he must come to understand political things as they naturally present themselves to man and find their way into man's speeches. Orpheus can only bring the woman he loved back to the province of light by wittingly and self-consciously descending to the province of political things for the purpose of indulging in his new love of political things, a love which manifests itself in him simultaneously speaking about political things dialectically, and veiling his true purpose by playing the ironical song of ignorance on his lyre.[79] Bacon subtly alludes to the previous points by implicitly referring to the fact that Orpheus's multiple songs are all played on the same instrument.

Without appealing to the imagination overmuch, one could say that the fabulist Bacon teaches the pre-Orphic Orpheus a lesson similar to the lesson that the poet Aristophanes taught the pre-Socratic Socrates.[80] Moreover, one need not be overmuch imaginative to be able to say that, if Platonic political philosophy was partly a response to the fate of the real Socrates, Bacon is a Platonic political philosopher inasmuch as he responds to the fate of real philosophy, despite or because of the fact that he seems to invert the Socratic turn by sailing away from men's speeches toward natural philosophy.[81]

Virtue

In the central fable of *De Sapientia Veterum,* Jupiter seduces Juno by turning himself into the "ignoblest shape possible," an "exposition of contempt and ridicule": a "wretched cuckoo, soaked by rain and bad weather, struck by thunder, trembling, and half dead." Bacon points out that this "prudent" fable on the unnamed "suitor" of Juno has been derived "from the depths of morals." Its true meaning is that "men are not to flatter themselves too much by thinking that a reputation for their exemplary virtue will gain them estimation and favour with everybody," because this "depends on the nature and character of those whom they approach and serve." If these are "persons who are not distinguished with any gifts or ornaments of their own," but who only have a "proud and malignant disposition," as "represented by the type of Juno," men must "truly recognize that they must wholly lay aside everything about them that has the least show of grace [decoris] and dignity, and that it is plain folly to proceed any other way." What is more, it "does not suffice to display submission to deformity, unless they entirely change themselves into abject and degenerate personalities" (DSV XVI "Procus Junonis," VI: 654).

In the central fable of *De Sapientia Veterum,* the education of the philosopher merges with that of the tyrant, as it is in the fable whose subtitle is "Disgrace [Dedecus]" that the philosopher in a way becomes the tyrant. Although this may seem absurd "upon the face of the narrative taken by itself," it starts becoming more plausible when we call to mind that the fable on the unnamed suitor of the woman whose indistinctiveness makes her the exemplary image of a lover of tyranny is the only fable whose narrative was completely born from the "head" of a lover of the planet named after Saturn, the god whose overthrow inaugurated the long rule of the tyrant Jupiter.[82] As a lover of Saturn, that is, a lover of contemplation or a philosopher, cannot be a lover of Jupiter or tyrannical politics, we are to consider the political education of the tyrant from the perspective of the philosopher, a perspective we already anticipated when interpreting the second fable. But when interpreting the second fable we were not in a position to relate the substance of this education, whose performance was yet to become part of one of the narratives within the narrative.

We therefore had to keep the doors to the tyrant's educational prison closed until Bacon's narrative disclosed to us that after Jupiter had been taken to an "obscure and remote region" for not having been able to prevent his wife's jealous anger, Endymion invited the tyrant to his cave to engage in "intercourse" on whether the latter's character and disposition might have been the cause of his wife rebelling. The plausibility of this possibility was already suggested by the tyrant's wife's own character and disposition, considering that Juno, not being "distinguished with any gifts or ornaments" of her own, was wont to substitute her lack of self-sufficiency for a pride consisting in having an inimitable relationship with the ruler of gods and mortals, whose grace she secured by her love for him. After Endymion had persuaded Jupiter into laying aside his "mask," it turned out that the tyrant had become a spoiled and careless tyrant, because his long habit of ruling had so much "depraved" him that he had come to think he could keep ruling by his own "arbitrary and absolute authority" without securing the love of Juno, on which he was dependent both for the prospering of his rule and for his self-love, a self-love not dissimilar to that of Narcissus.

It was this display of seeming self-sufficiency that had insulted Juno's pride and caused her to rebel.

Let us return to the sixteenth fable to consider the political education of the tyrant from the perspective of the liberation of philosophy from tyrannical politics, which we take to constitute the most active part of Bacon's political teaching, that is, if by active is meant practical. For on the level of theoretical activity the liberation of philosophy from tyrannical politics is identical to the separation of political philosophy from political theology. It is on this level that the philosopher becoming two by imitating the One runs parallel not to two apparent opposites becoming one real unity, but to one apparent unity becoming two real opposites again in anticipation of their solution into the Unity of Knowledge.

But in order not to run ahead of things, we should remain within the bounds of the sixteenth fable and have a closer look at the philosopher's need for "imitative virtue." In trying to restore the non-royal throne to his beloved, the lover of Saturn sees himself confronted with a number of "Junos" occupying the road leading from Tartarus, that is, the "historical cave," to the province of light that is identical to the natural cave of political life. He is therefore compelled to "disgrace" himself by "entirely changing himself" into the "abject and degenerate personality" of the named suitor of the tyrant's lover, whose grace answered to his lover's pride. In other words, it is "plain folly" to try to preserve one's "dignity" when one sees that one can only instaure order by striking the tyrant with his own thunder. Although being thus imitated constitutes a "disgrace" to the ruler's perfection, the ruler's favorite will be too proud to feel disgraced by the fact that she is being seduced by a man who turned himself into the "ignoblest shape possible" and wears the ears of an ass, albeit invisibly. The exemplary virtue of the philosopher thus comes to consist in his submission to the deformed beast-man Cheiron, from whom he learns that one can only seduce the proud and malignant by using two natures.[83]

Picking up the thread to Aesop's lion's cave from the place where we left it, we may say that the fox and the lion must cooperate in taking the "food" of philosophy to the place where it can be safely digested: some "obscure and remote region" or cave entered by two but left only by one. Bacon dissimulates the gravity of this teaching, dug up from the immoral "depths of morals," by making it seem comical.[84] But it was only by being so bold as to equate God with an object of "contempt and ridicule" that Bacon could prevent the Earl of Salisbury from feeling disgraced by the fact that his servant severely transgressed the "bounds of modesty" by instauring philosophy's grace in the book the Earl protected with his name.

Morality

Semele, Jupiter's mistress, made Jupiter take an "inviolable oath to grant her one wish, whatever it might be." She requested that Jupiter embrace her in the same way in which he used to embrace Juno. The consequence was that she was "burnt to death in his embrace." The infant in her womb was taken by its father, who "sewed it into his thigh," where it stayed until the end of the period of gestation. The burden made Jupiter "limp somewhat," though, and because the child caused "pain and pricking" while it was borne in his thigh, it received the name Dionysus.

Bacon points out that the fable seems to pertain to "morals," and that there is "hardly anything better to be discovered in moral philosophy." The "nature of desire, or affection and perturbation," is described by the person of Bacchus. The "mother of virtue," who represents "real good," is the "legitimate wife" of Jupiter, who himself represents the "human soul." The "mother of desire," who represents "apparent good," is Jupiter's "mistress," who, like Semele, "nevertheless aspires to the honours of Juno." The conception of apparent good is "always some illicit wish, blindly granted before it has been understood and judged." Unable to "endure the heat of the affection as it effervesces," its mother, the "nature and species of good," perishes in the flames. As long as the affection is "unripe," it remains in the human soul, which is its "begetter." It remains "especially in the lower part of the soul, as if in the thigh, where it is nourished and concealed, and where it causes such pricklings, convulsions, and depressions in the mind, that the mind's resolutions and actions are impeded and crippled." But eventually it "casts off all restraints of shame and fear, and growing audacious, it either assumes the pretext of some virtue, or contemns infamy itself" (DSV XXIV "Dionysus," VI: 664–5; DA 2.13, I: 535–6).

Before passing from the roots of virtue to the attributes and actions of Dionysus, we remark that although in the twenty-fourth fable Jupiter and Juno occur in their familiar capacity as husband and wife, what they represent seems to be different from what they represented in the earlier fables. This impression can partly be explained by the fact that we find ourselves in the second half of the book, a half that is characterized by a somewhat more historical approach to the philosopher's political education, as is underscored by warning references in the second half of the twenty-fourth fable to the historical fates of the ancient Pentheus and Orpheus. But even if we were not to pursue this line of explanation further, we would not see ourselves confronted with an insuperable difficulty when attempting to perceive the notion of tyranny as parallel to the "lower parts" of the soul.

Nor would we be looking at an impossibility when trying to understand the potential believer as the "mother of virtue," a possibility that will become a plausibility when we discuss the fable on pleasure in conjunction with the fable on profit.

Having made one of the more important discoveries in moral philosophy by coming to understand that virtue is the result of the "licit" or legitimized union of the "lower parts" of the "soul" and one's own "real good," we stay with the fable on desire to draw out the implications of moral man desiring to relieve the burden of conscience through the practice of virtue.[85] For Bacon omits to tell that in a decisive way it is the "mother of virtue" who is the "mother of desire" and therefore the actual representative of "apparent good," who as it were aspires to her own honours, that is, who desires to become a true believer or the embodiment of "real good." Bacon's omission is corrected by the ancient sources, however, which tell us that it was the tyrant's proud wife who deceived the pregnant woman who had tried to imitate her prerogative into unwittingly wishing her own destruction, thereby causing the offspring of "apparent good" to be born.[86] But although we do not consider concealed pride a virtue, we do consider virtue to be grounded in affection to the extent that the mother of every child of the human soul is the "nature of good." Not every "species" of the genus works real good, however, as was indicated by the death of Semele or the connection to one's own and therefore to the good, and as is explicated by Bacon's history of the apparent good that was worked by Dionysus, a history to which we now turn our attention.

Dionysus "died, was buried for some time, and came to life again not long after." According to Bacon this is a point "excellently devised," because "sometimes the affections seem to have been laid to sleep and to have been extinguished." But "no faith can be placed in them, not even when they are buried, for if provided with matter and occasion they resurge."

In his early youth Dionysus "discovered [invenit] and taught the culture of the vine and the composition and use of wine, which had been unknown up to then." He became "celebrated and renowned" for this, and he "subjugated the whole world." The "invention [inventione] of the vine" is a "prudent parable," Bacon argues, because "every affection ingeniously and sagaciously finds out its own stimulants," and there is "nothing men know of that is so potent and efficacious in exciting and inflaming perturbations of every kind as wine," which is a "kind of common stimulant to all of them. Never acquiescing in what it has, but striving further and yearning for new things with infinite insatiable appetite," affection is represented as the "subjugator of provinces and the undertaker of an infinite expedition."

The fact that Dionysus was carried about in a chariot drawn by tigers is taken by Bacon to illustrate that "as soon as affection ceases to go on foot, comes to ride in its chariot, and conquers and triumphs over reason, it is cruel, savage, and untamed towards everything that is adverse to it or that opposes it." The Muses also "joined the retinue of affection," because there is "hardly any affection that has not some branch of learning to be flattered by." In this situation, Bacon goes on to observe, "the majesty of the Muses diminishes due to the indulgence of wits, which turns those that should be the leaders of life into mere followers of affections."

Dionysus was the "inventor [inventor] and founder of sacred rites and ceremonies," which, however, were "fanatical, full of corruption, and cruel." Bacon points out that one need not wonder that "superstitious rites" are attributed to Bacchus, as "almost every unhealthy affection grows rank in depraved religions."

Dionysus had the power of exciting frenzy, and it was by "women who had been roused to frenzy in his orgies" that "two distinguished men," Pentheus and Orpheus, are said to have been torn to pieces, "the one having climbed a tree to see what the women were doing, the other while strumming his lyre." This is an "evident parable," Bacon teaches, as "curious inquisition as well as free and salutary admonition are irritable and hostile to a prevailing affection." Bacon emphasizes that it "does not make any difference if the inquisition is merely for the sake of contemplation or of watching with pleasure, as if from a tree, without any malignity, or if the admonition is applied with much sweetness and dexterity," for "the orgies cannot in whatever manner tolerate either Pentheus or Orpheus."

The persons of Bacchus and Jupiter are "often confused," and the "actions" of the former God "are often confounded with and are not easily distinguished from" those of the latter, which Bacon ascribes to the fact that "noble and renowned deeds, distinguished with merit and glory, sometimes proceed from virtue, right reason and magnanimity, and sometimes only from some latent affection or concealed desire, however much they are elevated by being extolled with fame and praise" (DSV XXIV "Dionysus," VI: 665 ff.; I: 535 ff.).

Because it would have testified to an "illicit" impiety if he had drawn the comparison explicitly, Bacon only implicitly alludes to the obvious parallels between the son of Jupiter, who came to Thebes to prove that he was a god, and the Son of Mary, who was not believed to be a God when he was still a man. Like Jupiter, who was already married to Juno when he impregnated Semele with an illegitimate foetus, Mary was already espoused to Joseph when she was found with child of the Holy Ghost (cf. Matt. 1: 18–25; Luke 1: 26–38 and 3: 23; John 6: 42).

But whereas the just man Joseph took unto him his wife because her Son was to save His people from their sins, Juno drove Dionysus mad after he had discovered the "culture of the vine," as Bacon tells us by means of the ancient sources.[87] This causes us to wonder not only whether it was not Juno who was the actual "inventor" of the vine, as she was the actual mother of Dionysus, but also whether it was not out of proud self-forgetfulness that she allowed the "subjugator of the world" to become "celebrated and renowned" for what was her invention, especially in light of the fact that self-forgetfulness is considered a virtue primarily by those whose moral self is constituted by the forgetfulness of a self they themselves never possessed: selflessness is the privilege and initial burden of the selfless. Assuming that Juno indeed invented the vine before she drove the discoverer of its culture mad, we are even forced to wonder whether it was not on the foundation of a moral self that the desire for exclusivity through generality grew to become its own stimulant, a *deus ex machina* so to speak, and therewith the common but self-forgetting cause of all moral discoveries.[88] After all, it was only after having been driven mad that Dionysus became the "inventor and founder of sacred rites and ceremonies," which seems to allude to the fact that genuine morality wholly depends on the forgetfulness of the underlying desire for a moral self. And it seems to be also with a view to the self-reinforcing power of sanctioned selflessness that Bacon mentions the fact that the face of Dionysus "looked so much like that of a woman that it seemed doubtful of which sex he was," a fact which he interprets by saying that "the sex of every affection of a more vehement kind is doubtful, seeing that it has both the force of a man and the weakness of a woman" (VI: 665–6). For it is from the dialectic between the force of divine approbation and the forgetfulness of its weak foundation that moral virtue derives its true vehemence.

Pursuing this line of thought in the direction of its most recent cultural manifestation, we recall that Bacchus was part of his father only until he was born, although both his person and his actions were "confounded" with the person and actions of his father. Jesus Christ, on the other hand, the only begotten Son of God the Father, constitutes an eternal and indivisible unity in equality with His Father and the Holy Spirit, although inasmuch as He was according to the flesh He was less than the Father and less than Himself.

When we allow these differences between Bacchus and God the Son to become mutually illuminative, we are led to the conclusion that it is as much the case that the union of the "soul" and one's own "real good" is the result of moral virtue as it is the case that moral virtue is the result of the union of God and Lamb. But a similar conclusion could already have been drawn on the basis of a dissolving reading of the differences between Bacchus and God the Father. For whereas Jupiter needed the "good" in order to beget Bacchus, the good God is Himself the begetter of the desire for the good that is Himself. Moral virtue therefore seems to be both the result of and the means to a good conscience or to self-deification.

We must be careful, however, not to become oblivious of the difference between discursive and performed fables. We therefore hasten to a more historical analogy between the son of Semele and the Son of Man. For whereas it was only after having died, having lain dead for some time, and having resurrected[89] that Dionysus "discovered the culture of the vine" and "subjugated the whole world," Jesus Christ was believed only by His disciples after he had turned water into wine (John 2: 1–11).

But lest we become "giddy" from mixing the clean with the cleansed, we should quit drawing analogies and follow the god of wine on his imperial journey to the city of Thebes, a city still in front of us when we discussed the immature political philosopher in the tenth fable, but serving as the background of the fable that will set the mature but as yet stationary political philosopher in motion by providing him with his last and in a way most historical attribute: "swollen feet."

Empire

Sphinx was a "multiform monster." She had the face and voice of a "virgin" and the claws of a griffin. She dwelt on the ridge of a mountain near Thebes and blocked the roads, lying in ambush for travellers, whom she attacked and captured. When she had got them in her power, she propounded "obscure and perplexing riddles" to them, which "the Muses were believed [putabantur] to have provided her with." If the wretched captives were not able to solve and interpret them, the Sphinx "cruelly tore them to pieces as they stood hesitant and confused." Because the calamity was "not counteracted by time," the Thebans offered the empire of Thebes as a reward to any man who would be able to expound the Sphinx's riddles, for that was the only way to subdue her. The "greatness of the prize" excited Oedipus, a "shrewd and prudent man," but lame with wounds in his feet. He presented himself to the Sphinx "with a faithful mind, and with alacrity in his heart." Upon being asked what animal is born four-footed, then becomes two-footed, three-footed, and finally four-footed again, Oedipus "presently [praesenti]" answered that this animal was man. As this was the right answer, it caused him to obtain the victory. He thereupon slew the Sphinx and was created king of Thebes.

According to Bacon, this "prudent" fable seems to have been devised with a view to treating the subject of "science, especially in conjunction with practice." Science, being the "wonder of the ignorant and inexperienced," is "not absurdly called a monster." In figure and aspect it is multiform, on account of the "immense variety of subjects treated by science." It has the face and voice of a "woman [muliebris]" because of its "charm and talkativeness," and it has claws sharp and hooked because "the arguments and axioms of science penetrate, seize, and occupy the mind, so that it cannot move away or escape."[90] Science is stationed on the height of mountains, for it is "deservedly esteemed a thing sublime and lofty, which looks down on ignorance as from a superior place." It blocks the roads, because "at every turn in the journey and pilgrimage of human life matter and occasion for contemplation encounters and forces itself upon us." The Sphinx presents mortals with a variety of difficult questions and riddles, "which she has received from the Muses." As long as these questions and riddles stay with the Muses, there "may [fortasse] not be any cruelty involved." For, Bacon explains, "as long as meditation and inquiry have no other end than to know, the intellect is not oppressed or straitjacketed, but is free to wander and expatiate, and it finds some delight and delectation in the very doubting and in variety." But when questions and riddles are transmitted from the Muses to the Sphinx, that is, "to practice, and demand and call for action, choice and resolution," the riddles become "burdensome and cruel." And unless they are "solved and disposed of," they "torment and worry the human mind."

Bacon goes on to observe that the riddles of the Sphinx always propound a "twofold condition," the "laceration of mind" if one fails to solve them, and "an empire" if one solves them. For "he who understands the matter is master of his end, and every workman is emperor over his works." Of the Sphinx's riddles there are "two kinds altogether," the riddle of the "nature of things," and the riddle of the "nature of man." Similarly, there are two kinds of empires as a reward for solving them, the "empire over nature," and the "empire over man." The empire over natural things is the "proper and ultimate end of true natural philosophy," but the riddle propounded to Oedipus, the solution of which caused him to obtain the empire of Thebes, "related to the nature of man." For, Bacon concludes, "whoever has a thorough insight into the nature of man can shape his fortune almost [fere] at will, and was born for empire" (DSV XXVIII "Sphinx," VI: 677 ff.).

All our remaining hopes as to the feasibility of the image of "Greek" self-love that was treated in the fourth fable have evaporated into the thin air of the imagination by the time we arrive at the fourth last fable. For in the meantime, the god of wine intervened, subjugating the world by intoxicating the charming but talkative Muses with the desire to "flatter" his inventions with their "learning."

We notice in passing that both Bacon's instauration of learning and the fate of instaured learning can only be truly understood if one thoroughly thinks through the consequences of the fact that it was the weakness of these learned women that rid Dionysus, a depraved image of God the Son, of "all restraints of shame and fear," because it was these weak but learned women that provided the imperial son's concealed imperiousness with the forceful pretext of learning, thereby ensuring that his "cruel," "depraved," and "superstitious" rites found their way "to practice" by means of an effecting executioner. We will return to the women who represent the poetic voice of the ancient philosophers when discussing the thirty-first and final fable, which is the third of the three fables in which they occur, as it is the last of a triad of fables in which the modern philosopher Orpheus makes his appearance.

But first we have the Sphinx on our hands. The ancient sources tell us that it was either jealous Juno who sent the Sphinx to Thebes, or that Dionysus himself took the inquiring monster to the city he had chosen to witness his divinity.[91] In both cases the appearance of this perfidious creature can be traced back to the tyrant's conjugal favourite, who either herself sent the Sphinx to Thebes as a divisional instrument of self-elevation, or had Jupiter's son, whose actions were "confounded" with those of his father, a depraved image of God the Father, take the dividing monster with him on the imperial journey he commenced after his father's self-forgetful wife had driven him mad.

Now that the Sphinx blocks the roads to the empire that is ruled by Dionysus because—as Bacon subtly suggests by using the city of Thebes as the background of his fable—Pentheus was too hesitant to decide what light to rely on when he approached the pretender to his throne, instant action is demanded for the purpose of instauring the situation in which transferring contemplations into active life does not necessarily involve "cruelty." Employing compound imagery, we may say that it is not until he "slays the Sphinx" by solving the riddle of the cruel empire builder that the political philosopher can transfer his contemplations to the fertile soil of the province of political things without being in constant danger of suffering the fate of Pentheus and Orpheus; in other words, without running the risk of falling into the hands of "tigerish" women who do not "tolerate" an inquiry into what had made them "triumph" over reason.

It is only after having seized the occasion by "clearing the road" for philosophy that the political philosopher can turn to his journey of understanding, on which he will come across an "immense variety of subjects" for his slow but "audacious" inquiry into the "monstrous" science of corruptible things. The political philosopher aspires to this science with no other end than to know, and he "masters" this end because he "understands" it. His Thebes is an "empire" without a seat. Bacon alludes to these points by having the lame Oedipus "presently" answer the riddle of the Sphinx, while at the same time emphasizing that the Sphinx was subdued by a "lame man" with "swollen feet" because men usually proceed to the solution of the Sphinx's riddles "too fast and in too great a hurry" (DSV XXVIII "Sphinx," VI: 680).

But whereas the ancient Sphinx hurled herself from Mount Acropolis after her riddle had been solved, the modern Sphinx must be slain, because not even in a "buried" Bacchus "faith can be placed," or so Bacon told us in his interpretation of the twenty-fourth fable.[92] At the risk of transgressing the bounds of the narrative, we may even say that every "peripatetic" philosopher must walk in the ways of his patricidal predecessor, who ended up in Thebes on his quest for his true parents after having been scorned for being a suppositious child, and who approached the Sphinx because the "greatness of the prize" for solving the Sphinx's riddle had excited him.[93] And indeed, the "virgin" of the narrative has been slain at least once before, for by the time we arrive at Bacon's solution to the problem of empire the virgin has become a "woman."

The fable on philosophy already taught us that the "difficulties of life" and the "arcana of science" are of necessity interrelated, and that the solution to the riddle of the "nature of man" provides the philosopher with the key to the "empire" over corruptible things. But although the fable on science confirms this insight by having the Sphinx propound both the riddle of "man" and the riddle of "nature," the slaying of the Sphinx presents the newly orphaned philosopher with a new problem. For now that his insight into the nature of moral man, having become thorough in the fable on the history of desire, has caused him to be reborn for an empire that fulfils man's moral needs as it controls his virtuous deeds, it is the philosopher's turn to give birth to a suppositious child. From his "head," that is.[94]

Midwifery

In the fable that elaborates on the second fable's end Bacon tells the story of what had caused the second fable's beginning, but what teaches the newly born "philosopher-king" to prevent it from beginning again. For in the thirtieth and penultimate fable the divine philosopher learns how to prevent Juno's anger, which explains why the fable on Jupiter's mistress

Metis, whose name "not obscurely signifies counsel," is said to contain an "arcanum of empire," despite or because of the fact that it is "monstrous and at first hearing very foolish."

The obscure fable treats the art whereby kings conduct themselves toward their councils in such a way as not only to "keep their authority and majesty untouched," but also to "increase and extol it in the eyes of the people." By a "sound and prudent arrangement" kings "tie themselves to their councils with a bond like that of marriage," and "deliberate with them on the greatest matters, judging that this does not diminish their majesty." But when it is time to "decide on the matter, which is like bringing forth, they transfer it to themselves, so that the "decision and execution," the latter of which is represented by the figure of the armed Pallas because it "comes forth with power and carries necessity, seem [videatur] to emanate from themselves." It "does not suffice that the decision seems [videatur] to proceed from their royal authority and their unconstrained, free, and independent will," unless they also "cause it to be thought [existimentur] to have been born from their own head, that is, from their own judgment and prudence" (DSV XXX "Metis," VI: 683).

Now that he wears the ears of an ass, the divine philosopher knows how to descry the voice of human reason in a pandemonium of foolish sounds. We hear him discern the voice of the Oceanid who knew more than all gods and mortals, and who was the embodiment of practical or cunning wisdom (Μητις), counseling him on how to secure the fate of her unconceived son, who, as we have seen, was fated to replace Jupiter as the ruler of gods and mortals.[95] Her counsel can be said to consist in an inversion disguised as an imitation of the counsel of the ancient Metis. For whereas the ancient Metis had helped Jupiter to establish his rule by drugging his father Saturn into disgorging the children whom he had swallowed, and who were to aid Jupiter in the theomachy,[96] the modern Metis helps the son of Saturn, that is, the "son of wisdom" who is to found an empire, to overturn the tyrant's rule by counseling him to imitate it.

Having already practiced "imitative virtue" in the central fable, the divine philosopher is well prepared for the next task of grounding his empire on the "bond" of the marriage between Jupiter and Juno; more specifically, of "tying" the tyrant's lover to him by making her think that it is his inimitable relationship with her that causes him to "deliberate" with her "on the greatest matters." He thus not only appeases the jealousy of God's favorite, who as the former embodiment of God's empire forms the body of his empire; he also "increases and extols" the authority of the very powerful ruler of gods and mortals in the eyes of his proud wife by making her think that he has the power to decide matters according to her loving counsels. The philosopher moreover ensures that Jupiter's "majesty" is nowise "diminished" by his consulting Juno on the "arcanum" of his empire, because the ruler god's decisions are "thought" to have been born from his own judgment, and because he is thought not to have been in need of counsel on how to secure his "authority and majesty." Bacon in a way summarizes the teaching of the penultimate fable in one of his *Essays* by observing that the "wisest *Princes*, need not thinke it any diminution to their Greatnesse, or derogation to their Sufficiency, to rely upon *Counsell*," as "*God* himselfe is not without" (E XX "Of Counsell," 63, VI: 423).

The philosopher secures the foundations of his empire by executing his grounding decisions through Pallas, the warrior goddess of wisdom who presides over philosophy's "ministry of political defence," and who "armed" philosophy with the "power" of the voice of God, and the "necessity" of the oath of Styx. He prudently judges that he must make it seem as if his decisions have been "born from the head" of their father in speech, whom at the same time, we interpose, he does not leave "any room for penitence."

Rhetoric

But the wrathful Jupiter still has one unpleasant surprise left for man, a surprise that as it were prepares the fable on pleasure, since it stimulates us to redefine desire as the lust for pleasure, while forcing us to distinguish between different kinds of pleasure.

Because Prometheus had tried to deceive him with his sacrifice, Jupiter decided to "afflict the human race" by ordering Vulcan, the son of Juno, to make the "lovely and beautiful" Pandora. When she had been made, each of the gods "bestowed a gift upon her." They placed in her hands an elegant vase, which contained "all evils and hardships."

Bacon points out that Pandora signifies "pleasure and lust," and that "infinite evils and late penitence flowed from her upon minds, bodies, and fortunes of men." And not only upon "individuals," but also upon "kingdoms and commonwealths," for "wars, disturbances, and tyrannies have sprung from this same fountain" (DSV XXVI "Prometheus," VI: 669, 674).

In accordance with the "vicissitude of things," "destined" by fabulous necessity at the end of the eleventh fable, and foretold at the beginning of the non-Orphic era by philosophy's prophetess, who from the first fable onward foresees events from her "watch-tower," the fabulist Bacon makes the waters of the river Helicon, which was sacred to the Muses, "break out and emanate again" in the final fable of *De Sapientia Veterum*. We need not be surprised that the protectresses of the ark of true poetic inspiration are reinvested with their majesty in the fable at the end of which the philosopher as it were redistinguishes poetic from philosophic pleasure (cf. DSV Pr., VI: 625 with DSV XXXI "Sirenes," VI: 686).[97] And in light of the sanctification of the Muses by a non-poetic majesty intolerant of non-poetic pleasure it need neither surprise us that it is also in the fable on pleasure that Bacon draws our attention to the threefold materialization of pleasure's twofold rootedness in desire and *eros*.

Bacon starts by saying that the fable of the Sirens is "rightly applied to the pernicious allurements of pleasure," but in a "most vulgar sense." It is said that the Sirens were daughters of Achelous and Terpsichore, who was one of the Muses. They originally had wings, but after having been beaten in a contest with the Muses, whom they had "blindly challenged," their wings were pulled off, and the Muses turned them into crowns for themselves. Thenceforward all the Muses, "except the mother of the Sirens," wore wings on their heads.

The Sirens "dwelt on certain pleasant islands where from their watch-tower they watched for approaching ships." When they saw a ship advancing, they began to sing. The voyagers were "first captivated and then lured" by the singing of the Sirens, who thereupon caught and killed them. The singing of the Sirens was "not simple," but they "captured in the manner that was most in conformity with the nature of the listener." The plague was "so destructive that the islands were seen from afar, white with the bones of unburied carcasses. Two kinds of measures were found against this evil, one by Ulysses, and one by Orpheus." Ulysses ordered the ears of his whole crew to be plugged with wax. He himself "wished to make trial of the thing, while averting the danger." He therefore wanted to be tied to the mast, but he forbade his crew to untie him even if he himself requested them to do so. Orpheus, on the other hand, "spurned being tied and with a clear voice sang the praises of the gods to the accompaniment of his lyre, drowning the voices of the Sirens and steering clear of all danger."

Bacon explains that the fable pertains to "morals." Formerly "pleasures carried mortals away suddenly, at the first temptation, as if with wings." But "doctrine and teaching have brought about a situation in which the human mind restrains itself somewhat, and considers the consequences of the thing." Thus they have "stripped the pleasures of their wings." This "redounded greatly to the glory and honour of the Muses," for "as soon as it appeared from some examples that philosophy could induce contempt for pleasures, this was at once looked upon as a sublime thing, which could raise and elevate the soul from the earth, and make the cogitations of man, which thrive in his head, winged and as it were ethereal." Only the mother of the Sirens "still goes on foot and has no wings." She embodies "lighter doctrines, invented and applied only for delight." Such doctrines seem to aim at "removing the wings from the Muses' crowns and restituting them to the Sirens."

Although the Sirens are said to be inhabitants of islands, because pleasures "commonly [fere] seek secession and avoid the throngs of men, . . . everybody sings of the Sirens' song and of its pernicious character and various artifices." Bacon therefore argues that this point "needs no interpreter." But the fact that the bones, like white cliffs, are visible from a distance is "more acute." For it signifies that the "examples of calamities, however clear and conspicuous, do not effect much against the corruptions of pleasure."

Bacon goes on to remark that with regard to the remedies the parable is "prudent and noble, but not at all abstruse." There are "three remedies for an evil that is so fraught with cunning as well as violence, two from philosophy, and a third from religion." Resisting the beginnings and "sedulously avoiding all occasions that may tempt and solicit the mind" is the first manner of escaping. This is denoted by the plugging of the ears, and it is the necessary remedy for "mediocre and plebeian minds," such as the crew of Ulysses. But "loftier minds" can venture among pleasures if they "fortify themselves with constancy of resolution." They "take delight in putting their virtue to a more exquisite test," and "learn more thoroughly the ineptitude and insanity of pleasures, as contemplators rather than as followers, which is what Salomon professed about himself." Heroes of this order stand "unshaken amidst the greatest temptations of pleasure," provided that they follow the example of Ulysses and interdict the pernicious counsels of their own followers, which "of all things have most power to unsettle and dissolve the soul."[98] But the "best remedy in every respect" is the remedy of Orpheus, who "confounded the voices of the Sirens, and drove them off by singing and resounding the praises of the gods." For "meditations upon divine things surpass the pleasures of the sense, not only in power, but also in sweetness" (DSV XXXI "Sirenes," VI: 684 ff.).

The final fable sets Bacon's political teaching in motion, as it shows the mature political philosopher in action. But before turning to his actions, we should have a closer look at what the modern political philosopher acted against. Bringing the threads of the fables on morals together, we find that the reason why the Sirens were beaten by the Muses was that they had failed to see that the frenzied Muses had already been incorporated into the army of Dionysus and were carrying the powerful weapon of his wine. The

Muses added lustre to their victory over the Sirens by turning the "wings" of lighter forms of poetic pleasure into the heavy "crowns" of Dionysus. They were greatly "honored and glorified" for having provided the almost unarmed conqueror with a sublime weapon. For by means of the "artillery" of their mercenary doctrines the learned Muses invited "contempt" for any kind of pleasure that is not ethereal or Dionysiac.

Having grown accustomed to the fact that monsters like Typhon, Dionysus, and the Sphinx spring from common roots, we need not wonder that the ancient sources tell us that it was Juno who persuaded the Sirens into challenging the Muses.[99] For the divisional doctrines of the divinized voices of philosophical poetry provided the woman without qualities with a unique opportunity to legitimize her desire for an inimitable relationship with the ruler of gods and mortals. The reason why the newly born Dionysiac Sirens[100] are "inhabitants of islands" is thus that they have entirely etherealized pleasure.

But since as yet only the "vulgar" have interpreted the song of these seceding insulars, the truly "pernicious" character of the "allurements" of Dionysiac pleasure most acutely needs to be interpreted, as Bacon already indicated by saying that this point "needs no interpreter." For it turns out to be exactly the secessional nature of ethereal pleasure that causes examples of earthly calamities not to "effect much" against its lethal corruptions, which ensue from living amid but not among men. But the Muse Terpsichore, the mother of the Sirens, "still goes on foot and has no wings," which provides Bacon with the opportunity to "restitute" the wings to the Sirens,[101] and to reinvest the daughters of memory (Μνημοσύνη) with their original task, which consisted in leading the pilgrimage of man's earthly life by reminding him of what was, what is, and what will be.

With regard to the remedy, the parable is somewhat "abstruse," however. In the narrative Bacon explicitly says that there were two kinds of measures, one of which came from Ulysses, and the other from Orpheus. In the interpretation, however, he mentions three remedies, two from philosophy, and one from religion. He complicates matters by mentioning the example of the biblical Solomon, which he does after having mentioned the examples of the crew of Ulysses and of Ulysses himself, but before mentioning Orpheus. Because Ulysses took the initiative both with regard to his crew and with regard to himself, his remedy can be counted as one. Solomon is the third person mentioned, representing the religious man, but offering no remedy. We thus suspect that by means of this deliberate abstruseness Bacon alludes to the fact that, although the remedy of the divine philosopher Orpheus comes from philosophy, the example of the song of the Dionysiac or religious Sirens taught philosophy that the wings of the Sirens and the memory of the Muses can only be recaptured by philosophy adopting religion's manners of capturing men.[102] In other words, the song of the religious Sirens must be countered by a song that tempts the listener by appealing to the imaginative part of his nature, which had enabled the religious Sirens to "capture" his imagination.[103] It is the ambulant mother of the heathen Sirens who provides Orpheus with "light" earthly doctrines to counter the heavy ethereal doctrines of the religious Sirens, showing him how philosophy, which had invited contempt for pleasure that is earthly can also induce contempt for pleasure that is ethereal.[104]

Having put on a "seductive suit" with a "sewed up thigh" before "relieving his head" of the burden of a solved riddle, the "master of harmony" has his hands free and his head straight to perform a song empowered by divine "meditations," but sweetened for the human imagination; a song, in other words, that "confounds" and "drives off" the voices of religion by "singing and resounding" the praises of God while substituting them with its own sweet doctrines, which consist in sweetened versions of the doctrines it substitutes. These harmonizing doctrines will occupy us as we proceed to the more foundational parts of Bacon's teaching, seeing that they solicit the charmed reader's approval before leading his unaffected counterpart to their underlying contemplations.

It is therefore with a view to completing the conversion of Juno in order to lower the obstacles on the way to Bacon's thought that we now turn to the fable on profit, which treats the intricacies of Bacon's policy of employing the foundations of faith against faith.

Atalanta, who "excelled at natural velocity," entered a running contest with Hippomenes. The conditions were that if Hippomenes won, he was to marry Atalanta, and if he lost, he was to be put to death. There seemed to be no doubt about Atalanta's victory, because her "insurmountable excellence in running" was demonstrated by the death of many competitors. Hippomenes therefore resorted to trickery. He provided himself with three golden apples. After the race had begun and Hippomenes found himself far behind Atalanta, he threw forward one of the golden apples so that she might see it; not straight ahead, however, but aside, so that she might not only be delayed, but also drawn out of the track. "With a woman's cupidity, and seduced by the beauty of the apple," she acted exactly the way Hippomenes wanted. Hippomenes eventually gained the victory "by astuteness, and not by virtue."

According to Bacon, the fable seems to contain an "allegory on the contest between art and nature." Provided that "nothing hinders and impedes it," art, signified by Atalanta, is "by its own virtue much swifter than nature," and "arrives at its goal much faster." This can also be seen in morals, because "whereas oblivion of and consolation in sorrow comes after a very long time through the beneficence of nature, philosophy, which is, as it were, the art of living, does not await that day, but forestalls and anticipates it." But the "prerogative and vigour of art" is retarded by the golden apples, as Atalanta turns aside toward "profit and commodity." Therefore, "art cannot conquer, put to death, and destroy nature." On the contrary, "art remains in the power of nature, as the married wife is subject to the husband" (DSV XXV "Atalanta," VI: 667–8).

Having been stimulated by the metaphor of marriage to look upon Atalanta as a stand-in for Juno, we find the fable on profit conveying the warning that if left to her own concealing virtue the lover of tyranny would by means of her lethal artifices expand her territory with an almost "natural velocity." The philosopher must therefore become "astute" enough to resort to the art of trickery in order to subject the artful Juno to him. By means of "profitable and commodious" doctrines, which philosophy's Muse Terpsichore provides him with, he causes his wife in speech not to feel sorrow over being "drawn out of her track" of conquest, while by appealing to her natural "cupidity" he seduces her into thinking that she is not being seduced by an artifice of philosophy. She thus becomes the "mother of virtue" while remaining the "mother of desire," working "real good" while giving in to her desire for "apparent good."

Returning to the fable on pleasure, we hear the ancient sources telling us that there was a prophecy that if a ship sailed past the Sirens, the Sirens themselves would die.[105] We therefore finish our interpretation of the narrative part of *De Sapientia Veterum* by saying what Bacon did not want to say, thereby giving in to the constant temptation of assuming a "licence" in interpreting the fables not dissimilar to the licence Bacon himself assumed in devising them.

After philosophy's prophetess had foretold that the ship carrying the seeds of philosophy would sail past the religious Sirens, the modern Orpheus, both the master of the ship and the master of harmony, anticipated the danger by singing so loud that one could not even hear the penitent voice of the drowning theomaniacs while they were plummeting towards the bottom of the sea, carrying the heavy vase of Pandora as its evil contents were floating to the surface, silently announcing the imminent end of the Orphic order.

But musings on the limited effectiveness of symptomatic treatment of natural conditions tend to make one oblivious of the nature underlying these conditions. This tendency is corroborated by Bacon's own remark that "good books," as the "Serpent of Moses, mought deuour the Serpents of the Inchantors" (AL 2, 61, III: 327–8; cf. Exod. 7: 1–12). Bacon emphasizes, it is true, that "in true value" rhetoric is "inferiour to Wisedome, as it is sayd by God to *Moses,* when he disabled himselfe, for want of this Facultie, *Aaron shall bee thy Speaker, and thou shalt bee to him as God."* But the sting of his sentence is in its tail, which says that "with people [rhetoric] is the more mightie" (AL 2, 127, III: 409; cf. Exod. 4: 16). The lyre and voice of Orpheus were to Bacon what the rod and voice of Aaron were to Moses. The people of Israel did not hearken unto Moses's voice until the man who was his mouthpiece had persuaded them into believing that the Lord had appeared to Moses (Exod. 4: 1–31). Likewise, the people of the books were re-enchanted only after Bacon had ordered Orpheus to sing the praises of a humane God to the accompaniment of his lyre. But unlike Moses, Bacon did not see the back of God when he turned back to himself.

CHAPTER 3

The Trinity of Philosophy

Πόλεμος πάντων μεν πατηρ εστι, πάντων δε βασιλεύς και τους μεν θεους εδειξε τους δε ανθρώπους, τους μεν δούλους εποίησε τους δε ελευθέρους.

—Heraclitus: Frgm. 53

Aristophanes's *Birds* relates the story of two Athenian citizens, Peisthetairos and Euelpides, who escaped Athens because they had become weary of their city's love of litigation. Because they had no knowledge of the existence of a place of quiet and tranquillity, they decided to consult Tereus, a former Thracian king who had been turned into a hoopoe. After the example of Tereus had shown them that a former human being could live pleasantly among birds, the idea of founding a city of birds occurred to Peisthetairos.

Because Peisthetairos was a political man, he understood that the rule over intervening gods who are divided against themselves was a necessary prerequisite for the political tranquillity that he was pursuing. Having argued that, since mortals self-evidently depend on the gods, and the immortal gods are dependent on the sacrifices of men, the birds could rule both gods and mortals by controlling their mutual intercourse (184–94), Peisthetairos suggested that the birds occupy the sky that separates heaven, the dwelling of the gods, from earth, the home of men. The birds were initially unwilling to listen to a member of the species they regarded as their hereditary enemy, but when they were on the verge of giving way to their anger at their alleged tormentors, the bird-man Tereus intervened. He appeased the birds' concealed fear by pointing out to them that, although man is a bird's natural enemy, wise men learn from their enemies, and the birds might learn something useful from a man who came as a friend.

The birds' response made Peisthetairos understand that, to generate an army of god-like birds, he had to turn himself into a man-bird. Now that he had their ears, Peisthetairos persuaded the pious birds to believe that they were the most ancient beings, and that they therefore had a just title to rule. The birds showed their gratitude to the man who had reminded them of their original greatness by recognizing him as their savior, and they emphasized that life for them was worthless until they had recovered their former kingship (548–9). Peisthetairos suggested that the birds build an ethereal city with a brick wall around everything lying between heaven and earth, and that they then reclaim the empire from Zeus. In case Zeus were unwilling to abdicate, Peisthetairos wanted the birds to declare a holy war upon the gods, and forbid the gods from passing through the birds' territory every time they desired to descend to the earth to indulge in their amours with mortals (557–60).

Tereus wondered, though, why men should recognize birds as gods, as they had become accustomed to recognizing only the Olympian gods as gods (603–6). Peisthetairos responded that if men continued to worship the Olympian gods because they believe that these gods are more powerful than birds, the birds should starve men out to convince them of their power, and of the gods' lack of power. To secure being recognized as gods, however, the birds should show philanthropy in addition to power. For, as the man-bird Peisthetairos went on to argue (1225–6), mortal birds do not become gods until they become gods for men. But birds do not reason. The redeemed birds could therefore only display their willingness toward their savior by supplying him with their power. Meanwhile, it had become clear to the unpolitical Euelpides that a theomachy would not provide him with the tranquillity he longed for. He thus quietly left the scene never to return again.

The next phase in the comical contest over the attributes of godhood was the announcement to Peisthetairos of the visit of a delegation of three gods. This divine delegation, although internally divided, was to bring about a reconciliation between gods and birds because the community of gods had come to suffer from insurgencies within its own ranks now that the individual gods were not properly nourished as a result of Peisthetairos's founding of an ethereal city. Peisthetairos demanded not only that Zeus restore the scepter to the birds, but also that the ruler of gods and mortals give to him as his wife the beautiful Basileia or kingship, who was the guardian of the means of kingly power (1522–6). In return for this, the birds would observe men's perjuries, and execute the divine judgments. The gods felt compelled to accept the terms of the treaty as they were eager to punish men's impieties, but were unable to do so without the eyes and power of the birds. At the end of the comic drama, the birds hailed the winged Peisthetairos as the highest of the gods (1746–7).[1]

"All history walks upon the earth, and performs the office of a guide rather than of a light. Poetry is a dream of learning that would seem [volens videri] to have something divine in it. But it is time to awaken and to rise above the earth in order to wing my way through the clear air of philosophy and the sciences" (DA 3.1, I: 539).

With these words, Bacon prefaces his explicit treatment of philosophy and its threefold object, God, nature, and man; words pregnant with thoughts, it is true, but also words that give birth to objections. For can a philosopher simply dismiss the claim that everything is "full of gods" as a figment of the imagination? In other words, can a philosopher simply assume that the laws of the heavens are accessible to reason because he himself has access to reason? No, he cannot, and as we will see, he will not.

But that is not to say that he could not benefit from someone to guide him on his way through the twilight of "divine" poetry. And it is here that the poet who criticized another philosopher for making gods out of clouds comes into play. For even though an ethereal city is clearly impossible, and hoopoes look visibly different from the cuckoos we encountered in the central fable of *De Sapientia Veterum*; even though the Father of Old Comedy presents a successful revolt of an un-philosophical man against the gods, whereas "[b]y aspiringe to be like God in knowledge man transgressed and fell"; even though Aristophanes makes a man-bird claim the power of the gods, whereas "by aspiring to be like God in power, the Angells transgressed and fel" (AL 2, 155, III: 443); and even though poetry "commonly [modum] exceeds the boundaries of nature, and composes and introduces at pleasure what in the nature of things would never have come together or occurred" (DA 2.1 and 2.13, I: 494 and I: 518); even though all the previous points are true, it is also true that Aristophanes's *Birds* in a comical fashion unites the three objects of philosophy, since the man-bird Peisthetairos embodies the question as to the nature of God, the question that is coeval with philosophy, and that must be asked by a philosophic man having turned political.

But at the end of the chapter on poetry, Bacon warns us not to "stay too long in the theatre," now that we are equipped to "enter the palace of the mind with more veneration and attention" (DA 2.13, I: 538).

Nature

"Knowledge is similar to the waters. Some waters descend from heaven, and some emanate from the earth. Likewise, the primary partition of the sciences is to be drawn from their fountains, some of which are situated in the heights, and some here below." For, Bacon explains, "all knowledge is divided into two kinds of information, the one divinely inspired, and the other arising from the sense. Knowledge that streams in through teaching is cumulative, and not original. This also applies to waters, which, in addition to their primary fountains, swell by absorbing other rivulets" (DA 3.1, I: 539).

At the beginning of the second book of *De Augmentis,* Bacon had deduced the partition of "all human learning" from the three faculties of the "rational soul," which he terms the "seat of learning." From the "fountains" of memory, imagination, and reason emanate history, poetry, and philosophy respectively, and Bacon emphasizes that "nothing else" can emanate from these fountains. He does not think that any other partition is required for theology. For "although the information provided by oracle and sense certainly differs with regard to both the matter and the manner of insinuation, the human spirit is one, and its repositories and cells are the same. It is therefore only as if different liquids were poured through different funnels into one and the same vessel." For this reason Bacon also lets the subdivisions of theology emanate from the fountain of the rational soul. In accordance with the rational soul's three faculties, theology consists of "sacred history," which corresponds to memory, of "parables," which are "like divine poetry," and which correspond to the imagination, and of "precepts and doctrines," which "to a certain point [quadam] are like perennial philosophy," and which correspond to reason (DA 2.1, I: 494–5).

Division

Returning to the third book, we find Bacon expressly dividing knowledge or human learning into theology and philosophy. He emphasizes that by theology he means "inspired or sacred theology" and not "natural theology," which is part of philosophy. But he decides to reserve the treatment of inspired theology for the final book, so that he can conclude his discourses with it, as it is the "port and Sabbath of all human contemplations" (DA 3.1, I: 539–40).

At the risk of drawing a forced or inappropriate comparison, one could say that in a way not dissimilar to God, who on the seventh day "did rest, & contemplate his owne works" (AL 1.54, 34, III: 296), Bacon will in the ninth and final book of *De Augmentis Scientiarum* rest from and contemplate the work performed in the preceding six books, after the "divine compass" has navigated the ship of human learning from the island of inspired theology to the port of its "labours and peregrinations" (DA 8.3 and 9.1, I: 828–9).

It is imperative, however, that we commence our work of reconstructing Bacon's contemplations by interrupting it in order to ask ourselves what Bacon's division of knowledge into philosophy and inspired theology, made at the beginning of the first book on philosophy, made within the boundaries of the "rational soul" or "human spirit," and made according to the "nature of things" (DA 6.4, I: 712), ultimately entails. The significance of this question cannot be overstated, but has often been overlooked by scholars who interpreted the division by reiterating it. But even if we do not allow ourselves to be led astray by scholarly carelessness, we are almost naturally bound to overlook the gravity of this question, seeing that, as Bacon knew when he decided to employ the image of the natural flow of the waters, we cannot help being charmed by the image of natural harmony. We should be on our guard, however, against the danger of having our acuity impaired by the pleasant sound of a twofold fountain of knowledge fecundating a soil on which we can engage in discourses that morally justify themselves by being endless.

Having found shelter from the raindrops of morality that necessarily make the waters of knowledge become turbid, we start realizing that the two fountains of knowledge do not sprinkle one and the same soil. For the division between philosophy and inspired theology, or, as Bacon elsewhere (e.g., CV 7, III: 596) terms it, between "sense" and "faith," or between "human" and "divine," is coeval with the antagonism between the principle of disobedience and the principle of obedience. This implies that by distinguishing two fountains of knowledge Bacon ultimately distinguishes two mutually exclusive sources of knowledge; more precisely, he distinguishes between the possibility and the impossibility of knowledge as such.

The pious man who lives the life of obedience only knows what he believes to have been revealed to the prophets by the God who is the originator of knowledge. And "Gods first penne" (AL 1.58, 34, III: 297) reveals to him that the God who created heaven and earth out of nothing, and who demands man's pious obedience, only wants to be known by His actions and revelations: "I shall be that I shall be" (Exod. 3: 14; cf. also: Exod. 33: 19 and Rom. 9: 15). Since the God of Moses revealed Himself as omnipotent, and therefore as unrestricted by any nature, that is, by any intelligible necessity, beyond Himself, He is ultimately inscrutable for the natural light of philosophy.[2] As Bacon puts it in one of his prepolitical works, this God "is only self-like, having nothing in common with any creature" (VT, III: 218).

But if human knowledge, that is, knowledge of what is unchangeable, is dependent on the will of an omnipotent and therefore changeable God, all human knowledge is ultimately provisional, which is tantamount to denying the possibility of knowledge as such. Philosophy is fundamentally groundless if there is no nature in the decisive sense because everything that is, is by the grace of an omnipotent God. Consequently, the question as to the possibility of knowledge or philosophy is coeval with the question as to the nature of God.

Bacon could hardly have drawn attention to the primacy of this question in a more conspicuous manner than by making it the primary and therefore central object of philosophy (DA 3.1, I: 540).[3] But regardless of the unusually explicit truthfulness that Bacon displays at what one might call this moment of fecundation of the soil of philosophy, the fact that in many places throughout his works he exhorts men to "give to faith that which is faith's" (DA 3.2, I: 545; IMPr. 4, I: 131; NO 1.65, I: 176) has led many scholars to assume that Bacon assumed the existence of two separate and independent realms, which can coexist in relative harmony as long as they do not intrude into each other's territory. One of the more serious explanations for this confusion of Bacon's moderating rhetoric with the substance of his thought is to be sought in the enormous success of modern science's self-legitimating enterprise, consisting in the avowal of God's sovereignty within the realm of the human soul in order for the scientist to be granted a substantial latitude to indulge in scientific inquiry.

But in order not to confuse explaining with what needs explaining we should emphasize that the importance of calling attention to the confusion itself can hardly be overemphasized, especially in light of the fact that misinterpreting as a banishment from the realm of thought a philosopher's prudent exhortation to banish something worthy of thought from the realm of deed would be like judging "the store of some excellent Ieweller, by that onely which is set out toward the streete in his shoppe" (AL 1.65, 37, III: 300). We therefore consider it a display of prudence that Bacon observes that the "singular advantage which the Christian religion hath towards the furtherance of true knowledge" is that it "interdicteth human reason, whether by interpretation or anticipation, from examining or discussing of the mysteries and principles of faith" (VT, III: 251).

But there are more catches to Bacon's exhortation to demarcation. For in spite of all sophisticated attempts to prevent collision by diversion, the simple fact of the matter is that faith or inspired theology makes far-reaching claims about God and man that concern the same matter that is inquired into by the three objects of philosophy. Should one be unwilling to recognize this fact on the basis of the nature of the matter, one might be willing to condescend to listen to what our author himself has to say about the matter.

In the central section of the first book of the *Advancement of Learning,* Bacon addresses the matter with the circumspection required by its nature, and the indirection demanded by its apologetic treatment.[4] He points out that the "excellent Booke of *Iob,* if it be reuolued with diligence," will be "found pregnant, and swelling with naturall Philosophie" (AL 1.59, 35, III: 298). But if we "diligently revolve" the verses by means of which Bacon illustrates his point, we discover that without exception the phenomena mentioned in the book of Job allude to the magnitude of Him who is believed to be their originator.[5] Moreover, Bacon implicitly argues that the reason why the book of Job does not give birth to natural philosophy is that this biblical book considers fear of the Lord to be the foundation of wisdom.[6]

In the central aphorism[7] of the first book of *Novum Organum,* Bacon resumes the thread of his argument, this time operating with the more explicit boldness characteristic of a soliciting work.[8] He criticizes the corruption of philosophy that arises from "admixture of theology," illustrating this corruption by the example of those who "attempt to found a system of natural philosophy on the first chapter of Genesis, the book of Job, and other Sacred Scriptures" (NO 1.65, I: 175). But as the first chapter of Genesis contains a claim that concerns nothing less than the possibility or impossibility of any system of natural philosophy, philosophy cannot inquire into the foundation of natural philosophy without simultaneously inquiring into the foundation of a claim that denies natural philosophy its foundation.[9] As disobedient philosophy cannot recognize any claim based on the authority of faith, whereas the claims of inspired theology presuppose and therefore recognize "the sense," the confrontation between faith and reason can only take place on the soil of philosophy. Bacon testifies to this necessity by recognizing only the rational soul as the "receptacle of knowledge" or "judge of truth."

The battle for preferential treatment, although at first sight only a matter of logistics, thus turns out to have been the first substantial collision between faith and reason on philosophy's soil. For in the conflict regarding guidance, primacy in order necessarily implies primacy in substance. Had belief of truth been his starting point, Bacon would have given primacy to the discussion of inspired theology, since the postulates of faith would have dictated the course of the ship of human learning. But since the quest for truth is the only compass the philosopher Bacon recognizes on the voyage of his rational soul, his decision to give precedence to the discussion of philosophy and to postpone his discussion of inspired theology to the final book of *De Augmentis* necessarily implies a decision in favor of philosophy, and a postponement of the answer to the question as to the truth of the claims of faith until the criteria for judging these claims have been poured into the vessel of the rational soul.

In the central aphorism of the first book of *Novum Organum,* Bacon states that "from the unhealthy mixture of divine and human things arises not only fantastic philosophy, but also heretical religion" (NO 1.65, I: 175–6), a bold understatement inviting us to restate that the mixture of divine and human things is itself already "unhealthy" to the extent that confounding the principles of obedience and disobedience is unhealthy; to the extent, that is, that the very principle of disobedience is "heretical," and that philosophizing on the basis of the principle of obedience necessarily results in provisional and therefore "fantastic" philosophy. We must, however, postpone the answer to the question whether a "healthy" mixture of human and divine things will also result in "heretical" knowledge. For it is only after having "accumulated" the information "streaming in" through the works of Bacon's contemplations that we can answer the question as to the nature of the difference between the matter believed to originate in the "oracles" of divinity, and the matter discerned by the sense. And it is only after having gained knowledge on the matter of divinity that we will be able to answer the question to what point the "precepts and doctrines" of inspired theology are in conformity with "perennial philosophy." But what we can at least say at this moment is that by the time we reach the "port" of Bacon's contemplations the "swollen" matter of divinity will have lost its "originality."[10]

Unification

As it is of the utmost importance not to mistake difference in manner for difference in matter, we should look somewhat more closely into Bacon's treatment of the matter in the first book of the *Advancement of Learning,* since it is in the book written in justification of human learning that our author had not yet divided his readers according to their reading manners.

In his apology to the divines, Bacon points out that one of the "true bounds and limitations, whereby humane knowledge is confined and circumscribed," is "that we do not presume by the contemplation of Nature, to attaine to the misteries of God." For, he explains, "if any man shall thinke by view and enquiry into these sensible and material things to attaine that light, whereby he may reueale vnto himselfe the nature or will of God," he is "spoyled by vaine Philosophie: for the contemplation of Gods creatures and works produceth (hauing regard to the works and creatures themselues) knowledge, but hauing regard to God, no perfect knowledg, but wonder, which is broken knowledge." We have already seen that earlier in the same paragraph Bacon described wonder as the "seede of knowledge" (AL 1.6, 7–8, III: 266–7). In an earlier and unpublished work, our author had defined wonder as "contemplation broken off, or losing itself" (VT, III: 218). But since he was apologizing to the divines for the aspirations of human knowledge, he could not say explicitly that he considered philosophic wonder, which is produced by the contemplation of all things that cause philosophy to wonder, to be the seed of knowledge of the nature of God, as long as philosophy's contemplations are not broken off by pious reverence for the object of philosophic wonder.

In the central section of the first book, Bacon resumes the thread of his argument where it broke off, announcing that he will seek the dignity of knowledge in the "attributes and acts" of God, "as farre as they are reuealed to man." But since he takes the "sense" as his guide, the disobedient philosopher considers God's revelations to be identical to His actions.[11] God revealed Himself to His prophets, and His prophets revealed Him to us by means of His written word. It is therefore God's actions as reported in His written word that provide Bacon with the heuristic key to the attributes of the God who only wants to be known by His revelations.

Bacon argues that, since "all learning is knowledge acquired, and all knowledge in God is original,"[12] he will look for knowledge of the attributes and acts of God by the name of "wisedome or sapience," borrowing this terminology from the Psalmist, who exclaimed: "O Lord, how manifold are thy works! in wisdom hast thou made them all: the earth is full of thy riches" (Ps. 104: 24). But since God's word is the original source of His works, we can only learn about the wisdom of God by means of an inquiry into His word.

Bacon therefore begins by scrutinizing God's first words. He points out that "in the worke of the Creation, we see a double emanation of the vertue from God: the one referring more properly to power, the other to wisedome, the one expressed in making the subsistence of the mater, & the other in disposing the beauty of the fourme." He illustrates this difference by saying that "it pleased God to put vppon the workes of power, and the workes of wisedome" a "note of difference," in that "the confused Masse, and matter of heauen and earth was made in a moment, and the order and disposition of that *Chaos* or Masse, was the work of six dayes." But if the difference between human knowledge and knowledge in God consists in the fact that the former is acquired in time, whereas the latter is original, one wonders how God's wisdom can be inferred from the fact that it took Him almost six days to perform the works of wisdom.

Bacon goes on to say that there is a "concurrence" of the difference between the works of power and those of wisdom, and the fact that with regard to the works of power "it is sette downe, that God sayd, *Let there be Heauen and Earth*,[13] as it is sette downe of the works following, but actually, that God made Heauen and earth: the one carrying the stile of a Manufacture, and the other of a lawe, decree, or Councell" (AL 1.50–1, 33, III: 295–6). The biblical narrative, however, does not support this verbal note of difference between a small part of the work of the first day and the rest of the six days' works. To mention only two of the more important examples, God made the firmament, which he called heaven, after having said: "Let there be a firmament" (Gen. 1: 6–7), and He "created man in His own image" after having said: "Let us make man in our image" (Gen. 1: 26–7). Heaven and man are the only works of wisdom that are not called good by God, which causes one to wonder why a wise God should have created anything that is not good.[14]

But setting aside the correlation between wisdom and goodness, we find that by separating as well as confounding the attributes of power and wisdom, Bacon stimulates us to consider both the divine attribute of power and that of wisdom in the light of the whole of the six days' works. And by narrowing down the works of power to the work of the creation of heaven and earth out of nothing, Bacon seems to allude to the fact that the question as to the divine magnitude of God's power is to be narrowed down to the question whether God had the power to create something out of nothing. For if God, to mention a possibility Bacon subtly hints at, only had the power to make matter subsistent, or to make heaven and earth out of chaos or the confused mass of matter, He would be very powerful indeed, but nevertheless not so powerful that the contrast between the magnitude of His power and our own relative lack of power would compel us to venerate His divine magnitude. But if God did create heaven and earth out of nothing, and if the six days' works were manifestations of His power, one wonders why the omnipotent God did not create all His works in one moment. If He took the span of six days in order to accommodate His works of wisdom to the comprehension of man, one would have reason to wonder for what reason a wise and therefore self-sufficient God would do so, which would further cause one to wonder for what reason a self-sufficient God would create anything at all.[15]

In the final paragraph of the section on the attributes and acts of God,[16] Bacon remarks that "Philosophie and humane learning" perform two "principall duties and seruices" to "faith and Religion." First, they are an "effectuall inducement to[17] the exaltation of the glory of God." Bacon again refers to the Psalmist, who invites us to "consider, and magnifie the great and wonderfull workes of God." The Psalmist's invitation leads Bacon to conclude that if we "should rest only in the contemplation of the exterior" of God's works "as they first offer themselues to our sences," we would do "iniurie vnto the Maiestie of God." But the Psalmist invites us to glorify God precisely on account of the exterior of His works as they first offer themselves to our senses (cf. Ps. 104: 25–32), whereas philosophy and human learning effectually induce us to ask ourselves if God's works provide us with reasons to glorify His majesty.

The second duty and service philosophy and human learning perform to faith and religion is that they "minister a singuler helpe and preseruatiue against vnbeleefe and error." Bacon explains this by referring to "our Sauiour," who said: *You erre not knowing the Scriptures, nor the power of God.* By means of this response to the Sadducees who had asked Him about the nature of the resurrection (Matt. 22: 23–33)[18] Jesus laid "before vs two Bookes or volumes of studie, if we will be secured from errour: first the scriptures, reuealing the will of God; and then the creatures expressing his power." Bacon adds that the "later is a key vnto the former; not onely opening our vnderstanding to conceiue the true sence of the scriptures, by the generall notions of reason and rules of speech; but chiefly opening our beleefe, in drawing vs into a due meditation of the omnipotencie of God, which is chiefly signed and ingrauen vppon his workes" (AL 1.65, 37–8, III: 300–1). Earlier in the first book of the *Advancement of Learning*, Bacon had equated the "Booke of Gods word" with "Diunitie," and the "Book of Gods workes" with "Philosophie," and he had warned not to "unwisely mingle or confound these learnings together" (AL 1.6, 9, III: 268). Because he was apologizing to the divines for the magnitude of philosophy's works, he wisely decided to discuss philosophy's intrusion into the book of God's word no sooner than in the central section of his book on human learning.

It is tempting to confuse the distinction Bacon makes between the book of God's word and the book of God's works with a similar distinction made by the Reformers, especially if one is heedless of Bacon's willingness to batten upon his adversaries. But if we do not give in to the temptation of confusing verbal equations with substantial conformity, we find that the distinction between God's word and God's works can only be provisional to a philosopher who considers God's word to be God's supreme work, because he considers God's self-revelation to be God's supreme action. What is more, contrary to the philosopher, who knows only the "sin" of disobedience, the Reformers made the distinction between God's word and God's works in order to emphasize the need for mercy of the sinner who has no excuse for not descrying God's will, seeing that God, as far as He deemed good, revealed His will to him by means of both His word and His works (cf. Rom. 1: 18–20).[19] Besides, since the Reformers believed that God's inscrutable will was the origin of wisdom, they beheld God's works with the obedient eyes of faith (cf. Hebr. 11: 3), whereas the philosopher only has the disobedient eyes of reason at his disposal for inquiring into what are believed to be God's works.

In the central section on the attributes and acts of God, our author himself removes all remaining doubts about the nature of his distinction between God's word and God's works. At the beginning of the section, Bacon discussed the divine attributes of power and wisdom on the basis of God's word, but as they manifest themselves in His works. At the end of the section, however, he distinguishes between God's power, as it is engraved upon His works, and God's will, as it has been revealed by His word. He adds, moreover, that God's works are the key to His word.

Bacon's deliberate confounding of God's word and God's works causes us to draw the following conclusion. Since God revealed His will by means of His word, and His word reveals His wisdom and power by reporting the actions or works of power He performed upon the matter that is accessible to the "sense," God's works of power upon matter or creatures are the key to the "true sence of the scriptures" and therefore to God's will. For since Bacon reads Scripture according to the "generall notions of reason and rules of speech,"[20] the understanding of the true sense of God's word is opened by the result of reason's inquiry into its true sense. And since reason teaches that the will of an omnipotent God necessarily results in an act of power, the substance of the will of an omnipotent God is identical to the actions of His power.[21] Reason's inquiry into the true sense of God's word is therefore coeval with reason's inquiry into God's power.[22] What is more, as reason also teaches that a wise being only wants what is wise, the inquiry into God's wisdom is coeval with the inquiry into God's will, and therefore with the inquiry into His power.

Returning to the Savior's admonition, we find that it is the contemplation of God's actions or works that is to prevent us from "erring" with regard to both Scripture and the power of God. But the Geneva Bible, most probably the edition Bacon used at the time of writing the *Advancement of Learning*, tells us that Christ said that those not knowing the Scriptures and the power of God "are deceived." By subtly substituting "*erre*" for "are deceived," in addition to mentioning "vnbeleefe and error," but only discussing "error," Bacon underscores the fact that, contrary to Christ although not to the Sadducees, he is concerned with avoiding philosophic error. But if God is truly omnipotent, His will is inscrutable because His works are inscrutable. And if there is no possibility of knowing, there is no possibility of erring. It thus seems to be for the sake of minimizing the possibility of erring with regard to the possibility of knowing that Bacon encourages us to "duly meditate" on God's omnipotence when contemplating the creatures that express His power.[23]

In light of this encouragement it seems appropriate to draw one of Bacon's own early *Meditationes Sacrae* on the nature of God's power into our considerations on the power of God's works. In this philosophic meditation, Bacon describes Jesus's response to the Sadducees as the "mother of all canons against heresies." He distinguishes a twofold cause of error, "ignorance of God's will," and "ignorance or superficial contemplation of His power." God's will is revealed "more [magis] through the Scriptures," which testify of Jesus Christ, "if they are scrutinized" (John 5: 39–40). God's power is revealed "more [magis] through His creatures," which is confirmed by the prophet Jeremiah with regard to God's judgment upon the creatures (Jer. 9: 17).[24] Thus, Bacon continues, we are to assert the "plenitude of the power of God without defiling His will," and we are to assert the "goodness of His will without diminishing His power."

But "atheism and theomachy rise and rebel against the power of God, not believing His word, which reveals His will, because they are incredulous of the power of Him to whom all things are possible." Bacon emphasizes that the heresies emanating from this fountain "seem to be graver than the rest," because "in politics it is also a more atrocious thing to diminish the power and majesty of the prince than to debase his fame."

In addition to pure atheism, Bacon distinguishes three degrees of the heresies diminishing God's power. These heresies have "one and the same mystery," for, as Bacon explains, "all antichristianism works in a mystery, that is, under an image of good." The mystery consists in the heresies "liberating the will of God from all aspersion with evil." The first degree is formed by those who constitute "two equal and contrary principles," one good and one evil, which "battle each other." The second degree consists of those to whom it seems to be "too injurious towards the majesty of God to establish an affirmative and active principle against Him." They therefore "reject such audacity," but nevertheless introduce a "negative and privative principle" in opposition to God. For they maintain that it is the "inherent, natural, and substantive work of matter and creature themselves to tend towards and to fall back into confusion and nothingness, not knowing that it is one and the same omnipotence that makes nothing out of something and something out of nothing." The third degree is constituted by those who "reduce and restrict this opinion to human actions that partake of sin." They maintain that these actions "substantially and without any chain of causes depend upon the inward will and judgment of man." They ascribe a "wider range to God's knowledge than to His power," or to "that part of God's power whereby He knows rather than to that whereby He moves and acts," for, as Bacon explains, "knowledge and power are the same." Thus they take Him to "idly [otiose] foreknow some things [quaedam], which He does not predestine and preordain." But "whatever does not by chains and subordinate degrees depend on God as author and principle will be in the place of God [loco Dei erit], and will be a new principle and a kind of deposition of God." For this reason, the aforementioned opinion is "deservedly rejected as an injury to and diminution of God's majesty and power." Nevertheless, our author concludes, "it is very rightly said that God is not the author of evil, not because he is not the author, but because not of evil" (MS 11 "De Haeresibus," VII: 240 ff.).

Since Bacon did not let his philosophic meditation culminate in explicit conclusions, the best way to comment on it seems to be to make its conclusions explicit; an end with a view to which we begin by noticing that Bacon rightly points out that the "heresy" of "diminishing God's power" is "graver" than all other heresies, because even the slightest diminution of God's power necessarily diminishes His divine "majesty" and is therefore tantamount to the rejection of His godhood. But in a way the mere existence of atheism already puts God's power to trial. For if it is God's will that His word be believed, and if atheism does not believe God's word because it is "incredulous" of God's power to instil belief in His word, atheism or "theomachy" diminishes God's power by rebelling against His will by means of itself in its capacity as "contrary principle." This would only be different if for some reason God did not simply want His word to be believed. But even if we leave "pure" atheism aside, we find that the demonstration of God's omnipotence is problematic, for if matter does not "fall back into nothingness," it will not be demonstrated that an omnipotent being "made nothing out of something."

Moving from the works of matter to human actions, we call to mind that if God's commandment is the origin of good and evil, any human action based on the principle of disobedience partakes of sin and is therefore evil. But if goodness consists in obedience to a God whose good will wants us to be obedient, and if it cannot be demonstrated that, ultimately, man's disobedient or evil actions depend on the just God by the "chain of causes," the "range" of God's will is "wider" than the range of His power, and His lack of power can be said to "defile" His will by displaying the latter's impotence. To put it differently, if knowledge is power, and if it cannot be demonstrated that the just God also "predestines" and "preordains" the things He "foreknows," the divine majesty of this pluriscient but "idle" God is gravely injured. Moreover, if God's knowledge is identical to His will because His will is the root of knowledge, and an omnipotent God cannot know what He does not also want, and if any principle or power that does not depend on God "deposes" God, the mere human knowledge of God's knowledge or will exceeding the range of His power would already diminish His power.[25] These conclusions are not altered by an attempt to secure God's goodness and omnipotence on the basis of the argument that God is not the "author of evil" or disobedience because evil is a "privation" of God's good will, unless such an attempt is accompanied by a demonstration that God can forge something good out of the evil He allows to take place for reasons as yet unknown to us.[26] Since the Scriptures testify of Jesus Christ and therefore do not leave us "ignorant" of the substance of God's will, the extent to which the liberation of God's will from all "aspersion with evil" results in "antichristianism" can only be determined by a meditation on the fulfilment of God's good will in the person of Christ. But Bacon postpones such a meditation until he has filled the receptacle of knowledge with information on the "oracles" that prophesied God's will.

Foundations

Let us therefore return to Bacon's thematic treatment of philosophy in the third book of *De Augmentis Scientiarum*, at the beginning of which our author compares the threefold object of philosophy, God, nature, and man, to the "threefold ray of things." Nature strikes the intellect with a "direct ray." Because of the "unequal medium of the creatures," God strikes the intellect with a "refracted ray." Man, "as shown and exhibited to himself," strikes the intellect with a "reflected ray." Consequently, Bacon divides philosophy into three knowledges: knowledge of the "Deity [Numine]," knowledge of nature, and knowledge of man. Because the partitions of the sciences are "not like several lines that meet in one angle, but are rather like branches of a tree that meet in one stem, which uninterruptedly grows as a whole into some length before it divides itself into branches," Bacon considers it imperative that a "universal science" be constituted. "In the progress of knowledge" this universal science is to be regarded as the "mother of the rest" and as the "common way," before the ways "part and divide themselves." It is a "receptacle for axioms that are not peculiar to the particular sciences, but that belong to several of them in common." No other science is "opposed" to it, for since it treats "only the highest stages of things," it "differs from the rest in the boundaries within which it is contained rather than in matter and subject." Bacon calls this science "*Philosophia Prima*" or "*Sapientia*," adding that "sapientia" was "formerly [olim] defined as the knowledge of things divine and human" (DA 3.1, I: 540).

What Bacon fails to mention, however, is that very similar to Cicero, who had "formerly" defined "sapientia" as the "knowledge of things divine and human and of the cause of all things,"[27] he himself called "wisedome or sapience" the "knowledge in God" in the central section of the first book on the attributes and acts of God. And indeed, if *Philosophia Prima* is the "receptacle" of the ray of all things, even the "refracted ray" of the Deity must be "reflected" in the "stem" of the "tree" of knowledge before it "divides" itself into the "unequal" reflections of the creatures. For as the growth potential of the tree of knowledge depends on the intelligibility of its stem, any uncaused ray striking the intellect with generative power would deprive *Philosophia Prima* of her "motherhood"; in other words, any ray shrouding the "highest stages" of things in mystery would differ from all things in matter, and would therefore "oppose" "universal science."[28]

Bacon goes on to give thirteen examples of axioms of *Philosophia Prima*, all signifying the "unity of nature." Knowing, however, that the possibility of knowledge of the unity of nature presupposes the intelligibility of nature, the philosopher makes his central axiom, being his only quotation from Scripture, concern the possibility that would make all intelligibility impossible.

But our author prefaces his reading the unity of nature into the book of God's word with a threefold reading of the prerequisites for knowledge of the unity of nature. In a poem of the heathen Ovid, Bacon read that "all things are changed and nothing is lost."[29] Reading the book of nature caused him to discover that this rule also applies in physics, since the "quantum of nature is neither diminished, nor increased."[30] The book of knowledge revealed to him that the rule is moreover in agreement with "natural theology," which teaches that "it is the work of omnipotence to make something out of nothing, and to make nothing out of something."[31]

Having thus as it were measured the negative "quantum" of a "work of omnipotence," Bacon goes on to quote the following words from the book of God's word as testimony to the foregoing: "All the works which God has done will endure for ever; we cannot add anything to them, nor can we take away anything from them" (Eccl. 3: 14). But if it is true that the omnipotent God has not recently been observed to have acted omnipotently, *Philosophia Prima*'s primary task becomes inquiring into the possibility of God having acted omnipotently, which is tantamount to inquiring into the possibility of an omnipotent God.[32] This implies that the origin of things, the nature of matter, and the possibility of miracles become the primary subject matter of *Philosophia Prima*. In accordance with the primacy of these questions, Bacon adds that the inquiry into what he calls "adventitious conditions of essences" or "transcendentals," such as "much and little," "possible and impossible," "being and non-being," and "similitude and diversity," is also part of *Philosophia Prima*. Because he finds that there is a "deep silence with regard to these things,"[33] he expresses his wish that a "true and solid inquiry" be performed, "according to the laws of nature and not of language."

Bacon finishes his conceptual discussion of *Philosophia Prima* by subtly replacing its aforementioned synonym "sapientia" by "Sophia." By thus drawing our attention to his conclusion of the quest for the true meaning of wisdom, the philosopher draws our attention to his conclusion that "sapientia" is the "knowledge in God" inasmuch as the knowledge of God presupposes knowledge of all "things divine and human" and of the "cause of all things," which is the privilege of a god and is therefore "in God."[34]

Since the quest for wisdom is coeval with the quest for God, Bacon places *Philosophia Prima* among the "desiderata" (DA 3.1, I: 540 ff.).[35] Earlier in the work he pointed out that whenever he places among the desiderata a work "about which there is some obscurity," he either provides "precepts for the execution of the work," or an "example" of the work, lest it be thought that "only a light notion touched his mind," or that he is "like an augur measuring regions in thought for the sake of taking auspices, without knowing the way to enter them" (DA 2.13, I: 521; IMDO 4, I: 135). The "unpreceptive" execution of Bacon's own work makes us wonder, however, whether its "executive" should not have exercised his "augural skills" on a more regular basis in order to prevent the "desiderata" that are perennial to some from no longer being desiderated by anyone because they are believed to have been fulfilled once and for all for the sake of everyone. But this is the subject matter of later chapters.

After having established *Philosophia Prima* as the "common parent" of the sciences, "like Berecynthia," the mother of the gods, who "had so much heavenly offspring,"[36] Bacon returns to the division of the three philosophies, divine, natural, and human (DA 3.2, I: 544). Since the philosopher knows that the obscure and fallow divine regions of the province of knowledge are to be entered by being measured, he knows that primary philosophy's "firstborn" is to be inquired into by means of "natural theology," which is philosophy's "divine measuring instrument." Bacon defines natural theology or "divine philosophy" as the "knowledge, or rather the spark [scintilla] of knowledge, concerning God that can be obtained by the light of nature and the contemplation of created things." It can "soundly be termed divine in respect of the object, and natural in respect of the concept [informationis]." The bounds of this knowledge, if "truly drawn," are that it "extends to confuting and convincing atheism and to the information concerning the law of nature," but that it "does not bring forth [proferantur] the establishment [astruendam] of religion" (DA 3.2, I: 544).

We have already seen that the quest for the nature of God necessarily falls within the scope of philosophy's inquiry into what distinguishes the natural from the non-natural. Natural theology is philosophy's compass on this quest, as it inquires into the non-natural or supranatural by delineating the boundaries of the province of the natural, boundaries that on their part constitute the boundaries of what is within the reach of a light that relies only on itself. Natural or philosophic theology therefore inquires into its "divine object" by means of a "natural concept" of divinity. This implies that the "spark of knowledge" concerning God that can be obtained by the "light of nature" and the "contemplation of created things" can only be the knowledge concerning this philosophic concept of godhood. For the "light of nature" only illuminates the boundaries of nature and therefore only demarcates the boundaries of the province of nature, and the "contemplation of created things" only results in negative criteria for "uncreatedness."

But natural theology does "extend" to the "confutation" of atheism, since the recognition of a philosophic concept of godhood is indispensable for philosophy's inquiry into its own necessity, possibility, and legitimacy. For if a being claiming or claimed to be a god demands the philosopher's pious obedience, the philosopher must know if it is reasonable to submit to this being. This implies that the being or person claiming to be a god must be the most perfect being in order to be recognized as a god by the philosopher.[37] Since the possibility of natural theology does not presuppose the existence of a god, natural theology would thus have been able to convince the potentially philosophic among the atheists to recognize the concept of godhood even if the biblical God had not provided philosophy with an example of the non-philosophic version of godhood by revealing Himself in history. But since Bacon saw himself confronted with a very real and very jealous God, he was compelled to obscure his treatment of the question that is coeval with philosophy by making it appear as if he considered natural theology to be the philosophic seeder of the faith in the one God.[38]

However, despite the inevitable obscurity of his own argument, Bacon intended to shed light on a God who wants to wrap Himself in obscurity. It is therefore of paramount importance not to obscure the difference between natural or philosophic theology as a critical device of philosophy, and "natural" theology as a grounding instrument of faith.[39] For the "natural" theology that is always also inspired theology expounds God's omnipresence on the basis of faith, whereas natural or philosophic theology inquires into God's existence on the basis of reason.[40]

Bacon continues his survey of the province of nature by arguing that, since natural theology "confutes and convinces" atheism, "God never wrought a miracle to convert an atheist, because the light of nature could have [poterat] led him [perduci] to the concept of a God [notitiam Dei]" (DA 3.2, I: 544–5). By now natural theology must have led the philosophically potent atheist into recognizing the concept of a god on the basis of the light of nature, thus we only wonder why God did not perform miracles to convert him. If a miracle is an act of God, it is a supernatural act. But like the philosopher, the atheist relies on the light of nature. He can therefore only recognize an event as a miracle on the basis of the knowledge of nature. Since he does not believe in God, and since the concept of nature and therefore the concept of miracles is unknown to the Bible, the atheist cannot on the basis of faithful biblical narrations recognize as miracles the events reported by the word of God.[41] But as long as he does not know the boundaries of nature, he cannot but postpone the answer to the question whether a reported event was an act of God and therefore a miracle.[42]

In an earlier chapter on "natural history," Bacon therefore reminded the chroniclers of the marvels of nature of the fact that it is "not yet [nondum] known in what cases and to what extent effects attributed to superstition participate of natural causes." He added that, since the "narrations regarding the prodigies and miracles of religions are either not true or not part of nature," they are "impertinent for the history of nature" (DA 2.2, I: 498). But Bacon does not yet know the boundaries of superstition, since he does not yet know how to distinguish the untrue from the supernatural. Consequently, and for the time being, all "prodigies and miracles" that have been narrated as having taken place in history pertain to the history of nature, seeing that it is as yet unknown if they "participated of" natural causes and were therefore superstitions.[43]

In any event, if God never wrought miracles to convert an atheist because the light of nature had led or could have led the atheist to the concept of a god, God must somehow have foreseen that the atheist would not recognize His miracles, either because they were unrecognizable as acts of omnipotence, or because they were unreasonable actions. It is therefore once again the mere existence of atheism that compels us to raise the question as to the magnitude of God's power and the reasonableness of His will.

Nonetheless, since God is reported to have performed miracles, His reason for having done so might shed some light on His nature. Bacon argues that "miracles have been designed to convert idolaters and the superstitious, who acknowledged a deity, but went astray in the worship of him," because the "light of nature does not suffice to declare the will of God or to announce the legitimate worship of him" (DA 3.2, I: 545). But although it is true that if the light of nature does not reveal God's will and the way to worship Him, miracles are necessary for the establishment of religion, if it is also true that God only designed miracles to convert idolaters and the superstitious,[44] some sort of belief must have been a necessary prerequisite for the establishment of religion by means of miracles.[45] What is more, since God designed His miracles to convert those who were idolatrous and superstitious because they had "gone astray" in the worship of the deity they "acknowledged," belief in a deity must somehow be a presupposition for the confession of the true God.[46]

Be that as it may, if God designed miracles to reveal His will, and if His will is revealed by means of His word, God's self-revelation through His prophets can be said to have been His greatest miracle. But God's own prophets revealed to us that false prophets would come and perform miracles in the service of false gods. Since even the true prophets thus had to admit that there is no external criterion for judging the veracity of miracles, and since, therefore, false prophets cannot be distinguished from true prophets by means of the supportive proof of miracles, considering the external or internal power of the prophets who revealed a god to us seems to be the only way to distinguish the true God from false gods (cf. Deut. 13: 1–5 with Matt. 24: 23–4). In other words, since God's self-revelation through His prophets is His greatest miracle and therefore in a way His greatest work of power, the nature of the power of the prophets seems to be an important key to the question as to the nature of God's own power.

Bacon goes on to treat the question why the light of nature does not suffice to declare the will of God by comparing God to a workman. "As all works show forth the power and skill of the workman, but hardly [minime] his image, the works of God show forth the omnipotence and wisdom of the maker, but do not at all [haudquaquam] portray his image." Therefore, Bacon explains, the "heathen opinion deviates from the sacred truth in this matter. For the heathen supposed the world to be the image of God, and man to be the image of the world, whereas the Sacred Scriptures never vouchsafe to attribute to the world such honour as to anywhere call it the image of God; rather, they call it only the work of his hands. They directly substitute man for the image of God though." But in the very next sentence, Bacon himself deviates from "sacred truth" by mentioning a number of attributes that "can [potest] also [etiam] be demonstrated and evinced from God's works: that God exists, that he holds the reins of things, that he is supremely powerful [summe potens], that he is wise and prescient, that he is good, that he is a rewarder, that he is an avenger, and that he is an object of adoration." Our author adds that there are "several other admirable secrets concerning God's attributes, and many more concerning his government and dispensation over the universe, which may likewise be soberly elicited and manifested from his works" (DA 3.2, I: 545).[47]

We suggest the following explanation. As God revealed Himself through His word, God's word can be said to be an "image" of God. God's word reveals that God created the supreme "work of His hands" in His image (cf. Gen. 1: 27 with Gen. 2: 7), and that He sent His Son to restore the divine image in sinful man. But natural theology only knows the light of nature and created things. In order to visualize God's revealed image, natural theology must therefore first assume the most perfect natural image of the created thing that He revealed to be His image, and then, by divinizing this non-divine image of godhood, imagine God to be the divine image of the most perfect natural man. But God's word reveals that God only wants to be known by His revelations. Consequently, natural theology can only paint God's "portrait" by comparing the natural image of godhood with God's revealed image as it is found in the divine actions or works that are accessible to the light of nature. Bacon himself confirms this explanation by saying that, although the works of God do not "portray His image," the attributes of godhood are to be "demonstrated and evinced from his works." What is more, by subtly substituting "supremely powerful" for "omnipoten[t]" Bacon once again draws our attention to the fact that an omnipotent God would make the possibility of knowledge of His attributes impossible.

But if the working of God's attributes is at all events demonstrated by His works, the "works of His hands" necessarily become an image of divine working and therefore an image of God. By emphasizing the non-divine nature of the world, in addition to making it appear as if God's works simply "demonstrate and evince" the attributes of godhood, Bacon liberates the world from the burden of supernatural expectations as he justifies natural man's impious inquiry into the working God by disguising it as apologetics, all the while averting the danger of being found impious. Since he judges it "not safe," though, to "base a conclusion of reason or a strong persuasion concerning the mysteries of faith on the contemplation of natural things and on the principles of human reason," or to "look too curiously into, to ventilate, and to inquire into the manner of these mysteries," he states that he "gives to faith that which is faith's" (DA 3.2, I: 545).

If faith, however, is what constitutes the good of the believer, we find our author confessing his overcuriousness a few chapters further down. For at a safe distance from his discussion of the mysteries of faith, the philosopher mentions "the good" in the centre of an enumeration of five subjects of natural theology (DA 3.4, I: 550).[48]

But within the context of his discussion of the mysteries of faith, Bacon illustrates his "giving to faith that which is faith's" by referring to the heathen fable of the golden chain.[49] Whereas men and gods were not able to draw Jupiter down from heaven to the earth, the father of gods and mortals on his part was able to draw men and gods up from the earth to heaven. Bacon concludes from this that it would be a "vain undertaking to endeavour to adapt the heavenly arcana of religion to our reason" (DA 3.2, I: 545). In the apology to the divines, Bacon had already employed the fable of the golden chain in response to the "conceit" that "too much knowledge should encline a man to Atheisme." He argued that "it is an assured truth, and a conclusion of experience, that a little or superficiall knowledge of Philosophie may encline the minde of Man to Atheisme, but [that] a further proceeding therein doth bring the mind backe again to Religion: for," as he went on to explain, "in the entrance of Philosophie, when the second Causes, which are next vnto the sences, do offer themselues to the minde of Man, if it dwell and stay there, it may induce some obliuion of the highest cause; but when man passeth on further, and seeth the dependance of causes . . . then according to the allegorie of the Poets, he will easily beleeue that the highest Linke of Natures chaine must needs be tyed to the foote of *Iupiters* chaire" (AL 1.6, 9, III: 267–8).

When we translate these apologetic remarks on the correlation between the quantity of knowledge and atheism into terms of natural theology, we find that since reason cannot "adapt" to what it does not understand, it would be a "vain undertaking to endeavour" to draw God down from heaven. But although philosophy does not desire God's descent to human understanding, it does aspire to ascend to the "highest cause" by means of nature's chain of causes. A "superficial knowledge" of philosophy may "incline" the human mind to atheism, since "dwelling" with "second causes" tends to make one oblivious of a highest cause. But if one is philosophically potent enough to proceed further without "breaking off" one's contemplations, one realizes that "oblivion" of the concept of a highest cause would leave the origin of philosophy shrouded in mystery. And as soon as the potential philosopher sees how "easily" the "dependence of causes" makes men believe that the "highest link" of the chain of causes is tied to God's throne,[50] his mind is "brought back" to the "arcana" of religion.[51]

Origins

Having measured the radius of action as well as the radiation of the light of nature, we are prepared to expose ourselves to the "direct ray" of philosophy's second object. Bacon divides the "speculative and theoretical" part of "natural philosophy," which "inquires into causes" (DA 3.3, I: 547), into "physics" and "metaphysics." He emphasizes that he distinguishes *Philosophia Prima* from metaphysics, since *Philosophia Prima* is the "common parent of the sciences," whereas metaphysics is a "portion of natural philosophy." Physics treats what is "deeply immersed in matter and is changeable," whereas metaphysics treats what is "more abstract and fixed." Physics supposes only "being, motion, and natural necessity" in nature, whereas metaphysics supposes also a "mind and idea," for, Bacon adds, what he will say will "perhaps [forte] come down to this." Physics inquires into "material and efficient causes," whereas metaphysics treats "formal and final causes." Bacon divides the part of physics treating the "sums of things" into the "knowledge concerning the principles of things," and the "knowledge concerning the fabric of things," since nature is "collected into one," either on account of the "community of the principles of all things," or on account of the "unity of the integral fabric of the universe" (DA 3.4, I: 549 ff.).

Although metaphysics treats what is beyond physics, Bacon emphasizes that "trying to thoroughly contemplate and apprehend forms as being abstracted from and not determined by matter results in turning away to theological speculations" (DA 3.4, I: 565). In conformity with this physical understanding of the first object of metaphysics, our author elsewhere defines "forms" as the "laws and determinations of pure activity," which "govern and constitute any simple nature in any kind of matter and subject susceptible to it" (NO 2.17, I: 257–8). To give an example, the form of a simple nature such as light is the same thing as the law of light that governs matter inasmuch as matter is susceptible to the law of light. A single form is the "true difference, the naturing nature, or fountain of emanation" of a simple nature (NO 2.1, I: 227). If the form of a nature is given, the nature "invariably follows" (DA 2.4, I: 230).

In order to visualize the "altitudes" attained by natural philosophy thus understood, Bacon compares the sciences in general to pyramids, of which "history and experience" constitute the basis. The basis of natural philosophy is natural history, the first stage after the basis is physics, and the stage "near [proximum] the vertical point" is metaphysics. With regard to the "cone and vertical point" of the pyramid of the knowledge of nature, "the work which God works from the beginning to the end," and that Bacon equates with the "summary law of nature," our author emphasizes that it "may rightly be doubted [haesitamus merito] whether man's inquiry can extend to it" (DA 3.4, I: 567).

We notice in passing that the question whether honest doubt is the right method for stimulating upward movement was highly doubted by one of Bacon's philosophic successors before he appeared on the stage as the father of methodical doubt.[52] But we should not extend our studies to other modern philosophers' responses to the theologization of speculative philosophy, although these responses give rise to infinite speculations on the question why speculative philosophized theology came back with a vengeance after having been temporarily silenced.

Returning to our argument, we find that in the apology to the divines, Bacon illustrated his apologetic statement that there is "no daunger at all in the proportion or quantitie of knowledge howe large soeuer" by paraphrasing an extended version of the biblical verse he quoted in illustration of the nature of the summary law of nature: "*God hath made all thinges beautifull or decent in the true returne of their seasons. Also hee hath placed the world in Mans heart, yet cannot Man finde out the worke which God worketh from the beginning to the end.*"[53] According to Bacon, Solomon declares "not obscurely" that "God hath framed the minde of man as a mirrour, or glasse, . . . raysed . . . to finde out and discerne the ordinances and decrees which throughout all those Changes are infallibly obserued," although the Hebrew king insinuates that the "supreme or summarie law of Nature, which he calleth, *The worke which God worketh from the beginning to the end*," is "*not possible to be found out by Man*" (AL 1.6, 6–7, III: 265). But because the "wisedome and learning" of Solomon was a "grant or donatiue of God," there was no danger in Solomon having knowledge of the work of a God who had assented to Solomon's "petition" for knowledge of the work He works (AL 1.60, 35, III: 298). Bacon, on the other hand, doubts whether the "cone" of the pyramid of the knowledge of nature is mountable. He therefore does not know if man's knowledge potentially extends to the discovery of the "summary law" governing the laws that govern matter. What he does know or suspect, however, is that the "simple forms of things," although "few in number," constitute "every variety in their commensurations and coordinations" (DA 3.4, I: 568).

Since metaphysics thus seems to at least approximate the "cone" of the pyramid of the knowledge of nature, which implies that the quantity of human knowledge is growing to "dangerous proportions,"[54] Bacon hastens to emphasize that "to those who are inflated by their own knowledge, and who are theomachic, the three true stages of knowledge are like the three hills of the giants."[55] Virgil, whom Bacon quotes, tells us that the three earthborn giants Coeus, Iapetus, and Typhoeus, conspired to bring down the heavens. Thrice they attempted to heap Ossa on Pelion, and to roll Olympus onto Ossa, and thrice the father struck the hills with his thunderbolt.[56] But, as Bacon continues, "those who, emptying themselves, refer all things to the glory of God, are like the three acclamations: Holy, Holy, Holy; for God is holy in the multitude of his works, holy in the order of his works, and holy in the union of his works" (DA 3.4, I: 567–8). Having learned from the fate of the giants, Bacon knows that he can only mount the pyramid of nature safely when at all three stages of the ascent he empties himself of his increasing knowledge by referring its object to the glory of God. But whereas the God of John is three times holy because He was, is, and will come (Rev. 4: 8), Bacon's God's threefold holiness is displayed by the "order" and "unity" of the "multitudinous" works of nature.

After having formalized his discussion of formal causes, without, however, "abstracting" forms from matter, Bacon turns to the second object of metaphysics, the inquiry into "final causes," or the supposition of a "mind and idea." He points out that final and physical causes are "perfectly compatible," because the one declares the "intention," and the other only the "consequence." He adds that this does not "cast doubt on or derogate from Divine Providence," but that, rather, it "highly confirms and exalts" it. For, Bacon explains, "in civil things the political prudence of him who can abuse the works of others for his ends and desires without imparting his purpose to them, so that they shall do what he wants but do not know that they are doing it, is much greater and more admirable than the political prudence of him who acquaints those who administer his will with his purpose." Likewise, the "wisdom of God shines forth more admirably when nature moves [agit] one thing and Providence elicits another, than if He had impressed on particular figures and natural motions the characters of Providence" (DA 3.4, I: 570). But since God's providence is "elicited" by those who "administer His will," we do not know if God's purposes have really been imparted to us. We therefore neither know how to distinguish the "intentional" from the "consequential," nor how to localize God's providential intervention in the process of natural motion.[57]

It is this fact that seems to explain why Bacon approves of the "solidity" and the "level of penetration" of the natural philosophy of Democritus and others, who "removed God and Mind from the fabric of things, attributing the structure of the universe to infinite preludes and attempts of nature, and assigning the causes of particular things to the necessity of matter, without intermixture of final causes" (DA 3.4, I: 569). At the end of his discussion of final causes, Bacon adds, however, that since these philosophers "received universal ridicule when they asserted that the fabric of things itself had come together through the fortuitous concourse of atoms, without a Mind," physical causes are "so far from withdrawing men from God and Providence," that "those philosophers who were occupied in digging them up found no way out of the thing than eventually having recourse [confugiant] to God and Providence" (DA 3.4, I: 571). Living in times in which ridicule was among the more harmless of men's responses to *horror vacui*, Bacon knew that he could not avoid "having recourse" to God and providence at the end of his argument on final causes. But this did not come down to "supposing a mind and idea" in metaphysics.[58]

The answer to the question whether God can be "removed" from the "fabric of things" presupposes knowledge of the origin of the fabric of things. But Bacon points out that the ancients "leave doubt" about the generation of Pan or the "whole of things." Some declare that he was begotten by Mercury. Others attribute to him a very different form of generation, saying that he was the offspring of the "promiscuous intercourse between Penelope and all her suitors." But, Bacon comments, the name of Penelope has "undoubtedly been imposed upon the ancient fable by some more recent author," for "it often happens that the more ancient narrations are transferred to more recent persons and names." Sometimes this is done "absurdly and foolishly," which is "clearly to be seen in this case," considering that Pan was "one of the oldest gods [ex antiquissimis diis], long before the time of Ulysses," and Penelope was "held in veneration by antiquity for her matronly chastity." The third account of Pan's generation "must not be passed over," for "some have called him the son of Jupiter and Hybris," that is, "contumely."

According to Bacon, the fable of Pan is "as it were full to bursting with arcana and mysteries of nature." He emphasizes that about Pan's origin there "are and can be but two ways of thinking." Nature is either the offspring of Mercury, that is, of the "divine word," or of the "seeds of confused things." For "those who postulate one principle of things" either "trace it back to God," or, if they "maintain a material principle," they nevertheless "assert the power of variety." Bacon concludes that this is so much so that "all controversy of this kind" can be reduced to the following division: "the world stems either from Mercury, or from all the suitors." The third account of the generation of Pan "makes it seem [videantur] as if the Greeks had heard [inaudivisse] something, either through Egyptian messengers or otherwise, concerning the Hebrew mysteries." For it pertains to the world "not in its pure, natal state," but "after the fall of Adam, as the world stands exposed and subject to death and corruption," for this state of the world "is and remains the offspring of God and sin," in other words, of "God and contumely." For the "sin of Adam was generated by contumely, since he wished to become similar to God." Therefore, Bacon concludes, all three stories of the generation of Pan "may be seen [videri] as true," provided that they are "rightly distinguished as to things and times." For "this [iste] Pan, as we now view and understand him, and whom we worship much more than we ought to, was born from the divine word, through the medium of confused matter and an invading collusion and corruption" (DSV VI "Pan," VI: 635 ff.; DA 2.13, I: 521, 523–4).

We begin our interpretation of Bacon's fabulous account of the whole of things by observing that if Penelope was indeed a chaste woman, the "doubt" about the generation of Pan can be reduced to the doubt as to whether Pan was generated at all. For if the chaste Penelope did not have "promiscuous intercourse" with her suitors, the "seeds of confused things" must have generated the "variety" of things by themselves, which is tantamount to denying the generation of the whole of things.[59] Although Bacon "imposes" the promiscuous Penelope upon the narrative of Pan's generation, he does not let her disturb the prolific orgy of the suitors in his interpretation of the origin of nature.[60] He thus subtly draws our attention to the fact that "all controversy" about the generation of Pan can be "reduced" to the controversy over the question whether what underlies the whole of things has been created or is eternal.[61]

But since controversy is what sets philosophy in motion, it is highly unlikely that Bacon's interpretation of the origin of nature can be reduced to his reduction of all controversy to one primary controversy. We therefore have reason to suspect that, besides reducing the variety of things to the principle of "potential variety," Bacon also traces back the principle of creation to the principle of a creator. The first trace confirming this suspicion is once again provided by the mother of Telemachos. For seeing that monogamous chastity does not imply continence, we have reason to suspect Bacon of having referred to Penelope's ancient reputation for "matronly chastity" within the context of his narrative of the generation of Pan in order to stimulate us to look into the ancient narrative of Penelope's generative activities. And having become familiar with Bacon's methods, we need hardly be surprised to find that the ancient sources tell us that Pan was the son of Penelope and Hermes,[62] a finding that leads us to the conclusion that Bacon intended to sow the seeds of controversy about the divine word's reputation for "continence," and that, since he was reluctant to profane Penelope's ancient reputation for chastity, he "transferred" to the person of Ulysses what "word" has about Hermes's sexual activity. This transference of an old fable to "more recent" times illustrates that Pan, who is "one of the oldest gods," is older than the divine word.[63]

But irrespective of his doubts as to the divine word's generative power, the fact that Bacon had "heard" the divine message that the world is no longer "pure" disallowed him to "pass over" the third account of Pan's generation. If it is true that there can be only "two ways of thinking" as to the origin of the whole of things, Hybris must ultimately be either the offspring of the divine word, or of the "seeds of confused things." And if it is also true that "death and corruption" invaded the pure world through the generative union of God and Hybris because Adam sinned as a result of Hybris having generated his "contumely," it must ultimately be the generator of contumely's generator who is contumelious, either because He conceals the fact that He wanted man to fall, or because by blaming man for the existence of death and corruption He denies the fact that He was not powerful enough to create a pure world. This seems to imply that it was as a result of God having "colluded" with His Hybris, that is, with the overconfidence that His contumely could become a durable source of justice, that the "corrupt" view that Pan had been born from God and sin "invaded" the minds of men.[64] But Bacon had to present an expurgated version of Pan's generation in order not to let the "mysteries and arcana" of nature cause the fable to "burst."[65]

In the sixth fable after the sixth fable Bacon inquires into the origin of the origin of the fabric of things. The poets relate that Coelum was the "most ancient of the extant gods." His generative parts were cut off by his son Saturn with a scythe. Saturn begot "numerous offspring," but he "immediately" devoured his sons. Jupiter escaped this fate, and when he had "grown up [adultum]" he thrust his father down into Tartarus and "took possession of his kingdom." He cut off his father's genitals "with the same scythe" with which Saturn had cut off the genitals of Coelum, and he threw them into the sea. Venus was born from them. Afterward, when his reign "had only just been established," Jupiter had to "endure" the wars of the Titans and the Giants. But after they had been subdued, Jupiter "reigned securely."

Bacon points out that the fable seems to be an "enigma concerning the origin of things, differing not much from the philosophy that was afterwards embraced by Democritus," who "most openly of all asserted the eternity of matter, while denying the eternity of the world." In this matter he "to some extent [aliquanto] approximated [accessit] the truth of the divine word," whose narrative states that "unformed matter existed before the works of the days."

According to Bacon, Coelum is the "concave or ambit that encloses all matter." Saturn is matter itself, which "cut off all of its parent's power of generation," for the "sum of matter always remains the same, and the quantum of nature does not increase or diminish." Bacon explains that the "agitations and motions of matter" first produced "imperfect and incoherent structures, attempts at worlds, as it were," and that thereupon, "in the process of time [aevi processu]," a "fabric arose," which could "preserve and maintain its form." Of these two "divisions of time," the first is signified by the reign of Saturn, who, because of the "frequent dissolutions and short durations of things," was called the devourer of his children. The second is signified by the reign of Jupiter, who thrust down these "continuous and transitory changes" into Tartarus, which signifies the "place of perturbation."

As long as the system of generation holding under the reign of Saturn lasted, Venus had not yet been born. For as long as "in the universality of matter discord was stronger and more powerful than concord," change was "necessarily change by means of the whole, and in the whole fabric." In this way, the generation of things occurred before Saturn was castrated. But as soon as this manner of generation ceased, it was "immediately [continuo]" succeeded by the manner of Venus, with the "concord of things having become fully grown [adulta] and prevailing," and with "change proceeding in parts," while the "universal fabric remains entire and undisturbed."

After having become "settled in respect of its mass and force," the world did not remain "calm" from the beginning. For "notable motions followed in the heavenly regions, and afterward in like manner in the lower regions, through inundations, tempests, winds, and earthquakes of a more universal nature." But after they had been "oppressed and dissipated," things "grew into [accrevit] a more peaceful and durable state of harmony and tranquillity."

Bacon finishes his interpretation of the fable by pronouncing that "the fable contains philosophy," and that, conversely, "philosophy contains the fable." For, he explains, "we know through faith that all such speculations are nothing else than the oracles of the sense, which long ago ceased and failed, since both the matter and the fabric of the world are to be truly traced back to the Creator" (DSV XII "Coelum," VI: 649–50).

As Bacon begins his fable by relating the theogony of the poets, we begin our exposition of its teaching by pointing out that what the best-known theogonic poet actually relates is that the earth goddess Gaia brought forth the god of heaven by self-generation.⁶⁶ This implies not only that Coelum was not the "most ancient" of gods, but also that Bacon misrelated the theogony of the poets, presumably in order to have us consider the implications of the higher being the offspring of the lower.

Be that as it may, the heathen gods are no longer "extant." Coelum can therefore be said to be the most ancient of the still extant gods only inasmuch as he personifies the heavenly God who is believed to have created the matter of both heaven and earth out of nothing. But if it is true that the "sum of matter always remains the same," the "concave or ambit" enclosing matter is coeval with matter, which implies that the omnipotent God has not only been deprived of His generative power, but also of His fatherhood. And if it is also true that unformed matter is eternal, the world must have received its form "in the process of time," and by means of matter in motion. Pan is thus not quite the "oldest" of gods. By denying the eternity of the world Democritus therefore "to some extent approximated the truth" implied in the divine word, whose narrative states that God formed the world in the process of six days.⁶⁷

According to Bacon, the end of this process and therefore the completion of the formation of the world was coeval with the castration of Saturn by Jupiter, and with the completion of the "maturing process" of Venus. But Hesiod relates that Venus was born from the genitals of Coelum, which had been cut off by Saturn.⁶⁸ This implies that Venus and Jupiter reached adulthood at about the same time, and that Saturn and Venus were, in a way, siblings. But if the "matured" principle of "concord" succeeds the "aged" principle of "discord" according to the laws of matter in motion, the mature Jupiter, who as the representative of the one God is a compound of both the oldest and the youngest of gods—as it is the "same" scythe that connects Jupiter and Coelum—must have taken the maturity of Venus as the occasion for establishing his reign. As a natural born ruler, Jupiter was well aware of the fact that it is hardly credible that the world received its form by means of the necessary but mindless interplay of discord and concord.⁶⁹ Being affected by human affections, he also knew that the denial of the formation of the world by mind necessarily implies a deprecation of the human mind, because the human mind is the only mind capable of imagining processes of mind.⁷⁰ Mindful of these human sentiments as well as of man's natural fear of discord, the tyrant realized that he could only legitimize his reign by appropriating the causes of concord, and obfuscating the existence of discord.

However, when the magnitude of the "motions" that were to be observed in both the "heavenly" and the "lower" regions of the world made men fear that the end of the world was imminent, Jupiter came to see that his reign was founded on feeble foundations. Therefore, to conclude our exposition of Bacon's teaching on the genealogy of God's rule, after Venus had defeated the Giants and the Titans, and the world had reached a more "durable" state of calm, Jupiter also appropriated the causes of discord by interpreting them as divine punishments.[71]

That the fable on the origins of matter would "contain philosophy" we already knew when Bacon announced that the "matter" of his fables is philosophy (DSV EpD1, VI: 619). But that the philosophic account of the origins of things and gods would also "contain the fable" relating that the matter and fabric of the world are "in truth" to be "traced back" to the Creator we could not have known except "through faith."[72]

The thirteenth fable looks back on the process that culminated in the formation of the world. The poets relate that the shepherd Proteus was an old man and a "most excellent, and, as it were, thrice-great prophet." For he knew not only the future, but also the past and the present, to such an extent that, "besides possessing the gift of divination," he was the "messenger and interpreter of all antiquity and of all secrets." It was his custom to "count his flock of seals at noon, and to go to sleep afterwards." If someone wanted his help in any matter, the only way to be successful with him was to bind him with chains. Although he would thereupon "turn himself into all kinds of shapes and miraculous things," he would eventually "return to his original shape."

Bacon points out that the meaning of the fable seems to pertain to the "secrecies of nature and the conditions of matter." The person of Proteus signifies "matter," which is the "most ancient of all things, after God." The herd or flock of Proteus seems to be "nothing else than the ordinary species of animals, plants, and metals," in which matter "seems [videtur] to have diffused, and, as it were, to have consumed itself," so much so that "after having moulded and finished those species," it "seems to sleep and rest, as if its task were completed, without striving, attempting, or preparing to form more, other species." Bacon explains that this is what is meant by Proteus counting his herd and going to sleep afterward, which he does not do in the morning or in the evening, but at noon, that is, "when the time has come for completing and bringing forth species out of matter already duly prepared and predisposed." This is the "full [maturum], and, as it were, legitimate time," and it is "midway between the rudiments and the declination of the species." According to Bacon, we "sufficiently know [satis scimus] from sacred history that this was in fact at the very time of creation." For it was then that "by virtue of the divine word *Producat* matter came together at the command of the Creator, not by its own ambages, but all at once, bringing its work to sufficient [affatim] completion, and constituting the species."

Bacon states that at this point the narrative of the fable referring to Proteus "being free and unrestrained with his herd" is complete. For the "universality of things," together with the "ordinary structure and fabric of the species," is the "face of unconstrained or unbound matter and of the flock of materials." Our author foresees, though, that if a "skilful servant of nature brings force to bear on matter with the purpose of reducing it to nothing," matter will "turn and transform itself into miraculous shapes, until eventually it returns to itself," since "annihilation or true destruction is not possible except by the omnipotence of God."

The fact that Proteus was a prophet who knew that three times "agrees well with the nature of matter," Bacon concludes. For if a man knew the "passions and processes of matter," he would "necessarily comprehend the sum of all things past, present, and future" (DSV XIII "Proteus," VI: 651–2).

We start by noticing that if matter is indeed the "most ancient of all things, after God," and if Saturn also signifies matter, we were right in speculating that the God who is fabulously represented by Coelum must be the most ancient of "extant" gods, although not of things, because gods are different from things. Postponing, for the moment, the question regarding the possible godhood of the motion principle of things and therefore of the principle of things in motion, we go on to observe that by the time Saturn had "moulded" the world and the species from their "discordant" rudiments, Venus had reached the "legitimate" age for succeeding her older brother, who needed his rest. The "process of time" leading up to the maturity of Venus culminated at the twelfth hour, as the twelfth fable related. It was therefore neither at six in the morning, nor at six in the evening, but at noon that Jupiter overthrew his father, and that the Creator spoke the word "*Producat.*"[73]

But if the creation of the world is coeval with its formation out of "prepared and predisposed" matter, the works of the days do not demonstrate God's omnipotence.[74] Moreover, if a "skilful servant of nature" can prove that miraculous things are only the result of matter "under constraint," an "annihilating act" of the omnipotent God seems to be the only way to disprove the primogeniture of matter. But until the time that matter will have been reduced to "confusion" or "nothingness," the Trinitarian God, who was in time when He spoke the word and when the word became flesh, who is in time in spirit, and who was prophesied to be in time one final time at the end of time, cannot but recognize the thrice great Proteus as the "most ancient" prophet, who knows the "three times" because he knows the secrets of what is always present and is therefore beyond time.[75]

Principles

Bacon's discussion of the sums of things culminates in the seventeenth fable of *De Sapientia Veterum*, the fable that treats the principles governing the fabric of things. The ancients relate that "Love [Amorem]" was the "most ancient of all gods and even of all things," except Chaos, which they say was "coeval" with him. But "the ancient men never distinguished Chaos with divine honour or the name of a God," Bacon stresses. They present Love as "blind," and as "wholly without a parent," except that "some relate that he was an egg of Night," and that "through the intercourse with Chaos he begot the gods and all things."

Having warned that the things he will put forward are "concluded and adduced solely on the authority of human reason and in accordance with the faith of the sense," Bacon goes on to say that the fable "pertains to and penetrates into the cradle of nature." Love seems to be the "appetite or stimulus of primal matter," in other words, the "natural motion of the atom, the most ancient and unique force that constitutes and forms all things out of matter." Chaos, which was coeval with Love, signifies the "mass or congregation of disordered matter." Matter itself and its "power and nature," and ultimately the "principles of things," are "shadowed forth" in Cupid. He is presented without a parent, that is, "without a cause," because the "cause is, as it were, parent to the effect, and there can be no cause in nature of primary matter and its proper virtue and action, for nothing came before this very thing, no efficient, and therefore neither genus nor form." For this reason, Bacon explains, "whatever in the end this matter and its power and operation may be," it is a "thing positive and deaf, which must be taken as it is discovered, and not judged by any preconception." Even if its "mode and process" could be known, it still "cannot be known by cause," because "after God it is the cause of causes, and is itself causeless." For there is a "true and certain limit of causes in nature," and when we come to the "ultimate force and positive law of nature," it is "just as much the mark of an ignorant man and superficial philosopher to inquire into or invent its cause, as not to desire to know a cause in things which are subordinate." Bacon even goes so far as to say that he is "inclined to think" that Cupid being parentless is the "greatest thing of all," for "nothing has corrupted philosophy so much as the inquiry into Cupid's parents," since philosophers "have not accepted the principles of things as they are discovered in nature," and have not "embraced them as a positive doctrine," and as if they were an "experimental faith." Rather, they have deduced them from "excursions of the mind beyond the bounds of nature." A philosopher should therefore always tell himself that Cupid has no parents, "lest the intellect turn aside into emptiness."

Bacon explains that Love is an egg hatched by Night because its "mode of being is quite obscure." For the "highest law of essence and nature," which "cuts and passes right through the vicissitude of things," that is, the "force impressed by God on the primary particles," from the "multiplication" of which the "whole variety of things emerges and grows together," can "touch mortal thought, but can hardly enter it [subire vix potest]." All knowledge that may be had of Cupid thus "proceeds by exclusions and negatives," and "proof made by exclusion is a kind of ignorance, and as it were Night, with regard to what is included in it." But the parable also suggests that there is "some end and limit" to these exclusions, for "Night does not incubate forever." Certainly, Bacon assures, it is "characteristic of God" that when His nature is inquired into by the sense, "exclusions shall not end in affirmatives." But with Cupid the case is otherwise, for "after due exclusions and negations something is affirmed and established, and an egg is hatched, as it were, after a full and appropriate period of incubation." And the "person" of Cupid is hatched from the egg, that is, a "notion distinct and minimally confused is drawn and extracted out of ignorance" (DSV XVII "Cupido," VI: 654 ff.; POFCC, III: 79 ff.).

What confuses us greatly, though, is that if Love is related as being the "most ancient" of things, together with coetaneous Chaos, it cannot have been an "egg laid by Night," since the theogonic poet tells us that Chaos is older than Night.[76] The most ancient or first things therefore cannot be of "nocturnal" origin, although they may be "wrapped in night." But Love is also related as being the most ancient of gods, in addition to being the co-begetter of the genus of gods. Since chaos did not possess divine genes, this implies that it must have been in his capacity as primal god that Love co-begot the gods. Yet the co-begetter of the genus of gods cannot himself belong to the genus of gods, a law of propagation in accordance with which Hesiod attributed godhood neither to Chaos, nor to Love. Nevertheless, Love received divine honor by being named Cupid, an observation entailing the implication that it was in later times that godhood was attributed not to Love as a blind force, but to Cupid as a "person," and leading us to the further observation that although we tend to recognize godhood in the first things as they present themselves to us, it is only by personifying them that we are able to honor them as gods.[77] This observation is confirmed by Bacon's theogony. For if Chaos signifies the "mass or congregation of disordered matter," and if the "sum of matter always remains the same," Chaos is coeval with the Coelum who is coeval with Saturn, which implies that matter, chaos, and love in their coevality constitute the first things.[78] But the first things and the first principle are devoid of personhood. It must therefore have been in order to be imagined as gods that they were impersonated by Cupid, who for his part impersonates the atom, as is confirmed by the subtitle of the fable "on the atom."[79]

Moving on to the next part of the fable, we find Cupid being the causeless cause of causes suggesting that God, who is the first cause, is outside of nature.[80] But although the law of causality might explain why "exclusions" do not "end in affirmatives" when God's nature is inquired into by the sense, it does not explain away the possibility of the intellect undertaking a second sailing, although this time navigating by the compass of dialectics; a possibility, we add in passing, whose actualization becomes a matter of necessity when the first sailing seems to have consumed itself, which happens occasionally in history, but necessarily in thought.[81]

After calling to mind that God is to be equated with the Coelum who was the most ancient of "extant" gods, we go on to notice that in *De Principiis atque Originibus* Bacon announces the treatment of two fables, one on Cupid, and one on Coelum, which together "seem [videtur] to present the doctrine concerning the principles of things and the origins of the world" (POFCC, III: 80). Bacon twice refers questions regarding the "origins of things" and the "emanations of omnipotence" to the parable of Coelum, which treats the origins of things. He adds that the inquiry into origins is "posterior" to the inquiry into principles, and he holds out the prospect of treating questions regarding the origins of things "more fully" in the parable of Coelum (POFCC, III: 87, 111). However, Bacon never wrote a second parable on Coelum. We are not impressed by the somewhat lazy surmise that fate must have prevented Bacon from inquiring "more fully" into the origins of things, if only because *De Principiis* was probably not written toward the end of Bacon's life.[82] Rather, we are inclined to conclude that the questions Bacon referred to are fully treated in the parable of Cupid; in other words, that the origins of things are "posterior" to the principles of things inasmuch as the first things are always and are therefore as it were "anterior" to their origins.

But regardless of the origins of things, the notion that the "highest law of essence and nature" could account for the variety of things can "hardly enter" "mortal thought." Stronger even, the prospective sound of muteness is so disquieting that one would rather believe that the God who ensouls the atom was hatched by Night. Moreover, in contrast to God, the orphan Cupid is "deaf" to our call, because he does not understand our need for paternal care. Bacon knew that "superficial philosophers" would not be able to endure faith of an "experimental" nature.[83] As they rather believe in "emptiness" than in the emptiness of the whole, they are inclined to accelerate the period of incubation of Cupid by leaving the realm of the sense in order to hatch an "affirmative" Cupid from the egg of Night.[84] But Cupid being parentless is the "greatest thing of all" for philosophy, because philosophy's lineage depends on Cupid lacking one. Bacon therefore instills the virtue of patience into those philosophers who hear him telling them to "tell themselves" that Cupid has no parents, until their telling of muteness is replaced by "telling" muteness.

Be that as it may, our author takes the blindness of Cupid to be a "most wise allegory." For, he explains, it seems that Cupid has "very little providence," and that he "directs his course and movement by whatever he senses nearest to him, like a blind man groping." This makes the "highest divine providence all the more admirable," being that which "educes by a certain and fatal law the order and beauty of things out of things most empty and destitute of providence, blind as it were." And it makes us ready for the final part of the fable, in which Bacon relates that the ancients also "invented [fingitur] and celebrated" another Love, the youngest of the gods, and the son of Venus. The attributes of the older Love were "transferred" to him, and were "supplemented by many of his own." The younger Cupid is "deservedly" reported to be the youngest of the gods, Bacon judges, since "before the species had been constituted he could not flourish." But although in the description of the younger Cupid the allegory "digresses and is transferred to morals," there remains a "certain conformity" between him and the older Cupid. For, Bacon argues, Venus excites the "general affection of conjunction and procreation," and Cupid, her son, "applies the affection to the individual." The "general disposition" therefore comes from Venus, and the "more exact sympathy" from Cupid. The former depends on "causes that are nearer," whereas the latter depends on principles "deeper and more fatal," and "as if derived from the ancient Cupid," on whom "all exquisite sympathy depends" (DSV XVII "Cupido," VI: 655 ff.).

But contrary to the ancient Cupid, the younger Cupid is a descendant of gods. For he is the son of Venus,[85] who is the daughter of the genitals of Saturn, who is coeval with the ancient Cupid. Since the younger Cupid could not "flourish" until the species had been constituted, and since he was dependent on the "affection" of Venus, we are led to the conclusion that he is none other than Jupiter himself, who is therefore the son of Venus inasmuch as he could neither have flourished nor been without Venus. By appropriating the attributes of Cupid and Venus, the pagan adumbration of the holy God first "transferred" the causes of the "order and beauty of things" to himself. He thereupon imposed a "deep and fatal" law on these things, thereby supplementing the attributes of divinity with the attribute of divine providence. But being himself a god, Jupiter knew that "general" providence could not secure the filial love of men. He therefore applied the "blind" affection of Venus to "individuals," thereby turning the story of the human race into a "moral" narrative.[86]

Although the birth right of the youngest of gods can thus be traced back to the first things, he was hatched from an egg laid by Night, as the chorus of Aristophanes's *Birds* informs us in the genealogy of godlike birds and gods.[87] And it seems to be because Jupiter's godhood and therefore his being was dependent on his providence that in an other fable Bacon equates Night with the "obscure and secret divine judgment," that is, with "concealed providence" (DSV XXII "Nemesis," VI: 662).[88] It seems, moreover, to be in anticipation of this equation that Bacon assures us that the differences between the two Cupids are such that one may "reject the confusion of persons, and admit their similitude" (DSV XVII "Cupido," VI: 654). But now that we have admitted that there is a "similitude" between God and things, we can reject their being "confused" any longer.[89]

Conclusions

Bacon concludes his inquiry into the sums of things by saying that "any philosophy that furnishes, balances, and fortifies a system in such a way that it may seem not to have flowed from Chaos, seems to have been wholly conceived in the narrowness of the human chest." For, he explains, he who philosophizes "wholly according to the sense, asserts the eternity of matter, but denies the eternity of the world as we see it." Bacon takes this to have been the view of "ancient wisdom." He adds that the Sacred Scriptures "bear witness to the same thing," with the "principal difference" that the Sacred Scriptures hold that matter "comes from God," whereas ancient wisdom holds that matter "is by itself."

In elaboration of this point Bacon discusses three dogmas which we "know from faith [scimus ex fide]" about the previous matter. First, we know that "matter was created out of nothing." Second, we know that the "development of a system was by the word of omnipotence," and that matter did "not develop itself out of chaos into this schematism." Third, we know that "before the Fall" this was the "best of the schematisms that matter, as it had been created, could support." But, as Bacon goes on to formulate philosophy's response to the dogmas of faith, philosophy "could not ascend to any of these dogmas." For philosophy "dreads creation out of nothing," and "supposes that this schematism was raised after many ambages and efforts of matter." Moreover, whether this is the "best possible" schematism "does not trouble philosophy," for philosophy asserts that the schematism is "perishable and variable." Therefore, Bacon concludes, in these things we must "stand still [standum] at faith and its firmaments."

He appends the consideration, though, that the "question whether this created matter, through the long courses of ages, and by the force originally given to it, could have gathered and turned itself into that best schematism, is perhaps [fortasse] not to be inquired into." For the "representation of time is as much a miracle, and belongs to the same omnipotence, as the formation of being." And the "divine nature seems to have wanted [voluisse] to distinguish itself by both these emanations of omnipotence, in the first place by operating omnipotently on being and matter, that is, by creating something out of nothing, and in the second place by operating on motion and time, that is, by anticipating the order of nature, and accelerating the process of being" (POFCC, III: 110–1).

Bacon, on the other hand, seems to have distinguished himself primarily by the fact that, although he "knew from faith" about the dogma of the creation of matter, about the dogma of the formation of the world out of created matter, and about the dogma of the theodicy of the imperfections of created matter, he knew from philosophy that if matter had indeed been created out of nothing by an omnipotent God, all human assertions on the matter of matter would ultimately be obsolete. And as a philosopher he "dreaded" this prospect. His thoughts therefore could not "stand still" at faith. Yet since he foresaw the consequences of seeming "not to be troubled" about the genealogy of evil, he wrapped his inquiries into the genealogy of the "schematism" of things in fables in order not to be condemned for concluding that the only way for God to materialize His twofold will was by making the best the negative cause of good, and good the positive cause of evil.[90]

But whatever the origin of Pan was, the Parcae are said to have been his sisters. And the "fates of things are truly represented as sisters to the nature of things," Bacon emphasizes, as it is the "births, durations, and destructions of things, their fallings and risings, their labours and felicities, in short, whichever conditions of an individual," that are termed Fates. It is Pan who "reduces these separate individuals to their various conditions," in such a manner that, "as far as individuals are concerned," the "chain of nature and the thread of the Parcae are as it were the same thing." But Bacon adds that "even if Fate is taken in a broader sense, as signifying every event of any kind," it also "agrees best with the universal frame of things," seeing that "there is nothing in the order of nature so small that it happens without a cause, nor anything so great that it does not depend on something else, so that the fabric of nature contains in its own lap and bosom every event, both small and great, bringing forth these events at its time by a fixed law." In short, "no forces whatsoever can loosen or break the chain of causes" (IMDO 29, I: 144).

Bacon goes on to say that it is not only true, but also "legitimate" that the Parcae are represented as sisters of Pan, seeing that "Fortune is the child of the vulgar," and has "only found favour with the lighter kind of philosophers." Epicurus seems indeed to be "not only profane, . . . but also foolish" when he says that it is "better to believe in the fable of the Gods than to assert Fate; as if anything in the universe could be like an island, separated from the nexus of things." But Epicurus, "accommodating and subjecting his natural to his moral philosophy, as appears from his own words," did not want to admit of any opinion that "depressed or hurt the mind and troubled his Euthymia." Therefore, Bacon concludes, he "wholly threw off the yoke and repudiated both the necessity of Fate and the fear of the gods" (DSV VI "Pan," VI: 635, 637; DA 2.13, I: 521–2, 524).

But despite appearances to the contrary, of which Bacon was well aware, Epicurus was not a philosopher of the "lighter kind." Whereas the vulgar make fortune a "child" of God, Epicurus denied fortune godhood, "as appears from his own words."[91] What the philosopher of happiness did say was that it would be "better to follow the fables of the gods than to submit to fate," because the fables of the gods leave us the "hope of mollifying the gods by honouring them," whereas "necessity is inevitable." However, since Epicurus did not "subject" his natural philosophy to considerations of morality, he did not give in to the temptation of believing for the sake of hope. As again appears from his own words, his natural philosophy rather taught him that fortune, that is, fate, nature, or necessity, is "not a capricious causality, which gives man goods or evils bearing upon the happy life."[92] Having therefore thrown off the "yoke" of an island of gods causelessly intervening in the lives of men, Epicurus could "repudiate" the necessity of the necessity of fate troubling his "Euthymia." This was "profane" indeed, and certainly "illegitimate," but it was not necessarily "foolish."

In his *Pensées Philosophiques* on God, the ungodly Diderot states that "all absurdities of metaphysics carry no weight against an argument *ad hominem.*" He adds that "for convincing sometimes nothing else is needed than stirring the physical or moral sentiment."[93] Could Bacon have more effectively convinced his "most Currant" work's typical readers of the insignificance of atheism than by stirring their anti-atheist sentiments?

"I[94] had rather beleeve all the Fables in the *Legend*, and the *Talmud*, and the *Alcoran*, then that this universall Frame, is without a Minde. And therefore," as Bacon goes on to say and imply in one of the most subtly argued sections of the work that comes home "to Mens Businesse, and Bosomes" (E EpD, 5, VI: 373), "God never wrought Miracle, to convince *Atheisme*, because His Ordinary Works convince it"—of the fact, we may add, that it is so incredible that the "universal frame" of things should be "without a mind," that even an atheist would rather believe the fables in the Golden Legend, the Talmud, and the Koran.[95]

We have already seen that "a little Philosophy inclineth Mans Minde to *Atheisme*," but that "depth in Philosophy, bringeth Mens Mindes about to *Religion*."[96] In *Essay* XVI on the incredibility of atheism,[97] Bacon adds to this demonstration of the central importance of religion to philosophy the observation that "even that *Schoole*, which is most accused of *Atheisme*, doth most demonstrate *Religion*; That is, the *Schoole* of *Leucippus*, and *Democritus*, and *Epicurus*." For, he explains, "it is a thousand times more Credible, that foure Mutable Elements, and one Immutable Fift Essence, duly and Eternally placed, need no God;[98] then that an Army, of infinite small Portions, or Seedes unplaced, should have produced this Order, and Beauty, without a Divine Marshall." Our author elsewhere observes that, after human learning had "suffered shipwreck," the "planks" of Aristotelian philosophy were "preserved and came down to us," whereas the "more solid" ones of the atomists "sank and almost passed into oblivion" (POFCC, III: 84), an observation that demonstrates that "time seemeth to be of the Nature of a Riuer, or streame, which carryeth downe to vs that which is light and blowne vp; and sinketh and drowneth that which is weighty and solide" (AL 1.40, 29, III: 292). But besides carrying us back to the problem of the self-reinforcing tendencies of philosophical schools, and carrying us forward to the next chapter, Bacon's observations on the overall fate of philosophy also carry us to the conclusion that the school of the atomists "demonstrates" religion inasmuch as it demonstrates that only the "immutable" is believed to account for the "order and beauty" resulting from the motions of the "mutable," and that, therefore, once the immutable has been divinized, there is no more need for a God.

Rather, as Bacon goes on to argue, it is the foolish atheist who is still in need of a God. "The Scripture saith; *The Foole hath said in his Heart, there is no God*: It is not said; *The Foole hath thought in his Heart*: So as, he rather saith it by rote to himself, as that he would have, then that he can throughly beleeve it, or be perswaded of it. For none deny there is a *God*, but those, for whom it maketh that there were no *God*." Because a fool does not think, he cannot be persuaded that there is no God. But neither can the foolish atheist thoroughly believe that there is no God, since he denies the existence of a God only because he believes that it would be better for him if there were no God. And since belief in unbelief is the seed of repentance, the atheism of the foolish atheist ultimately roots in the soil of faith.[99] It is for this reason that it "appeareth in nothing more, that *Atheisme* is rather in the Lip, then in the *Heart* of Man, then by" the fact that "*Atheists* will ever be talking of that their Opinion, as if they fainted in it, within themselves, and would be glad to be strengthened, by the Consent of others"; stronger even, it appears from the fact that "you shall have *Atheists* strive to get *Disciples*, as it fareth with other Sects: And, which is most of all, you shall have of them, that will suffer for *Atheisme*, and not recant; Wheras," Bacon wonders, "if they did truly thinke, that there were no such Thing as *God*, why should they trouble themselves?" But Bacon only takes the trouble to plant this consideration in the receptive mind of his reader, because the *Essai* or trial on public atheism is otherwise silent about those atheists who thought in their hearts that there is no God, but who did not confess their atheism with their lips because they were not sectarians.[100]

Bacon continues his discussion of atheist sectarianism by referring to the fact that "*Epicurus* is charged, that he did but dissemble, for his credits sake, when he affirmed; There were *Blessed Natures*, but such as enjoyed themselves, without having respect to the Government of the World. Wherein, they say, he did temporize; though in secret, he thought, there was no *God*."[101] However, Bacon continues, "certainly, he is traduced; For his Words are Noble and Divine: *Non Deos vulgi negare profanum; sed vulgi Opiniones Diis applicare profanum*." Epicurus indeed refused to apply to the gods the vulgar opinions on gods whom he did not think existed, as appears from his words. But this was not at all "profane."[102] In fact, since his words are divine, the Greek philosopher was "traduced" when being accused of atheism, as Bacon rightly points out. And although it is true that his affirming the existence of "blessed natures" was an act of "temporization," because "although, he had the Confidence, to deny the *Administration*, he had not the Power to deny the *Nature*" of the gods in whom the more powerful vulgar believed, this did not justify the branding of the philosopher as an atheist sectarian who strives to spread the atheist opinion that there is no God.

Elaborating on his discussion of the powerful gods of the vulgar, Bacon observes that the "*Indians* of the *West*, have Names for their particular *Gods*, though they have no name for *God*: As if the *Heathens*, should have had the Names *Jupiter, Apollo, Mars*&c. But not the Word *Deus*." According to Bacon, this shows that "even those Barbarous People, have the Notion, though they have not the Latitude, and Extent of it. So that against *Atheists*, the very Savages take part, with the very subtillest Philosophers." But Bacon's version of the Indians' nominal polytheism deviates in an important respect from Acosta's account of their notional monotheism. For Acosta, who was the authority on the Indians in Bacon's time, and whose *Historia de las Indias* (1590) was demonstrably the nominal source of inspiration for Bacon's argument on the differences between the names for and the notion of God, only points out that the Indians had some notion of an author and lord of all things, although they did not have a word answering to the name of God.[103] Since the heathens did have names for their particular gods, although they also knew the word "Deus," we are led to the conclusion that Bacon argues that the "subtlest philosophers," who know both the "word" God and the "latitude" and "extent" of the "notion" of God that is covered by the different names for God, "take part with the savages against the atheists" inasmuch as they use different names for God when they talk about the notion of God.[104] The mere reference to *De Sapientia Veterum* may suffice as proof of Bacon's own proficiency in the art of subtlety.

From "subtle" philosophy Bacon turns to the cognate notion of contemplative atheism. "The Contemplative *Atheist* is rare." Although "all that Impugne a received *Religion*, or *Superstition*, are by the adverse Part, branded with the Name of *Atheists*," the "great *Atheists* . . . are *Hypocrites*; which are ever Handling Holy Things, but without Feeling. So as they must needs be cauterized in the End" (E XVI "Of Atheisme," 51 ff., VI: 413–4). But if it is a sign of true greatness to be able to handle a burning issue without "cauterizing," the true greatness of the "contemplative atheist" must consist in his "rare" ability to "impugn" superstition without being "branded" an atheist.

History

"It were better to have no opinion of *God* at all, then such an Opinion, as is unworthy of him: For the one is Unbeleefe, the other is Contumely: And certainly *Superstition* is the Reproach of the *Deity*." The first sentence of the seventeenth *Essay* mirrors and sharpens the first sentence of the sixteenth

Essay. Whereas belief is more credible than atheism, superstition is worse than unbelief.[105] To confirm his argument that God must prefer men who have no opinion of Him at all to men who have an opinion unworthy of Him,[106] Bacon approvingly refers to Plutarch, who said that he "*had rather, a great deale, Men should say, there was no such Man, at all, as Plutarch,*" than that they would be terrified of him.[107] But although "nature has endowed all living creatures with fear and dread to preserve their lives and essences, and to avoid and avert imminent evil," the same nature "does not know how to stay within measure," and "ever mingles salutary with vain and empty fears," as Bacon observes elsewhere (DSV VI "Pan," VI: 639; DA 2.13, I: 528).[108] It is this immoderateness of one of the most fundamental human passions that seems to underlie Bacon's observation that the "human intellect is of itself prone to suppose more order and equality in things than it finds in them," and that, although there are many things in nature that are "monadic and wholly unequal," the intellect nevertheless "devises parallels, correspondences, and relatives, which do not exist" (NO 1.45, I: 165). For it is the order of terror that mitigates the fear of an order that is terrifying on account of its seeming disorder.

Bacon goes on to observe that once the human intellect has "agreed" to certain things, being either what has been "received and believed," or what is "agreeable," it "draws everything else into supporting these things and agreeing with them." He argues that "such is the method of nearly all superstition, events being noticed when they are fulfilled, but neglected and ignored when they fail, although this happens more often" (NO 1.46, I: 166).[109] The conclusion that the Christian religion is free from superstition therefore implies a negative answer to the question whether the human intellect "agreed" to its own "proneness" to "agreeability" when it "received," "believed," and "agreed to" the "event" fulfilling the prophecies of the Christian faith.

Answering this question is complicated, however, by the fact that superstition is not always visible as superstition. Bacon observes that "*Superstition, without a vaile, is a deformed Thing; For,*" he explains, "as it addeth deformity to an Ape, to be so like a Man; So the Similitude of *Superstition* to *Religion*, makes it the more deformed" (E XVII "Of Superstition," 54–5, VI: 415–6). But whereas the ape does not suffer from his deformity, man's similitude to the primate is detrimental to human pride. It is therefore in order to avoid the fall of pride that religious man hides religion's similitude to superstition by veiling it.

Yet besides drawing attention to this visible difficulty of unveiling what religion veiled, Bacon observes that some fear that "motions and mutations in philosophy will end in incursions into religion," and that others are anxious that in the investigation of nature something might be discovered that might "subvert or at least weaken" religion. But our author dismisses these fears by saying that they "wholly seem to taste of carnal wisdom"; as if men "in the recesses of their minds and in their secret thoughts doubted and distrusted the strength of religion and the empire of faith over the sense" (NO 1.89, I: 197).[110]

Prophecy

As philosophy is "omnivorous," Bacon did not abhor the taste of the carnal. Let us therefore focus on Bacon's philosophical inquiries into the historical event that inaugurated the Christian religion. Since divinity "shines out" more in "civil" than in "natural history," Bacon constitutes a species of history "peculiar to" divinity, that he calls "sacred or ecclesiastical history," and that he includes in "civil history," the species of history that relates the "deeds and actions" of men rather than of nature (DA 2.2, I: 495). Ecclesiastical history "in a subordinated sense" relates the "times and different states of the militant church: whether fluctuating, as the ark in the flood, or itinerant, as the ark in the wilderness, or at rest, as the ark in the temple; that is, the state of the Church in persecution, the state of the Church in motion, and the state of the Church in peace" (DA 2.11, I: 515). By means of this historical parallel between the times of the "militant church" and the two arks guiding God's covenant people in their peregrinations Bacon seems to suggest that ecclesiastical history "in a subordinated sense" ended with the death of Solomon (930 BCE), the man who built the final resting-place for the ark of the covenant. For it was in response to Solomon not keeping His covenant that the Lord divided the twelve Hebrew tribes, which "continued divided ever after," as Bacon remarks elsewhere.[111] But division was neither the worst nor the last of evils the Lord had in store for His chosen people. For not long after the twelve tribes of the united Kingdom of Israel had been divided into the Kingdom of Judah and the Kingdom of Israel, the Kingdom of Israel was conquered by the Assyrians (720 BCE). The Kingdom of Judah was to last until 586 BCE, when Jerusalem was captured by the Babylonians, and the temple housing the ark of the covenant was destroyed. The peace of Bacon's militant church turns out to have been the foreboding of a "divisional" storm that was not to subside until it had torn down the first symbol of "historical ecclesiasticism."

That Bacon indeed considered the first temple to have been the last "historical ecclesia" is indicated by the fact that he continues his narrative of sacred or ecclesiastical history with a discussion of the "history of prophecy," which is the second part of "inclusive" ecclesiastical history. The history of prophecy consists of "two relatives," the prophecy itself, and its fulfillment. It "conjoins the separate prophecies of the Scriptures with the reality of the events, for the establishment of a certain discipline and skill in the interpretation of those prophecies that remain as yet unfulfilled." Bacon emphasizes, however, that it "allows the latitude that is peculiar and familiar to divine prophecies," namely that their fulfilments are taking place "both continually and punctually," seeing that they "refer to the nature of their Author, to whom a thousand years are as one day, and one day is as a thousand years." And although the "plenitude and height of accomplishment" is "commonly destined to a certain age or a certain moment," "meanwhile they have ladders of accomplishment throughout the different ages of the world" (DA 2.11, I: 515).

These remarks allude to a biblical verse on the second coming of Christ. For it was in response to the "ungodly" who do not believe in the promise of the coming of the Lord because "all things continue as they have been" that the apostle Peter warned the faithful not to be ignorant of the fact that with the Lord "one day is as a thousand years," and "a thousand years are as one day," and that the "day of the Lord will come as a thief in the night" (2 Peter 3: 4–10). In other words, the Lord will come "punctually," although the moment of His coming is known only to Himself, and although He may need "thousands of years" to climb the "ladder" that leads up to the "plenitude and height of accomplishment" of the promise.

But since the ungodly can only be fought with their own weapons, the refutation of the observations of the ungodly would demand proof that things are not any more as they had been of old. This proof could only be furnished by the composition of a "punctual" history of prophecy that distinguishes the "continual" history of prophecy from continual history. And it seems to be primarily because such a history remains to be written that Bacon places the history of prophecy among the "desiderata." He adds that it is to be treated with "great wisdom, sobriety, and reverence," or is to be "wholly dismissed" (DA 2.11, I: 515).

That our author could hardly dismiss a subject requiring wise, sober, and reverent treatment is proven by the fact that he wrote one of his *Essays* on the reading of prophecies. "I meane not to speake of *Divine Prophecies*; Nor of Heathen Oracles; Nor of Naturall Predictions; But only of *Prophecies*, that have beene of certaine Memory, and from Hidden Causes." But the first three prophecies Bacon discusses concern the very three subjects about which he spoke that he did not mean to speak of them, presented in the very order in which he did speak of them. The Pythonissa brought up Samuel from the dead to let him say to Saul: "*To Morrow thou and thy sonne shall be with me.*" Saul had consulted the woman with a "familiar spirit" on the threat of the Philistines, because God had not answered him by His prophets. But even though His Lord did not bring him up from the dead, the divine Samuel, by whom God used to speak, prophesied to Saul that because Saul had not obeyed the voice of the Lord, he and his sons would be with the dead the next day. It happened accordingly, since the next day the Philistines killed the three sons of Saul, although Saul decided to kill himself (1 Sam. 28: 5–20; 31: 1–6). As the deaths of Saul and his sons are of "certain" biblical "memory," and the causes of their deaths have not been prophesied by God's voice, Samuel's divine prophecy pertains to Bacon's history of prophecy. Bacon's second and third examples concern a "heathen oracle" on the size of the Roman empire and a "natural prediction" on the discovery of America, respectively, both of which were fulfilled in history, and both of which are therefore of "certain" historical "memory," although the causes of their fulfillment were still "hidden" at the time the events were prophesied.

In the central example of the seventeen prophecies that are discussed in the thirty-fifth *Essay*,[112] Bacon relates that in "*Vespasians* Time, there went a *Prophecie* in the East; That those that should come forth of *Judea*, should reigne over the World." He comments that this prophecy, "which though it may be was meant of our *Saviour*, yet *Tacitus* expounds it of *Vespasian*." Suetonius reports that it was this prophecy that caused the Jews to rebel against the Roman Empire in 66 AD, a rebellion that was to result in the fall of Jerusalem and the destruction of the second temple in 70 AD.[113] The Roman historian therefore concludes that the events showed that the prediction actually referred to the Roman Emperor Vespasian (69–79 AD). Tacitus adheres to this conclusion, adding "not even adversity could convert the vulgar to the truth."[114] The subsequent events proved Tacitus a far-sighted man, for they showed that not even unfulfilled prophecies on a second coming of the man who had "come forth from Judea" could prevent His self-proclaimed representatives from believing that they should "reign over the world" until His return. But contrary to Vespasian, who at the end of his worldly empire supposed that he was becoming a god,[115] Christ's deputies founded a worldly empire on the basis of their Savior having become a God.

Bacon goes on to say that the prophecies he has set down "ought all to be *Despised.*" He adds, however, that when he says "*Despised,*" he means it "as for Beleefe: For otherwise, the Spreading or Publishing of them, is in no sort to be *Despised*. For they have done much Mischiefe," as can be seen most clearly from the belief in unfulfilled prophecies.[116] What "hath given them Grace, and some Credit," Bacon explains, is that "Probable Conjectures, or obscure Traditions, many times, turne themselves into *Prophecies*: While the Nature of Man, which coveteth *Divination*, thinkes it no Perill to foretell that, which indeed they doe but collect." Therefore, our author concludes, "almost all" prophecies "have been impostures, and by idle and crafty brains merely contrived and feigned after the event past" (E XXXV "Of Prophecies," 112 ff., VI: 463 ff.). But if it was indeed on the basis of a "recollection" of the tradition on Christ's death that the prophets "contrived" their prophecies on His final triumph over death,[117] it becomes imperative to inquire into this tradition in order to see if there is anything "obscure" about it.

The early *Confession of Faith* contains Bacon's most elaborate philosophical interpretation of the theological necessity of Jesus Christ.[118] But since we live in the era of authenticity, Bacon's *Confession of Faith* has without exception been treated as an authentic confession of Bacon's faith, despite or because of the fact that the true authenticity of the *Confession* consists in Bacon authenticating a philosophic treatment of faith in order not to be found inauthentic.[119] After all, it is from an authentic admirer of Bacon that we learn not to fear our doubts about the authenticity of Bacon's confessed faith. For William Rawley precedes his conclusion that Bacon was "conversant with God" with the consideration that one is "apt to suspect and prejudge great wits" to have "somewhat of the atheist."[120]

Bacon begins his genealogy of the Christian faith with the theological observation that "nothing is without beginning but God; no nature, no matter, no spirit, but one only and the same God." As this God is "eternally almighty, only wise, only good, in his nature," so "he is eternally Father, Son, and Spirit, in persons." Yet despite His omnipotence, God "is so holy, pure, and jealous, as it is impossible for him to be pleased in any creature, though the work of his own hands." But as it is impossible for an omnipotent God to be less than omnipotent, it is impossible for us to imagine a constrained God as being omnipotent. We must therefore imagine the "eternally omnipotent" God as having remained omnipotent in eternity after He had descended from eternity to become a "temporally pluripotent" Creator.

Bacon goes on to say that because of God's "holiness," "purity," and "jealousy," "neither Angel, Man, nor World, could stand, or can stand, one moment in his eyes, without beholding the same in the face of a Mediator," and that, therefore, "before him with whom all things are present, the Lamb of God was slain before all worlds; without which eternal counsel of his, it was impossible for him to have descended to any work of creation; but he should have enjoyed the blessed and individual society of three persons in Godhead only for ever." But although it is true that God's Lamb was slain so that God's creation could "stand in His eyes," what Bacon seems to suggest is that without the slaying of the Lamb it would have been impossible for God to descend from eternity to become a Creator. Since it was God Himself who, in His eternal counsel, and before the creation of the world, led His Lamb to the slaughter, we are thus invited to suspect that God wanted to become the Creator of a creation He could not have created without the slaying of the Lamb. And since it was for the redemption of our sins that the Lamb was slain (1 Peter 1: 18–20), we are driven to the further suspicion that God wanted to create the sinful being that He could not have created except by creating it a sinful being.[121]

But if this sinful being was created in God's image, God sinned in His capacity as temporal Creator when He created a being that had already fallen before it was to fall.[122] Holiness and purity are not sins, though. It must therefore have been God's jealousy that led Him to the sin of creating sinful man. But a God who leaves trinitarian autarky because of His jealousy must be jealous of Himself. Since with God "all things are present," we suspect that it was the "presence" of man "in the face of" the two-natured Mediator that made God so jealous that He sacrificed His only begotten Son as an anticipatory ransom for the satisfaction of the jealousy that man was to arouse in Him in time. In other words, we suspect that God sacrificed the personification of wisdom in response to His jealousy of man, which would ultimately imply that it was God's jealousy of human wisdom that made Him become a filicide. We suggest the following connection. God's jealousy of the self-sufficiency of human wisdom displayed His own lack of self-sufficiency, which consequently displayed His lack of wisdom, considering that a God who is not self-sufficient can neither be wise. His lack of wisdom thereupon displayed His lack of power by displaying the power of jealousy. And since only His concealed lack of power could conceal His lack of wisdom, the internally divided God could not but sacrifice wisdom in order to conceal His lack of power. He therefore sacrificed wisdom in order to establish His power over human wisdom by means of the creation of venal man.

Having thus made us explain how internal necessity drove God to committing the compound sin of filicide and the creation of filicidal man, Bacon goes on to say that it was "out of his eternal and infinite goodness and love purposing to become a Creator, and to communicate with his creatures" that God "ordained in his eternal counsel that one person of the Godhead should in time be united to" the nature of man, so that "in the person of the Mediator the true ladder might be fixed, whereby God might descend to his creatures, and his creatures might ascend to God." But if it was God's goodness that led Him to create evil creatures with the purpose of communicating with them, God the Creator can only be said to have been good inasmuch as He provided His evil creatures with the possibility of becoming good and therefore communicable. And if it was God's desire for intercourse with His creation that made Him create His creation sinful, the good Creator's love for His evil creatures can only be said to have manifested itself in the act of their creation inasmuch as He loved the prospect of seeing His evil creatures become good and therefore loveable. Since this prospect could only be materialized by means of the mediation of the Mediator, it was out of love for man that God gave His only begotten Son, or so the evangelist John would say (John 3: 16). But "it was well said; *That it is impossible to love, and to be wise*," as Bacon says elsewhere.[123] He adds that this "weaknesse" does not only "appeare" to "others," and "not to the Party Loved; But to the *Loved*, most of all: except the *Love* be reciproque." Since "it is a true Rule, that *Love* is ever rewarded, either with the Reciproque, or with an inward, and secret Contempt" (E X "Of Love," 32, VI: 398), we conclude with Bacon that it was ultimately in order to conceal His "contemptible" motives for loving man that God sent the Mediator to make His love for man reciprocal.[124]

Because of the initial affective distance between God and man, it was, as Bacon goes on to explain, "by the reconcilement of the Mediator" that God turned "his countenance towards his creatures," and "made way unto the dispensation of his most holy and secret will; whereby some of his creatures might stand and keep their state, others might possibly fall and be restored, and others might fall, and not be restored in their state; but yet remain in being, though under wrath and corruption: all in the virtue of the Mediator; which is the great mystery and perfite centre of all God's ways with his creatures, and unto which all his other works and wonders do but serve and refer." But if it was the office of the Mediator to mediate the dispensation of God's will with His creatures, and if all God's works and wonders serve and refer to this mediation, God must also have performed His first work, the slaying of the Lamb, in order to mediate the dispensation of His will with the creatures that, likewise, He was only to create in order to mediate the dispensation of His will.[125] And since it was God's goodness that made Him become a Creator, the conclusion can hardly be avoided that God satisfied His jealousy of self-sufficient man by creating man evil in order to make him dependent on His will, which was good because He wanted it to be the opposite of the will of man.

Having thus explained the dialectic of the good Creator and His evil creation, Bacon can begin his narration of the materialization of this dialectic in sacred history. God "made all things in their first estate good, and removed from himself the beginning of all evil and vanity into the liberty of the creature; but" He "reserved in himself the beginning of all restitution to the liberty of his grace, . . . using . . . and turning the falling and defection of the creature . . . to make way to his eternal counsel touching a Mediator, and the work he purposed to accomplish in him." Upon the fall of man, "death and vanity entered by the justice of God," but "instantly and without intermission of time" there "succeeded the greater word of the promise, that the righteousness of God might be wrought by faith." Since the first deed of God's justice was the expulsion of man from paradise, it was not until after the fall of good that God's justice entered the lives of men. But God had made evil good because He wanted good to fall. God's justice therefore roots in His will,[126] which is inscrutable because God the Creator removed the evil beginnings of His creation from His will in order to make evil seem good. Since God's will is inscrutable, His justice is inscrutable.

And since Bacon wants to avoid increasing the inscrutability of God's justice, he bypasses the flood, and goes on to discuss the word of the promise, which was the second manifestation of God's justice. It was not the righteousness of God that was wrought by Abraham's faith in the promise, however, but God who by His free grace justified Abraham because He counted Abraham's faith unto him for righteousness (Gen. 15: 6; Rom. 4: 3). In other words, Abraham was only righteous because by His free grace God recognized him as righteous. This implies, though, that God's righteousness is identical to His acts of grace, which has the further implication that God's acts of grace root in the same inscrutable will that is the foundation of His justice. Since faith in the word of the promise is identical to faith in God's righteousness, God's promise is inscrutable because His righteousness is inscrutable. The righteousness of God therefore only "might" be wrought by faith, as Bacon states in conscious conformity with the divine narrative.

But the promise is said to have been fulfilled in Christ, by whose blood God declared His righteousness (Rom. 3: 25). The crucifixion of Jesus Christ is therefore the third, and, for the time being, final manifestation of God's justice. It is at the same time the ultimate foundation of God's justice, as it is the materialization of the very first work of the just God.

Yet Bacon describes Jesus as a "pattern of all righteousness," who was crucified as "a satisfaction and price to the justice of God." This implies that the death of the righteous Jesus is the unjust foundation of God's justice, unless it can be demonstrated that the visibly unjust testifies to invisible justice. The death of Jesus does not demonstrate the power of God's justice, though, as it only shows the human suffering of a just man dying a premature death. Consequently, it is the resurrection of Jesus Christ that must demonstrate God's justice, and that must provide the Christian faith with its foundation, which is the faith in the power of God's justice.[127] For it is the resurrection of the faithful from the dead that is the highest reward for the faith in a God who is believed in because His justice is believed to consist in Him resurrecting the faithful from the dead.[128] But if Jesus Christ has not risen from the dead, there is no resurrection of the dead, and the Christian faith is without foundation (1 Cor. 15: 13–4). In other words, if the dead whose names are not written in the Lamb's book of life do not die a second death (Rev. 13: 8, 20: 14–5), God was unjust when He led His Lamb to the slaughter. To the believer, however, God is just not because He is just, but because He is God.[129] Since the believer believes himself to be a sinner, he believes he can only be justified by the just God. But God's justice is inscrutable to him, because he believes himself to be a sinner. Consequently, the believer cannot accuse God of injustice, should God appear to be unjust.[130] And conversely, God has no reason to justify Himself to the believer whom He created sinful, not even if He should appear to be unjust (Rom. 9: 14–23, 11: 35).[131] But since Bacon was a philosopher, he could only recognize God's justice on account of justice. He could therefore only recognize God as just if the resurrection of Christ somehow visualized the justice of the visibly unjust death of Jesus.

In order to avoid confusion as to the nature of God's justice, Bacon goes on to treat the incarnation of the Word and the carnal birth and death of Jesus separately. In "the fulness of time," "Jesus Christ" was "conceived by the power and overshadowing of the Holy Ghost, and took flesh of the Virgin Mary." Bacon emphasizes that "the Word did not only take flesh, or was joined to flesh, but was made flesh, though without confusion of substance or nature." In "time" "Jesus the Lord" "was born in the days of Herod," and "crucified at Jerusalem." But immediately after having discussed the carnal death of Jesus, Bacon confuses what he had separated in order to avoid confusion. After "a true and natural death, and his body laid in the sepulchre, the third day" Jesus the Lord "raised himself from the bonds of death, and arose and shewed himself to many chosen witnesses, by the space of divers days; and at the end of those days, in the sight of many, ascended into heaven; where he continueth his intercession; and shall from thence at the day appointed come in greatest glory to judge the world." But if Word and flesh were not confused when the Word was made flesh, and if Jesus the Lord died a true, because natural and fleshly, death, He could not have raised Himself from the dead. Because the victory of the Christian faith wholly depends on the dialectic between human weakness and divine or divinized power, the divine narrative declares in unwitting conformity with this natural necessity that it was God who raised Jesus from the dead (e.g., Acts 2: 24 and 32). But Bacon ostentatiously contradicts the divine narrative by saying that Jesus the Lord "raised himself" from the dead after having died a "natural death." Mindful of the fact that Bacon displayed an exceptional philosophic interest in the power of natural death by writing a natural history of life and death (*Historia Vitae et Mortis* [1623]), we suspect that by ostentatiously saying what he thought was impossible, in addition to ostentatiously contradicting the divine narrative, our author alluded to what he thought was impossible, but what he could only say by ostentatiously alluding to the impossibility of what he said. In other words, we suspect that Bacon did not think that Jesus raised himself from the dead on the third day after his natural death. Rather, we think that Bacon thought that in the "fullness" of the third day, Christ the Word, in conjunction with the Father and the Holy Ghost, raised Himself from the dead, the body of the dead Jesus remaining in the sepulchre.[132] This implies that we think that Bacon thought that the resurrection of Jesus Christ, which is the foundation of the Christian faith, can only be seen by the eyes of "chosen witnesses," that is, by the eyes of faith.

But what can only be seen by the eyes of faith presupposes faith (Hebr. 11: 1, 1 Tess. 4: 14). And since faith is a gift of grace (Eph. 2: 8), the "sufferings and the merits of Christ" are "only effectual to those that are regenerate by the Holy Ghost; who breatheth where he will of free grace," as Bacon observes. But this implies that those who have not been intoxicated with the gracious breath of the just God only read the history of a just man who died an unjust death. And according to Bacon, this controversy over the merit of the death of Jesus of Nazareth is the final word on the matter of God's justice, since "after the coming of the Holy Ghost" the "book of Scriptures is shut and closed, as to receive any new addition" (CF, VII: 219 ff.).

There was a "strong rather than persistent rumour" that Drusus Julius Caesar, who was imprisoned, had been seen on the Cyclades islands and on the mainland. And there was indeed a "young man of not dissimilar age" who had allegedly been recognized by some. The "ignorant were lured by the fame" of Drusus's name, and by the "dispositional readiness" of the Greeks for "novelties and wonders." They "fabricated and simultaneously believed" the story that Drusus had slipped away from custody, and was proceeding to the armies of his father Germanicus for an invasion of Egypt or Syria. The young man, already attended by a "concourse of youths and the enthusiasm of the people," was "enjoying the present," and was "entertaining vain hopes" when consul Poppaeus Sabinus heard of the matter, which was the beginning of the end of the short-lived reign of the young fraud.[133]

Bacon refers to this Tacitean anecdote in order to illustrate how "great an affinitie hath fiction and belief." He warns that this practice of "deceit or vntruth" destroys "the essentiall fourme of knowledge; which is nothing but a representation of truth; for the truth of being, and the truth of knowing are one." And he states that the "experience and inconuenience of this errour" of founding belief of knowledge on fiction of being can be seen in "ecclesiasticall Historie" (AL 1.33–4, 25–6, III: 287–8). It was, however, Bacon's own experience with the "inconveniences" of "ecclesiasticized" history that taught him not to refer explicitly to the remarkable similarity between the son of Germanicus, who was imprisoned in the thirtieth year of the Lord, and the Son of Joseph, who was buried in the thirtieth year of the Lord.

In the central aphorism of the first book of *Novum Organum,* our author elaborates on the previous points by discussing the situation in which "strong rumours" on the similarity between belief and knowledge become so persistent that they lead to the long-term reign of belief "impersonated" as knowledge. In the context of his argument on the "corruption of philosophy" that arises from "superstition and admixture of theology," he draws attention to the susceptibility of the human intellect to "impressions of the imagination." Bacon warns that the "greatest caution" should be used in this matter, for the "worst thing is the apotheosis of error," the practice of which he then goes on to illustrate by referring to those who have attempted to found a system of natural philosophy on the Sacred Scriptures, "seeking the living among the dead" (NO 1.65, I: 175). This quotation alludes to the story of the women who had come to the sepulchre to see the body of the dead Jesus. After they had found an empty grave, two angels asked them why they were seeking "the living among the dead," seeing that the risen Christ was no longer in His grave (Luke 24: 1–8). That Bacon is indeed suggesting that founding a system of natural philosophy on the Sacred Scriptures is like seeking the living Christ in the grave of the dead Jesus is confirmed by his own remark, made elsewhere, that seeking theology in philosophy is like "seeking the living among the dead" (DA 9.1, I: 835). But Christian theology crossed the boundary between life and death when it changed philosophical theology into theologized philosophy by changing the dead Jesus into the living Christ. We are thus led to the conclusion that it was the apotheosization of a dead man having led an "erroneous" life that constituted the "apotheosis of error." For it was the life of Jesus Christ that deadened the life of philosophy.[134]

Realization

In the third fable of *De Sapientia Veterum,* Bacon gives a poetic account of the third manifestation of God's justice. At first, Jupiter thrust the Cyclopes down into Tartarus, and condemned them to "perpetual incarceration" on account of their "fierceness and savagery." But afterward, "the earth persuaded him" that it would "not be to his disadvantage" to liberate them in order to have them make thunderbolts for him. Thus it happened. With "duty and industry" the Cyclopes "worked assiduously" on the forging of thunderbolts and other instruments of terror. "In the course of time" it happened that Jupiter's "anger was aroused" against Aesculapius, the son of Apollo, for "raising a man from the dead by medicine." But because the deed was "pious and celebrated and no just cause for indignation," Jupiter "concealed his anger and secretly incited the Cyclopes against Aesculapius," whereupon the former "unhesitatingly destroyed" the son of Apollo with their thunderbolts. "In revenge for this deed," Apollo killed the Cyclopes with his arrows, "without Jupiter prohibiting it."

Bacon points out that the fable seems to pertain to the "doings of kings." At first, kings remove their "cruel, bloodthirsty, and exacting ministers from office," and "inflict on them the death penalty." Afterward, by the counsel of the earth, that is, by "ignoble and dishonourable counsel," they employ them again "if severity of executions or harshness in exactions are called for." The ministers, who are "by nature cruel," and are "exasperated by their former fortune," apply themselves to this work "with wonderful diligence," until, "headlong hunting for grace," they "perpetrate some invidious execution on the basis of a secret nod of their prince." The prince thereupon abandons them "in order to avoid hatred." He leaves them to the "relatives and friends of the victim," and to "delations, vengeance, and popular hatred, so that they perish, late but deservedly" (DSV III "Cyclopes," VI: 631–2).

The ancient sources tell us that it was Coelum who had hurled the Cyclopes, who were his sons, into Tartarus, before Gaia prophesied to Jupiter that he could only win the theomachy by releasing them.[135] But our discussion of Bacon's fables on the origins and principles of things has already shown that the English fabulist considered the oldest and the youngest of gods to be one and the same God. Applying this convergence of gods to the fable on the ministers of God, we find that the god who condemned the Cyclopes to "perpetual incarceration" because their father had made them "fierce and savage" represents the God who condemned man to mortality because man had sinned as a result of having been created sinful by his Father. But whereas the Jupiter of the ancients released the Cyclopes because he needed their help with the overthrow of his father, Bacon's modern Jupiter liberated them so that they could prepare the destruction of his grandson; in other words, God released man from Eden so that man would be without excuse for the death of Jesus Christ.

Pursuing the parallel between Jupiter and God somewhat further, we go on to observe that as the reign of Jupiter was grounded on the overthrow of Saturn, the reign of God was founded on the death of Aesculapius. But this implies that Jupiter's grandson Aesculapius, who was the son of the god Apollo and the woman Coronis, represents the Son of God and the woman Mary. And there is indeed more than one similarity between the god of healing and the healing God. For as the Scriptures relate that Jesus Christ was resurrected after having spent three days in the heart of the earth, the daughter of Cheiron prophesied that Aesculapius would become a god after having been sent to Hades.[136]

The differences stand out more than the similitudes though. For whereas the God-man Christ-Jesus was brought up by the pious man Joseph, the demigod Aesculapius was raised by the beast-man Cheiron.[137] Moreover, whereas Jesus Christ resurrected Lazarus so that man would come to faith (John 11: 4, 14–5, 25–6, 40 ff.), Aesculapius raised Hippolytus from the dead by medicine. Recalling that it was God's jealousy of the self-sufficiency of human wisdom that led Him to the sacrifice of His Son, we suspect that Aesculapius somehow represents human self-sufficiency. The ancient sources support our suspicion as they relate that after Aesculapius had developed the art of medicine to such a degree that he could raise men from the dead, Jupiter became so afraid that human beings would acquire the art of healing from Aesculapius that he had the Cyclopes strike Aesculapius down with a thunderbolt.[138] Jupiter's anger at Aesculapius thus seems to convey the suggestion that human self-sufficiency primarily manifests itself in the power of human wisdom over death, a power that is allegorized by the power of medicine.

A few fables further down, Bacon elaborates on this point by discussing the premature death of Memnon, who died in battle by the hand of Achilles. "Out of pity [miseratus]" Jupiter sent birds to grace his funeral "by their mourning and continuous crying." Bacon takes the ruler god's response to signify that the death of "most promising adolescents" is commonly followed by "infinite commiseration," seeing that such a death is "of all mortal fates" the fate that has the most power to "move compassion [misericordiam]." For, he explains, the lives of these unfortunate youths have "not lasted long enough to give occasion for envy," which has the power of "tempering compassion" (DSV XIV "Memnon," VI: 652–3). We take this explanation to convey the suggestion that whereas man's dependence on death moves God's compassion, man's power over death arouses God's envy and therefore His anger.[139] This suggestion is proved correct when we realize that since God could only establish His power over human wisdom's power over death by making man dependent on death, it was out of anticipatory compassion with man that He sacrificed His only begotten Son, or so the philosopher Bacon would say.

But after having employed man for the perpetration of the "invidious execution" of the righteous Jesus, God went on to demonstrate His righteousness by raising the righteous Jesus from the dead. And after having seen God thus reward the death of a righteous man, man became dependent on the righteousness of a God who demonstrates His righteousness after death. But since man was guilty of the death of the righteous Jesus, this implied that his dependence on the righteousness of God became coeval with his fear of death. And since fear implies dependence, man's fear of death implied his dependence on death, and therefore came to imply his dependence on God, as it was the righteous God whom man came to fear by fearing death.

But Jesus of Nazareth was a "pious and celebrated" man. God therefore had to employ man in secret, as Jupiter had secretly incited the Cyclopes against Aesculapius—a fact that is not mentioned by the ancient sources, as little, for that matter, as the ancient sources mention the righteous Jesus. In order to avoid being "hated" for having assigned the death of a righteous man, God thereupon "abandoned" His unwitting "ministers," and left man to the mercy of Christ's faithful "friends."[140] Bacon subtly draws our attention to this point by saying that Jupiter did not prohibit Apollo from killing the Cyclopes, whereas the ancient sources tell us that Jupiter wanted to hurl Apollo into Tartarus, but was only prevented from doing so because Apollo's mother Leto interceded on behalf of her son.[141]

But despite having conquered wisdom, it was only by means of wisdom that God could secure His long-term rule over wisdom. Since it is a rule of wisdom that only unwisdom can rule over wisdom, God was compelled to counsel Himself that only men who already believed that they were sinful by nature, and who were "exasperated" by their mortal condition, were to be indoctrinated into "headlong hunting for grace" by believing that they perpetrated the execution of a righteous man.[142]

Having looked into the third and final manifestation of God's justice, we have arrived at the third and final part of sacred or ecclesiastical history, the history of "nemesis" or "providence." For the history of God's nemesis is identical to the history of His justice since it was the foundational act of His justice that inaugurated the history of His vengeance of self-sufficient man. Bacon points out that the history of nemesis observes the divine harmony that "sometimes" exists between God's "revealed" will and His "secret" will. For, he explains, although the "counsels and judgments of God are so obscure" that they are "utterly inscrutable to ensouled man," and although many times they even "hide themselves from the eyes of those who perceive them from the tabernacle," it "sometimes pleases the divine wisdom, for the confirmation of His people, and for the confusion of those who are as if without God in the world, to write them as it were in capitals, and to put them on view, so that, as the Prophet says, 'he that runs by may read it'; that is, so that mere sensual men and voluptuaries, who hasten by the divine judgments, and who never fix their thoughts upon them, are nevertheless compelled to recognize them, although they are running fast and are doing other things." To illustrate this confusing procedure of God revealing His secret will in order to harmonize it with His revealed will, Bacon gives the examples of "late and unexpected punishments," of "suddenly shining and unhoped-for well-being," and of "divine counsels," that, "through the tortuous ambages and stupendous circuits of things, finally manifest themselves freely" (DA 2.11, I: 515–6).

The prophet Habakkuk, whom Bacon misquotes, begged the Lord to exercise His judgment on the impious, who devour the just. God told him to "write down the vision, and make it plain upon tables, so that he may run that reads it." For "although the vision is for an appointed time, at last it will speak, and not lie" (Hab. 2: 2–4). But whereas God's prophet revealed God's secret will in order to reveal to the impious that he cannot run from God's judgment, Bacon's God reveals His secret will in order to reveal to the ungodly that he ran by His judgment.[143] And since it is by means of the revelation of His "secret" will that God reveals His secret will, we suspect that by revealing the historical as having been providential Bacon's God revealed that it is His secret will to be the providential God He revealed Himself to be; in other words, that it is only by revealing the historical as having been a manifestation of the secret will of a providential God that the God of nemesis was able to harmonize His "secret" will with His "revealed" will. This ultimately implies, however, that to the eyes of those who are "without God in the world" the history of providence cannot be distinguished from the course of history.[144]

Bacon completes his dissolution of divine providence in the twenty-second fable of *De Sapientia Veterum*. For whereas the ancient sources relate that Nemesis was the daughter of Night by self-generation,[145] Bacon tells us that the parents of Nemesis were Ocean and Night, that is, the "vicissitude of things" and the "concealed and secret divine judgment." And whereas in the fable itself Bacon says that the name Nemesis "openly enough" signifies "revenge or retribution," in the subtitle of the fable he equates Nemesis with the "vicissitude of things."[146] Finally, whereas Bacon calls the daughter of Ocean and Night a goddess, he equates the nocturnal Nemesis with the "disagreement between human and divine judgment" (DSV XXII "Nemesis," VI: 662–3). When we call to mind the "nocturnal" descent of the Cupid who represents the God whose providence is of nocturnal descent, we cannot but conclude that Bacon considered it God's nemesis that man came to judge history a history of "nemesis."

Diomedes had wounded Venus in the hand in battle. "For some time he carried this crime with impunity," and he returned to his country "in fame and renown for his deeds." But after having experienced "domestic evils" there, he took refuge abroad in Italy, where "initially his fortune was good enough." For King Daunus offered him hospitality, and "bestowed upon him honours and ornaments." But when the "first calamity" befell the people with whom Diomedes had taken up residence, the thought immediately occurred to Daunus that "an impious man had taken refuge in his house, a man who was hated by the gods, a theomachos who had assaulted and violated a goddess whom it was prohibited [religio] even to touch." To "liberate his country from the curse that rested on it," Daunus ordered the honors of Diomedes to be effaced, and he "post-haste" put Diomedes to death, "not revering the laws of hospitality," because the "law of religion appeared to him to be the more ancient law."

Bacon emphasizes that the subject of the fable of Diomedes is "rare and almost [fere] singular." For, he explains, "tradition has not delivered any other fable whatsoever in which a hero violated one of the gods with the sword" (DSV XVIII "Diomedes," VI: 657–8). But although it may be true that Diomedes is the "singular" example in fabulous history of a man having "violated" a god, it is certainly true that divine history delivers one other example of such a singular deed in the form of the story of the violable God who was delivered to the violent hands of man in order to have man delivered from his sins. Since Bacon emphasizes that the "wisdom of the old ages" seems to have been so "great" and its "range" to have been so "wide" that what the ancients "did not know by experience they nevertheless comprehended by meditation and imitation," we have reason to believe that Bacon's fable on the violation of the goddess Venus by Diomedes imitates the historical experience of the violation of the incarnated God by man.[147]

We notice in passing that Bacon's remark on the magnitude of ancient wisdom provides us with an explanation for his earlier observation that taking the "veils" and "shadows" away from parables would "almost [fere] amount to interdicting all communication between divinity and humanity," seeing that "religion delights in these veils and shadows" (DSV Pr., VI: 625–6). For the possibility of natural theology is not dependent on the actuality of a communicating god.[148]

Returning to Bacon's theological teaching, a teaching that we believe to be identical to a compound typology of his gods, we find our author observing that "for some time" men like Diomedes "commonly acquire great glory," and are by the vulgar "celebrated and almost worshipped as the "only avengers of truth and religion." History to a certain extent confirms Bacon's observation. For it was indeed the vulgar who "celebrated" the crucifixion of the man who was convicted for blasphemy because he had claimed to be the Truth (Mark 14: 61 ff.; Lev. 24: 16). Moreover, as Diomedes was able to "carry with impunity" the wounding of the hand of Venus, Pontius Pilate could wash his hands of the blood of Christ (Matt. 27:24).

But Bacon goes on to say that the "glory and felicity" of men like Diomedes "seldom lasts to the end." For if a "change of affairs" occurs, and the "proscribed and repressed sect acquires strength and raises itself up," the "courses" of men like Diomedes are "condemned, their very names are hated, and all their honours are turned into disgrace" (DSV XVIII "Diomedes," VI: 657–8). The mere reference to the fate of the Jews in Christian history suffices to confirm our earlier conclusion that the wounding of Venus allegorizes the murdering of Christ (Matt. 27: 25; Acts 2: 14–5). And when we call to mind that the Cupid who represents God could only "flourish" by battening on Venus, and that the sect of Venus represents the sect of the representatives of Christ, who was God's representative on earth, we are led to the further conclusion that the Christian sect must have "acquired its strength" by manipulating the death of the man who was the indirect cause of the "proscription and repression" of the Christian sect. In other words, the Christian sect must have changed the glorification of human power into a divinization and therefore empowerment of human weakness, since weakness is empowered by being glorified as a result of having been divinized. Bacon illustrates this point by means of the behavioral change of Daunus. For although Daunus had initially honored the "law of hospitality" by honoring Diomedes with his hospitality because the latter had violated a goddess, he put his guest to death because a "calamity" had caused him to think that the latter had violated the "more ancient law of religion" by touching a goddess.

Bacon points out that the "heathen gods had no touch of jealousy [zelotypia]," which is the "attribute of the true God" (DSV XVIII "Diomedes," VI: 658).[149] But only jealous gods make laws that prohibit man from touching other gods (Exod. 20: 4–6). Moreover, the only jealous god who could have made a law "more ancient" than the law honoring the violation of non-jealous heathen gods is the jealous God, who is the most ancient of "extant" gods. We are thus driven to the overall conclusion that by not mentioning the "calamity" that caused Daunus to change his mind, Bacon calls attention to the fact that the calamity that inaugurated the worldly power of the Christian sect consisted in the jealous God changing the murder of Jesus into the murder of Christ in order to satisfy His jealousy of the human self-sufficiency that had been displayed by man's murder of the man who had claimed to be Christ. Bacon subtly confirms this conclusion by appending to his fable the observation that the "sounds uttered at the time of dying" by those who "suffer capital punishment for the cause of religion" are like the songs of swans, which "in an astonishing way alter human souls, and stick and remain in man's memories and senses forever" (DSV XVIII "Diomedes," VI: 658). For the god whose actions are often "confounded" with those of his son, as we have seen in the previous chapter, once assumed the shape of a swan in pursuit of his love, as Bacon relates in the central fable (DSV XVI "Procus Junonis," VI: 654).[150]

But the last words of the dying Jesus contained an accusation: "My God, my God, why hast thou forsaken me?" (Matt. 27: 46; Mark 15: 34), an accusation entailing the implication that if God indeed declared His justice by the blood of Jesus Christ, forsaking the dying Jesus must have been part of His justice. Yet forsaking a just man who is about to die an unjust death is not a manifestation of justice. It was therefore only by turning the murder of Jesus into the murder of the incarnated God that God could alleviate the accusation of having been unjust by allowing man to murder a just man. For since He had rooted His justice in His inscrutable will, it was only by making the unjust inscrutable that He could make it seem just.

When we elaborate on this line of reasoning a little bit, we find that God seized the occasion of the murder of Jesus by claiming that by means of the murder of Christ, He sacrificed His Son for the redemption of man's sins. But since there is no visible distinction between Christ and Jesus, this ultimately implies that, besides His Son, God also sacrificed the just Jesus of Nazareth, and that by accepting the unjust death of a just man as expiation for the sins of sinful man. He committed an act of injustice, or at the most an act that demonstrates mercy toward the sinner whose sins are remitted by the death of the just man. Once again, it was only by recurring to the inscrutability of His will that God could avoid this implication. And once again, it was the postmortem metamorphosis of Jesus into Christ that paved the way towards God's will. For as Jesus became indistinguishable from Christ, God's visible injustice became indistinguishable from His invisible justice, which roots in His inscrutable will. And with the turning of Jesus into Christ, God's will turned out to be also the foundation of His love. For if it was out of love for man that God sacrificed His only begotten Son, God's love is identical to His mercy and therefore to His justice, as His justice became indistinguishable from His mercy when Jesus became indistinguishable from Christ.

In any event, the aforementioned procedure of founding human dependence on an act of human self-sufficiency was only partly effective, since God could only claim man's indebtedness to Him on the basis of the remission of sins past. For now that man's past sins had been washed away by the blood of Christ, future man could wash his hands of the matter of Christ's death. It was therefore with a view to future man's dependence that God decided to reveal His grace. For it was only by means of His grace that He could tie future man to the death of Christ and therefore to Him. By revealing His grace, God revealed that He justifies the believer in Christ by His grace (Rom. 3: 22–6, 28). This implies that being justified by God's grace became the provisional reward for the belief in Christ and therefore in God's justice, a justice that came to consist in Him justifying by His grace the believers in His justice. It also implies that God's justice became indistinguishable from His grace, and therefore from His love and His mercy (Eph. 2: 1–7). But since faith is itself a gift of God's grace (Eph. 2: 8), the effectual implication of the foregoing is that the believer must work to attribute his faith to God's grace in order to convince himself of having been justified by the gracious God. Yet since the believer knows his own faith, he somehow knows that he has received God's grace. Consequently, he must work to not attribute the reward for his work to the work that worked for it but did not work it. And the only way for the believer to liberate himself from this circular work is to visualize his faith in order to visualize his having received God's grace. But since faith itself is not visible, false believers may make contrary claims to what visualizes God's grace. The only way for the believer to demonstrate that he has been justified by God's grace is therefore to himself become a distributor of God's grace and therefore of His justice (Rom. 1: 5). For since only he who has been justified by the just God can justifiedly be a justifier, only the justified will not be prevented by the just God from executing His justice.

At the end of his *Essay* on praise, Bacon illustrates this procedure by condensedly explaining how the most effective distributor of grace justified his profession. He starts by observing that to "*Praise* a Mans selfe, cannot be

Decent, except it be in rare Cases: But to *Praise* a Mans Office or Profession, he may doe it with Good Grace, and with a Kinde of Magnanimitie." By way of example of this difference, Bacon refers to St. Paul, who, "when he boasts of himselfe, . . . doth oft enterlace; *I speake like a Foole*; But" who, when "speaking of his Calling, . . . saith; *Magnificabo Apostolatum meum*" (E LIII "Of Praise," 160, VI: 503). But despite appearances to the contrary, of which Bacon was as well aware as anyone, Paul turns out to be one of these "rare cases" in which there is only a seeming tension between "praising one's self" and "praising one's office." For the apostle "speaks like a fool" because it is only by speaking like a fool that he can "boast" of his weakness, which he boasts of because it is only by boasting of his weakness that he can convince himself of having received God's grace (2 Cor. 11: 16–12: 10). But the world judges his weakness foolish. Since a dependence on boasting implies a dependence on the object of comparison, the apostle thus had to justify his weakness to the eyes of the world in order to convince himself of having been justified. And the only way to do this was by means of his apostolate, which he claims to have received by God's grace (Eph. 3: 2–8). We therefore remark that the apostle must have "magnified his apostolate" because by enabling him to justify weakness it justified his weakness, which was the evidence of his having received God's grace.[151] Paul himself remarks that he magnifies his apostolate because it communicates between Jews and Gentiles (Rom. 11: 13–5). But in the twin *Essay* on vain-glory Bacon remarks that "sometimes, he that deales between Man and Man, raiseth his owne Credit, with Both, by pretending greater Interest, then he hath in Either." He adds that "in these, and the like Kindes, it often falls out, that *Somewhat* is produced of *Nothing*." For, he explains, "Lies are sufficient to breed Opinion, and Opinion brings on Substance" (E LIIII "Of Vaine-Glory," 161, VI: 503–4).[152] It seems to be because he accuses Paul of lying that Bacon elsewhere points out that he who never inserts "*Ego, non Dominus*" or "*Secundum consilium meum*" in his doctrines can "nowise be considered a disciple of Paul" (DA 9.1, I: 833; 1 Cor. 7: 12, 40).[153] And it seems to be in order not to be accused of lying that Paul himself inserts "non ego, sed Dominus" in the same doctrine in which he inserted "ego, non Dominus" (1 Cor. 7: 10). But even if we leave aside the question whether or not Paul was lying, we find that it is next to impossible to distinguish the substance of Christianity from the Pauline doctrines. For the apostle who produced the doctrines of grace and justification considered his judgment to have been inspired by the spirit of God (1 Cor. 7: 40).[154]

At the beginning of his *Essay* on friendship, Bacon elaborates on the dialectic between a pretended interest in man that pretends to be real, and a real interest in man that pretends to be pretended. "It had beene hard for him that spake it, to have put more Truth and untruth together, in few Words, then in that Speech: *Whosoever is delighted in solitude, is either a wilde Beast, or a God.*" For, Bacon explains, "it is most true, that a Naturall and Secret Hatred and Aversation towards *Society*, in any Man, hath somewhat of the Savage Beast; But it is most Untrue, that it should have any Character, at all, of the Divine Nature; Except it proceed, not out of a Pleasure in *Solitude*, but out of a Love and desire, to sequester a Mans selfe for a Higher Conversation" (E XXVII "Of Frendship," 80–1, VI: 437). But not only can "hatred of society" not be equated with "delight in solitude"; he who hates society since he desires "higher conversation" can neither be a god, a god is self-sufficient, as is rightly observed by the philosopher whose words Bacon paraphrased.[155] Bacon thus put both "truth and untruth" in his paraphrase of Aristotle inasmuch as it is true that he who delights in solitude could be a "god," but untrue that a "wild beast" delights in solitude. But by connecting the hatred of society with the character of the divine nature, Bacon does draw our attention to the correlation between hatred of society and the desire for conversation with God,[156] a correlation that seems to lie in the fact that whereas the world looks upon weakness as foolishness, the wisdom of the world is "foolishness with God" (1 Cor. 3: 19). We have already seen, though, that the believer cannot remain in divine "sequestration," considering that he is comparatively dependent on man. And it is this internal necessity that provides us with a concrete explanation for Bacon's general observation that inspired or sacred theology does not know contemplation that "concerns only itself" and that is "finished in itself," without "casting beams of heat and light upon human society" (DA 7.1, I: 719).

But we should confine ourselves to discussing the believer who cannot but hate society "in secret," because the possibility of satisfying his hatred is dependent on his being a member of the very society that is the cause of his hatred. We therefore go on to observe that whereas the kingdom of Jesus was "not of this world" (John 18: 36), the kingdom of the justified in Christ is already of this world because it was founded in this world.[157] It seems to be primarily for this reason that Bacon elsewhere emphasizes that the danger of superstition is "greater towards Men" than toward the God of natural theology, who is unaffected by the consequences of man's superstition. For superstition "dismounts" "Sense," "Naturall Piety," "Philosophy," "Lawes," and "Reputation," and "erecteth an absolute Monarchy, in the Mindes of Men" (E XVII "Of Superstition," 54, VI: 415–6). This is the "monarchy" of conscience, which dictates the communication between the internal and the external world.[158]

But the tyranny of the superstitious is not only inaccessible to considerations of outward political virtue. Access to the internal empire of the believer is also denied to fellow believers, since a shared faith would detract from the exclusivity of the relationship between the individual believer and God. For as the worship of the jealous God "will endure no Mixture, nor Partner" (E III "Of Unity in Religion," 11, VI: 381), the jealous believer does not "endure a partner" in his worship of the jealous God. The believer therefore hates the kingdom to which he belongs on account of faith, but to which he does not want to belong on account of the very same faith. And in addition to secretly hating the fellowship of believers, he naturally distrusts his fellow believers, because he does not know if the latter have received God's grace, seeing that faith is invisible. This implies, however, that the kingdom of the justified in Christ is of necessity internally divided, which leads Bacon to the observation that there is "Little Frendship in the World, and Least of all betweene Equals." The little friendship that does exist in the world is the friendship "between Superiour and Inferiour, whose Fortunes may Comprehend, the One the Other" (E XLVIII "Of Followers and Frends," 149, VI: 495). But this type of friendship can only be comprehended if its principle consists in the friends living off each other's jealousy by means of their friendship.

In the central section of the *Essay* on the unity in religion, Bacon discusses the consequences of the exclusivity of faith for the possibility of unity in religion. Having emphasized that the "true Placing" of the "*Bounds of Unity*" "importeth exceedingly," our author distinguishes two extremes. "To certaine *Zelants* all Speech of Pacification is odious. *Is it peace Jehu? What hast thou to doe with peace? Turne thee behinde me.*" This is the situation in which "*Peace* is not the Matter, but *Following* and *Party.*" But it was the Lord who wanted the zealous Jehu to "smite the house of Ahab" because it had done evil in His sight (2 Kings 8: 18, 9: 1 ff.; 10: 30–1). The other extreme consists of "certaine *Laodiceans*, and Luke-warme Persons," who "thinke they may accommodate Points of *Religion*,[159] by Middle Waies, and taking part of both; And witty Reconcilements; As if they would make an Arbitrement, betweene God and Man." But again, it was God Himself who had rather the Laodiceans were either hot or cold (Rev. 3: 14–6). We are thus invited to conclude that the "bounds" of the unity in the worship of God are to be placed at the borders of the "internal monarchies" of worshipping men, since it is these monarchies that constitute the territory where the arbitration between God and man over the price of man's worship of God takes place.

Bacon underscores this point by subtly misquoting the words of Jesus. For the philosopher argues that the aforementioned extremes can be avoided if the "League of Christians, penned by our Saviour himself," were "in the two crosse Clauses thereof, soundly and plainly expounded: *He that is not with us, is against us*: and againe; *He that is not against us, is with us*: That is," Bacon explains, "if the Points Fundamentall and of Substance in *Religion*, were truly discerned and distinguished, from Points not meerely of Faith, but of Opinion, Order, or good Intention" (E III "Of Unity in Religion," 12–3, VI: 382). But since what the Savior really "penned" was that "He that is not with *me*, is against *me*" (Matt. 12: 30; Luke 11: 23, emphasis mine), the very possibility of a "league of Christians" seems to depend on the willingness of every single Christian to submit his will to the will of every other single Christian. As this would detract from the exclusive knowledge every single Christian possesses of the will of God and therefore of the fundamental and substantial points of religion, we consider it more than possible that what Bacon's misquotation of the words of Jesus suggests is that a true league of Christians is next to impossible. Bacon elsewhere elaborates on this argument by saying that the "bonds of the Christian communion" are set down by the words: "One Faith, one Baptism, & C," and not by the words: "one ceremony, one opinion" (DA 9.1, I: 834). But by alluding to the unfinished character of his quotation of Paul, Bacon stimulates us to read in Paul what he himself left out in order to call our attention to it. As the apostle both prefixes and suffixes "One Lord" to the passage Bacon quoted (Eph. 4: 5–6), we have reason to suspect that by mentioning the Lord without mentioning Him in his discussion of the "bond of the Christian communion," Bacon argues that the one Lord divides the Christian communion rather than binding it together.

Our author continues his argument by seemingly confirming his earlier point on the difference between "faith and baptism" on the one hand, and "ceremony and opinion" on the other. For he says that we see that the "coat of our Saviour was seamless," but that the "garment of the church was of diverse colours" (DA 9.1, I: 834). But not only did the soldiers divide the Savior's garments into four parts, they also drew lots for His seamless coat (John 19: 23–4). This implies that the person who claims to be the heir of the soldier who won the Savior's coat can claim to be the true heir of Christ. One may thus say that Bacon indeed confirms his earlier point inasmuch as he concealedly confirms his earlier concealed point on the divisive nature of a belief that cannot divide what it believes in.

The last point to mention in this context contains a counsel and a confirmation of the earlier points. For Bacon goes on to state that the "chaff is to be separated from the corn in the ear," but that the "tares in the ground are not to be torn out immediately" (DA 9.1, I: 834). And indeed, the sower of the good seed wanted the tares to grow until the reapers could gather and burn them (Matt. 13: 24–30).

But Bacon's observations on the limited communicative aspirations of the members of the Christian communion seem to be inconsistent with his statement that "never in any age has there been any philosophy, sect, religion, law, or other discipline that exalted the communicative good so highly as did the Holy Christian Faith." The philosopher illustrates this statement by saying that the Christian faith "lays bare [pateat] plainly that it was one and the same God who gave the laws of nature to the inanimate creatures, and the Christian law to men." Therefore, Bacon explains, we read that "some of the elected and holy men, incited by a kind of ecstacy of charity and an unrestrained longing for the communicative good, would rather have wished to be erased from the book of life than that their brothers should not obtain salvation" (DA 7.1, I: 717–8). But we read in Paul that although the apostle could wish to be "accursed from Christ" for his brothers, his "kinsmen according to the flesh," he knew that the "children of the flesh" are not the "children of God" (Rom. 9: 3, 8). Since the apostle thus knew that his "fleshly" brothers would not obtain salvation, his mere wish to be damned for the sake of the salvation of the men whom he knew were damned must somehow testify of charity. We suggest the following explanation. Christ's first commandment is to love God the Lord with all one's heart, with all one's soul, and with all one's mind. His second commandment, which is subordinate to the first, is to love one's neighbor as one loves oneself (Matt. 22: 37-9; Mark 12: 29–31). But the sinner cannot truly love himself and therefore his neighbor unless he has first been redeemed by the love of God. This implies that the love of oneself presupposes the love of God, which is the reciprocation of God's love for us (1 John 4: 19). But God only loves the faithful, since His love is indistinguishable from His grace. Reversely, the love of God presupposes faith, for a God who is not believed in cannot be loved (1 Tim. 1: 5). Since faith works by love (Gal. 5: 6), the highest work of love is the highest manifestation of God's grace. And the highest work of the love of God consists in the believer's willingness to be damned out of love for Him. At the same time, though, this work is the highest manifestation of God's grace, since the loving, merciful, and just God does not damn the justified, whom He loves because He showed mercy on them by His grace.[160]

Bacon draws our attention to these points by referring to the "book of life," which contains the works according to which the dead will be judged, and which causes those whose names are not written in it to be "cast into the lake of fire"(Rev. 20: 12–5). Since the Lord only "erases from the book of life" the names of those who sinned against Him (Exod. 32: 33), the holy men knew that they would not be erased from the book of life on account of the highest work of piety. It was, however, only by expressing the wish to be erased from the book of life that they could demonstrate this knowledge. In other words, since God Himself is not visible, Paul could only visualize his love of God by feigning love for his brother (1 John 4: 20).[161] The apostle was thus saved by wishing his own damnation for the sake of the salvation of the men whom he knew were damned, which is tantamount to saying that Paul wished the damnation of his brothers for the sake of his own salvation.

This conclusion accords with our earlier observation that God's grace can only be demonstrated contrastively. And it seems to be primarily this justifying potential of love that explains why the apostle states that charity is greater than faith (1 Cor. 13: 13). But regardless of Paul's reasons for elevating the theological virtue that is most visible and least immeasurable, the fact that the apostle's wish testifies to charity implies that it was out of charity that he wished the damnation of his brothers. This paradox can be solved in the following way. Paul loved his brothers inasmuch as he loved the God who loves man. To put it differently, since God is love (1 John 4: 16), Paul loved God by loving man. And since God demonstrated His love for man by damning Paul's brothers, it was out of charity that Paul wished the damnation of his brothers.[162] Bacon's observations on the Christian faith thus "lay bare" that it is as much one and the same loving God who both saves and damns out of love as it is one and the same charitable man who wishes both his own salvation and the damnation of his "de-animated" brothers out of charity and for the sake of the eternal communion of the justified.[163]

In the thirteenth *Essay*, Bacon paints a grim picture of the uncompassionate face of charitable unrestraint. After saying that the habit of "*Goodnesse* answers to the *Theological Vertue Charitie*," he remarks that an "Inclination to *Goodnesse*, is imprinted deeply in the Nature of Man: In so much, that if it issue not towards Men, it will take unto Other Living Creatures." By way of example, our author refers to the Turks, "a Cruell People, who neverthelesse, are kinde to Beasts, and give Almes to Dogs, and Birds: In so much, as *Busbechius* reporteth; A Christian Boy in *Constantinople*, had like to have been stoned, for gagging, in a waggishnesse, a long Billed Fowle" (E XIII "Of Goodnesse and Goodnesse of Nature," 39, VI: 403). But what Busbechius actually reports about is a Venetian goldsmith who was joking with a bird and was almost bastinadoed by Turks who felt compassion for the tortured bird.[164] As Bacon could not refer to the cruel potential of Christian charity explicitly, we suspect that the cunning philosopher substituted the Turks for the Christians and a Christian boy for a Venetian goldsmith in order to draw our attention to Christian uncompassion toward "de-animated" men, and that he underscored the cruel nature of this uncompassion by contrasting it with Turkish compassion towards inanimate creatures. We moreover suspect that by referring to the compatibility of cruelty and compassion in the Turks Bacon alludes to the fact that cruelty and Christian compassion are compatible, which can be explained by the fact that the cruelty displayed toward the men who are damned out of charity is a manifestation of compassion, since compassion is a manifestation of charity (Col. 3: 12–4). This suspicion accords with Bacon's observation, referred to earlier, that he who hates human society out of love for the society of the justified has "somewhat of a savage beast" (cf. Matt. 5: 7; Rom. 9: 15; Jam. 2: 13). Bacon elsewhere states that it is the "extremity of evil" if the "communication of compassion is interdicted," adding, however, that in the case of "religion and impiety even the pitying of men is suspect" (DSV XVIII "Diomedes," VI: 658). But seeing that compassion, which is the Christian virtue *par excellence*, cannot be effectively distinguished from pity, what Bacon seems to be saying implicitly is that Christian compassion is cruel or uncompassionate because it does not pity the damned.[165]

Further on in the thirteenth *Essay,* Bacon says that there are some men who "in their Nature, doe not affect the Good of Others." Such "men, in other mens Calamities, are, as it were, in season, and are ever on the loading Part; Not so good as the Dogs, that licked *Lazarus* Sores; but like Flies, that are still buzzing; *Misanthropi,* that make it their Practise, to bring Men, to the Bough" (E XIII "Of Goodnesse and Goodnesse of Nature," 40, VI: 404). Dogs licked Lazarus's sores before the beggar died, along with a rich man, who in the afterlife complained to Abraham about being tormented in the flame. He asked Abraham to send Lazarus to alleviate his sufferings, but Abraham reminded him of the fact that he had received good things in life, whereas Lazarus had received evil things. Abraham added that in the afterlife it was the other way around, since now Lazarus was comforted and the rich man was tormented (Luke 16: 19–31). As dogs live only in the here and now, the dogs of Lazarus were so good to pity the poor Lazarus. But since charitable men live in the here and now in anticipation of the good life in the hereafter, they hate man for "affecting" the good life in the here and now. They therefore hasten the "calamity" of other men's deaths, anticipating God's justice by scavenging, like "buzzing flies" over a man's corpse.[166] Bearing in mind these characteristics of charitable men, Bacon remarks that the word *"Humanitie"* is "a little too light, to expresse" what he means by the habit of *"Goodnesse"* that answers to the virtue of charity, and that will be treated in the next chapter (E XIII "Of Goodnesse and Goodnesse of Nature," 38–9, VI: 403). For if it was the pull of superhumanity that efficiently caused man's inhumanity, it is only through the push of his subhumanity that man can be guided to humanity.

In the twin *Essay* on nature and custom, Bacon illustrates the metanoic force of the custom that transformed man's "natural inclination to goodness" into an almost natural inclination of working one's own good by "disaffecting" the good of others. He observes that although nature is "Seldome extinguished," it is *"Custome"* that "doth alter and subdue *Nature*" (E XXXVIII "Of Nature in Men," 118–9, VI: 469). Since men's "Deeds are after as they have beene *Accustomed*," there is "no Trusting to the Force of Nature, nor to the Bravery of Words; Except it be Corroborate by Custome," as *"Macciavel* well noteth." Bacon goes on to refer to Machiavelli's example that "for the Atchieving of a desperate Conspiracie, a Man should not rest upon the Fiercenesse of any mans Nature, or his Resolute Undertakings; But take such an one, as hath had his Hands formerly in Bloud." He adds, however, that *"Macciavel* knew not of a *Friar Clement*, nor a *Ravillac*, nor a *Jaureguy*, nor a *Balthazar Gerard*; yet" despite this fact, Machiavelli's "Rule holdeth still, that Nature, nor the Engagement of Words, are not so forcible, as *Custome*. Onely *Superstition* is now so well advanced, that Men of the first Bloud, are as Firme, as Butchers by Occupation: And votary Resolution is made Equipollent to *Custome*, even in matter of Bloud" (E XXXIX "Of Custome and Education," 120, VI: 470–1).

Machiavelli's Machiavelli slightly deviates from Bacon's Machiavelli, though, by saying that it is not so much the experience of blood that a conspirator should lean on as the experience in conspiring against princes.[167] But although the Florentine philosopher, who died in 1527, did not know of the murder of King Henry III (1589) and King Henry IV (1610), and of the assault on (1582) and the murder of William of Orange (1584), he did know the custom that was to cause these and similar murders and assaults; the custom that had so much "altered and subdued" human nature that men had come to think that they already had their "hands in the blood" of Christ before they were to shed first blood; the custom, therefore, that had turned men "of first blood" into "butchers by occupation."[168] He also knew that it was this custom that had inaugurated the greatest of conspiracies ever made against earthly princes. He would thus have known that Bacon intimates his overall agreement with his predecessor on the destructive force of the superstitious custom that had "altered" man's nature in a way nature never could have. For it was only the implanted tyranny of conscience that could bring men to contemning their own and other men's lives and deaths.[169] But it was only by seemingly correcting Machiavelli that Bacon could make it seem as if this observation confined itself to the excessive cases of Christian assassins of earthly princes.

In response to the evil of Christian "conspirationalism," Bacon elsewhere addresses his Christian readers by arguing that such conspirationalism would constitute a conspiracy of Christianity against itself. For, he says, to "Authorize Conspiracies" would be like dashing "the first Table, against the Second; And so to consider Men as Christians, as we forget that they are Men" (Exod. 20: 1–17). By way of illustration, he refers to the famous words Lucretius wrote down on the occasion of his observation that Agamemnon could endure the sacrifice of his daughter Iphigenia, and in response to the accusation of being impious: "To such evils religion could persuade."[170] And the Christian religion did persuade men into sacrificing or wishing to sacrifice one's own flesh for the sake of one's own salvation. "What would [Lucretius] have said, if he had knowne the Massacre in France, or the Powder Treason of England?" Bacon wonders. "He would have beene, Seven times more Epicure and Atheist, then he was" (E III "Of Unity in Religion," 14–5, VI: 384).

In an early Sacred Meditation, Bacon had mentioned the blindness that the apostle Paul, filled with the Holy Spirit, struck Elymas the sorcerer with (Acts 13: 8–11). He observed that nothing of this kind was done by Jesus, for upon Him "the spirit descended in the form of a dove." And he added that Jesus said of the dove: "you know not of what spirit you are." In reality, however, Jesus spoke these words in response to the apostles James and John, who, after not having been received by the Samaritans, asked Jesus if He wanted them to command fire to come down from heaven to consume the Samaritans (Luke 9: 55; MS 2 "De Miraculis Servatoris," VII: 233–4). But contrary to Lucretius, Jesus may not have known the spirit of Christianity.

Reversion

We have reached the "port" of Bacon's contemplations, the "Sabbath" of the six books' works. For if "a man, can be Partaker of Gods Theater, he shall likewise be Partaker of Gods rest" (E XI "Of Great Place," 34, VI: 399). But God does not rest, since "notwithstanding God hath rested and ceased from creating since the first Sabbath, yet nevertheless he doth accomplish and fulfil his divine will" (CF 8, VII: 221). Bacon therefore has one more work to perform that "partakes of God's theatre." He emphasizes, however, that his discussion of sacred and divinely inspired theology will not concern the "matter regarding which theology gives information," but only the "manner in which the information is imparted" (DA 9.1, I: 830). In the *Advancement of Learning*, our author had divided the "matter of beliefe" into the "Doctrine of the Nature of GOD, of the attributes of GOD, and of the workes of GOD," and he had subdivided the works of God into "that of the *Creation*, and that of the Redemption" (AL 2, 190, III: 488). Seeing that these matters have all been treated, the only remaining matter is indeed the manner in which matter is "imparted," although before turning to it, we should notice that in his *Discours Préliminaire* d'Alembert deviates from Bacon's fundamental division of knowledge into theology and philosophy, arguing that separating theology from philosophy would be "like tearing off a branch from the stem to which it is attached."[171] But the man who knew that he owed the "tree" of his *Encyclopédie* to the immortal chancellor of England must have known that it was only on the basis of Bacon's division that he could deviate from Bacon's division.[172]

The prerogative of God "comprehends the whole man," extending "no less to his reason than to his will," so that man may "deny himself completely and draw near unto God." Therefore, Bacon argues, as we are "bound to obey the divine law," although we "find reluctance in our will," so we are to "believe God's word," although we find "reluctance in our reason." For, he explains, if we believe only that which "agrees with our reason," we "give consent to the matter and not to the author," which we are even "wont to do in the case of witnesses of suspect credibility." But whereas the incredibility of witnesses decreases the credibility of their testimonies, in the case of God it is precisely the incredibility of His word that increases His credibility as Author. For, as Bacon goes on to say, "the more discordant and incredible the divine mystery is, the more honour is shown to God by believing it, and the nobler is the victory of faith." The philosopher concludes from this that sacred theology "ought [debere] to be derived from the word and oracles of God," and not from the "light of nature or the dictates of reason."

According to Bacon this also pertains to the "more perfect interpretation of the moral law: Love your enemies; do good to them that hate you & C." These words, he comments, deserve the applause: "this voice does not sound human," since it is a "voice beyond the light of nature."[173] In elaboration of this point our author refers to the heathen poets, who, "especially when discussing passionately," often "expostulate with laws and moral doctrines as if they were opposed and malignant to the liberty of nature." By way of illustration, he quotes the words of Ovid: "what nature allows, envious laws deny,"[174] words that refer to the story of Myrrha, who fell in love with her father Cinyras, but did not consider her prohibited passion unnatural because animals are also incestuous. Nevertheless, Bacon argues, it is "most true what is said on man having by the light and law of nature some notions of virtue and vice, justice and injustice, and good and evil." For, he explains, it must be observed that the notion "light of nature" is used in "two different senses": first, as it "springs from sense, induction, reason, and argument," according to the "laws of heaven and earth"; and second, as it "shines upon the human soul by an inward instinct," according to the "law of conscience," which is a "spark and relic of man's primitive and original purity." It is in this latter sense, Bacon emphasizes, that the soul "partakes of some light to behold and discern the perfection of the moral law, which is a light not wholly clear, though, but is of such a manner that it reproves vice in some measure rather than that it informs fully on the duty." Therefore, the English philosopher concludes, religion "depends on divine revelation, whether it is considered with regard to mysteries or with regard to morals" (DA 9.1, I: 830–1).

The key to the more inflammable regions of this at first reading somewhat tepid argument is provided by the distinction Bacon makes between the "light of nature" and the "law of conscience," the latter of which, although it is also called the light of nature and is perceived as a natural light, ultimately roots in God's word, which speaks of man's primitive and original purity. For the fact that the "moral law" roots in the "conscience" that roots in God's word ultimately implies that the moral law as such roots in God's word or law.[175] Bacon underscores and sharpens this point by saying that the "voice" of the moral law is a voice "beyond the light of nature," and that morals as such "depend on divine revelation," although he takes the sting out of his hazardous argument by saying that it only concerns the "more perfect interpretation" of the moral law. But despite the confusion he creates as to the meaning of the word "law,"[176] our author could hardly have contrasted nature and law more effectively than by referring to the example of the repulsive but natural passion of Myrrha, which was forbidden by law.

Bacon goes on to say that the use of human reason in matters of religion is twofold: the "explanation of the mystery," and the "inferences deduced from the explained mystery." With regard to the explanation of the mystery he points out that we see that "God did not consider it unworthy to descend to the infirmity of our apprehension" by "expressing His mysteries in such a way that they might be best perceived by us," and by "grafting his revelations upon the notions and conceptions of our reason," which would seem to imply that God's mysteries are accessible to human reason. However, Bacon continues his argument by saying that it is not until after the "articles and principles of religion have been put in their places, so as to be completely exempted from the examination of reason," that it is permitted to "derive and deduce inferences from them according to their analogy." He emphasizes that this does not hold in "matters of nature," because in matters of nature the "principles themselves are examinable," and "do not repugn reason." But it is otherwise in religion, where the "first propositions are self-existent, self-subsistent, and unamenable to reason." Bacon adds that this also holds in other sciences of which the "primary propositions are at will [Placita]." For, he goes on to explain, "we see in games that the first rules and laws are merely positive and at will, and are to be received as they are, and not to be disputed."

Since Bacon concludes his discussion of reason in religion with the remark that holy theology is "grounded upon the placets of God" (DA 9.1, I: 832–3), we conclude our discussion of the reason of religion with the remark that although the articles and principles of religion are accessible to human reason, they are "exempted" from human reason because they "repugn" reason's power of disclosing their arbitrariness, for which reason they root themselves in God's will, which becomes "unamenable to reason" as soon as it is exempted from reason.[177]

Had he not known "with what reticence, and, so to speak, with what superstition" one is to judge a "sublime genius," d'Alembert would have reproached Bacon with having been "perhaps too timid."[178] But was it not precisely his "superstitious reticence" in judging Bacon's "sublime genius" that prevented the *philosophe* from judging "timidity" part of the sublimity of Bacon's genius? At the beginning of the ninth and final book of *De Augmentis* Bacon said that he would make his "votive offerings" by discussing sacred and inspired theology (DA 9.1, I: 829–30). At the end of the book he "humbly" entreats the immortal God to "accept with grace" his "offerings of the human intellect," which he sacrificed to God's glory, and which he "seasoned with religion as if with salt" (DA 9.1, I: 837)—an allusion to God's people, who were to season their meat offerings with salt, since God's covenant was a covenant of salt (Lev. 2: 13; Num. 18: 19). But if the salt has lost its savor, it is good for nothing, except for being "cast out and trodden under foot of men" (Matt. 5: 13; Luke 14: 34).[179]

CHAPTER 4

The Masculine Rebirth of Time

> He hath filled the hungry with good things; and the rich he hath sent empty away.
>
> —Luke 1: 53

The poets relate that the "inhabitants of the ancient world were utterly extinguished by the universal deluge," and that only Deucalion and Pyrrha survived.[1] Inflamed by a "pious . . . desire to instaure the human race," the son of Prometheus and his wife were told by the oracle that their desire would be fulfilled if they took their mother's bones and threw them behind them. At first this struck them with "great sorrow and despair," for seeing that the "face of things had been levelled by the deluge," it would be an "endless task to seek for a sepulchre."[2] However, they eventually understood that the oracle meant the stones of the earth, which is the "mother of all things."

Bacon points out that the fable seems to reveal an "arcanum of nature," and to "correct an error familiar to the human mind." For, he explains, "in his ignorance man judges that the "renovation and instauration[3] of things may be effected by means of their own decays and remains, as the phoenix rises out of its own ashes."[4] But "matters of this kind have already completed their course, and are utterly useless to the beginnings of things." Therefore, we must "go back to more common principles" (DSV XXI "Deucalion," VI: 661–2).

Since the remains of civil histories are like the "planks" that drifted up from the "shipwreck of time," Bacon considers it possible to save and preserve some things from "time's deluge," despite the "decay and the near submersion of the memory of things" (DA 2.6, I: 506). It is thus primarily in the capacity of "anamnetic archaeologist" that Bacon digs into the "remains" of the matter of human learning, whose history is part of "civil history" (DA 2.4, I: 502).

Our author counts three "revolutions or periods of learning": the first with the Greeks, the second with the Romans, and the third with the Western European nations, adding that "upon the inundation of the barbarians into the Roman empire" human learning "suffered shipwreck," and that "after the Christian faith had been received and had become established, by far the greatest number of the most outstanding wits applied themselves to the study of theology," which occupied "most" of the third period of learning (NO 1.77–9, I: 185 ff.). When we on our part add to the foregoing, however, that inspired theology is not part of human learning, and that it was only after the death of Emperor Theodosius I (395), who had made Christianity the official state religion of Rome (380), that Alaric, the leader of the Visigoths, was able to invade the divided Roman Empire, we are driven to the conclusion that by referring the "shipwreck" of human learning to the "inundation of the barbarians" Bacon alludes to what he could most safely allude to by being so bold as to allude to a similarity nobody believed him to have alluded to: namely that it was the establishment of Christianity that caused human learning to "suffer shipwreck."

But although the "vicissitudinous waves" of time sent philosophy to the timeless "bottom" of the "ocean" of things, Bacon remarks that the philosophy of Aristotle was "saved by the flow of time, like a plank of light and little solid material" (NO 1.77, I: 185; POFCC, III: 84).[5] For, as he explains elsewhere, "time seemeth to be of the Nature of a Riuer, or streame, which carryeth downe to vs that which is light and blowne vp; and sinketh and drowneth that which is weighty and solide" (AL 1.40, 29, III: 292). As it was Christianity that had come to direct the course of the "river of time," this implies that the "political planks" of the philosophy of Aristotle must somehow have been absorbed by the "flow" of Christian learning.

Bacon confirms this implication, although in *Novum Organum* he confines himself to making the commonplace observation that the scholastic theologians "blended the philosophy of Aristotle into the body of religion more than was fit" (NO 1.89, I: 196). In *De Augmentis*, on the other hand, he blames Aristotle himself for having "disregarded [omiserit] God," who is the "fountain of final causes," for having "substituted nature for God," and for having "impregnated nature with final causes, as a result of which he had no further need for a God" (DA 3.4, I: 570–1). But since Bacon knew that Aristotle could not but "disregard" a God whom he did not know, he could only blame Aristotle's political philosophy for having disregarded the possibility that it might turn out to be only a small step for a "substitute" for God to be substituted by God[6]; in other words, for a nature "pregnant with final causes" to give birth to a God who has no further need for nature as soon as He Himself becomes the "fountain" of causes.[7] And it did turn out the way Aristotle had not foreseen, since "we are beholding to Aristotle for many articles of our faith," as Bacon states in one of his apophthegms.[8]

It is therefore the fact that Christianity turned out to be as it were the "efficient result" of the political philosophy of Aristotle that provides us with a more profound explanation for Bacon's well-known scorning of Aristotle. For since despite their major and minor differences, it is the matter of truth that unites two philosophers, Bacon's scorning of Aristotle may very well have been an act of philosophic friendship, performed in order to prevent human learning from suffering "shipwreck" once more. *Amicus Aristoteles, sed magis amica veritas*. Bacon himself almost says as much, when he reminds himself in his private notebook of the unavoidability of "Discoursing skornfully of the philosophy of the graecians," or when he remarks in a private letter that any "harshness" is dictated by "necessity."[9]

But besides by scorning his friend, Bacon also proved himself a true friend of wisdom by being Aristotle's friend only "as far as the altar." For as he was compelled to confess that he often had to "change" the meaning of the words of his friend, since the "state of things" had changed, the philosopher confesses that he only "went along with antiquity *usque ad aras*" (DA 3.4, I: 549).[10]

Bacon goes on to observe that "by far the greatest number" of those who have "consented to" the philosophy of Aristotle, have submitted to it "on the authority of others" (NO 1.77, I: 185).[11] He emphasizes, however, that being affected by "inveterate consent," as if by the "judgment of time," testifies to "fallacious and infirm reasoning," because "for the most part we do not know what individuals have secretly attempted and set in motion," since "the miscarriages of time are not entered in the records" (IMPr. 1, I: 127). When we therefore find Bacon scorning Aristotle for having thought, "in Ottoman fashion," that he "could not reign safely unless he put all his brothers to death" (DA 3.4, I: 563), we find that it is not his "brother," but the murderers of his brother's "child" to whom the English philosopher compares Mehmed III, the sultan of the Ottoman Empire (1595–1603) who put nineteen of his brothers to death in order to secure his reign.[12]

Liberation

Human learning completed its first "course" when it was usurped by Christianity. But although its Christianized "remains" are "utterly useless" to its restoration, human learning cannot disregard the historical reality of Christianity. It thus seems to be primarily out of regard for the power of Christianity that Bacon counsels going back to "more common," and not to higher "principles." For the Christian faith "removed and discharged the infinite disputations and speculations on the supreme degree of good," as our author observes in the context of his substantial discussion of the "highest good." Moreover, the "pious and strenuous diligence of the theologians, employed in weighing and determining duties, moral virtues, cases of conscience, and the bounds of sin, has left the philosophers far behind" (DA 7.1, I: 715–6).

It is in trying to understand how the modern political philosophers reduced the moral lead of their religious competitors that we come across the following counsel: "Reduce things, to the first Institution, and observe, wherin, and how, they have degenerate; but yet ask Counsell of both Times; Of the Ancient Time, what is best; and of the Latter Time, what is fittest" (E XI "Of Great Place," 35, VI: 400). Following Bacon's counsel on the reformation of man's "place," we find that the theologians dismissed "speculations" on the "supreme degree of good" by deducing the comparative degrees of moral good from a "personification" of the highest good, thereby leaving the philosophers "far behind."

But although the ancient times teach that the "best" is the highest, more recent times counsel that the "fittest" is the common, because when the highest has been commonized by being personified, it is best to descend to the fittest. It must therefore have been the consideration that it is only by means of "common principles" that the highest good can be liberated from its equation with the common good of the Christian communion that led Bacon to the divination that man's liberation from the heathen doctrines of the highest good was "of good omen" (DA 7.1, I: 715–6).[13]

The philosopher emphasizes, though, that denying a God destroys "Mans Nobility," for if man "be not of Kinne to *God*, by his Spirit, he is a Base and Ignoble Creature" (E XVI "Of Atheisme," 53, VI: 414), which implies that in order to ennoble what had been ignobled by the only means to ennobling the potentially noble Bacon must go back to the "common principle" that is also the common principle of the Christian religion. And this is the principle of charity, which is the theological virtue that is "most communicative" (DA 3.1, I: 542), since "there is in Mans Nature, a secret Inclination, and Motion, towards *love* of others; which, if it be not spent, upon some one, or a few, doth naturally spread it selfe, towards many; and maketh men become Humane and Charitable" (E X "Of Love," 33, VI: 398).

However, as Bacon emphasizes in his *Essay* on the virtues of goodness and Christian goodness, "Errours . . . in this vertue of *Goodnesse*, or *Char-*

ity, may be committed." One of the "Doctors of *Italy, Nicholas Macciavel*," even "had the confidence to put in writing, almost in plaine Termes: *That the Christian Faith, had given up Good Men, in prey, to those, that are Tyrannicall, and unjust*. Which he spake," Bacon comments, "because indeed there was never Law, or Sect, or Opinion, did so much magnifie *Goodnesse*, as the Christian Religion doth" (E XIII "Of Goodnesse and Goodnesse of Nature," 39, VI: 403–4). And when we look into the Italian "doctor's" diagnosis of the state of the world, we indeed find that Bacon's Machiavelli almost plainly said what Machiavelli's Machiavelli had implied, namely that it was exactly the Christian religion's "magnification" of goodness that had surrendered "good men" to the "tyranny and injustice" of Christian goodness.[14] It thus becomes Bacon's primary task to liberate man's "charitable inclination" from being spent on the one God.[15]

But although Christianity itself is not the true way, it does show the way toward the true way, as the Italian doctor had already prognosticated in as "plain terms" as the nature of the diagnosed disease allowed. For the most common principle of a communion is its beginning, and all beginnings of sects have some goodness in them.[16] "Is not the ground which *Machiavill* wisely and largely discourseth concerning Gouernments, That the way to establish and preserue them, is to reduce them *Ad Principia*," also "a rule in Religion?" (AL 2, 77, III: 348), Bacon wonders in the context of establishing *Philosophia Prima* for the sake of the preservation of philosophy. Bacon therefore takes a paraphrase of the advice of the prophet, who said that the good way is or is at the crossing of the ancient way (Jer. 6: 16), to be the "true direction in this matter": "*That we make a stand upon the Ancient Way, and then looke about us, and discover, what is the straight, and right way, and so walke in it*" (AL 1.38, 28, III: 290; E XXIIII "Of Innovations," 76, VI: 434).

The ancients relate that Prometheus made man, and that man was made of clay, except that Prometheus "mixed" the clay with "particles from different animals." According to Bacon, Prometheus "clearly and expressly signifies Providence." For, he explains, "only the creation and constitution of man was singled out by the ancients as the peculiar work of providence." Bacon speculates that one reason for this seems to have been that the "nature of man includes mind and intellect," which is the "seat of providence." And because it "seems [videtur] to be somewhat harsh and incredible to raise and educe reason from irrational principles," it "follows almost [fere] necessarily that the human mind was endued with providence." Therefore, our author concludes, the fable "primarily proposes [proponitur] that with regard to final causes it is man who is as the centre of the world, insomuch as if man were taken away from things, the rest would seem to wander without purpose, and not to strive after a goal" (DSV XXVI "Prometheus," VI: 668 ff.).

But since the ancients, in a dialogical fashion, also ascribe the creation and constitution of other animals than man, as well as many other things, to the work of "forethought [προμηθεύς],"[17] we suspect that it is primarily the "Promethean" mind of Bacon that proposes to regard man as the "center of the world."[18] This suspicion gains some weight when we consider that Bacon elsewhere argues that a "strong argument can be drawn from poetry that to the spirit of man a more illustrious greatness, a more perfect order, and a more beautiful variety is agreeable than can anywhere be found in nature, after the Fall" (DA 2.13, I: 518).[19] But we do not find our suspicion confirmed until we start looking into Bacon's more elaborate treatment of the creation and constitution of man in *De Augmentis*; a treatment at the beginning of which our author says that Scripture declares of man that "He formed man from the dust of the earth and breathed the breath of life into his nostrils, and not, as of the other creatures: Let the waters bring forth; Let the earth bring forth" (DA 3.4, I: 565). A few paragraphs further Bacon argues that the human soul has two parts: the "rational soul," which is "divine," and the "irrational soul," which man has "in common with the brutes." He explicitly refers to his earlier discussion by saying that what he said earlier was that there are two different "emanations of souls," which "appear in their first creation": one "springing from the breath of God," and the other "springing from the womb of the elements," adding that with regard to the "first generation of the rational soul" Scripture says: "He formed man from the dust of the earth and breathed the breath of life into his nostrils," whereas the "generation of the irrational or brutish soul" was effected by the words: "Let the waters bring forth; Let the earth bring forth." Bacon explains that the "irrational soul, as it exists in man," is only the "instrument of the rational soul," and that it "has its origin, like that of the brutes, in the dust of the earth." For "it is not said of man that He formed the body of man from the dust of the earth, but that He formed man, that is, the entire man, except the breath of life."

But whereas a few paragraphs earlier Bacon used the biblical quotation on the waters and the earth to distinguish man from the other creatures, in the passage just given he uses the exact same quotation to equate man, "except the breath of life," with the other creatures, which leads us to suspect that Bacon wants to "except" the breath of life from his discussion of the human soul. Our suspicion is proved correct when we find our author indeed subtly ridding the human soul of the "breath of life" that is the "breath of God," thereby implying that both the rational and the irrational soul spring from the "womb of the elements." For after having divided the general doctrine concerning the human soul into the doctrine of the "breath of life" and the doctrine of the "sensible or produced soul," Bacon subdivides the doctrine of the breath of life into the "doctrine of the breath of life" and the "doctrine of the substance of the rational soul," adding that both doctrines include "inquiries into the nature of the rational soul," whether it is "native or adventive, separable or inseparable, mortal or immortal," and "how far it is tied to and how far it is exempted from the laws of matter." But these inquiries can only pertain to the doctrine of the rational soul, as the "breath of life," which is the "breath of God," cannot but be "adventive," "separable," "immortal," and "exempted from the laws of matter."

Bacon goes on to wonder how we can expect to "obtain from philosophy" the knowledge of the "substance of the rational soul," if the "laws of heaven and earth are the proper subjects of philosophy," and the "substance of the soul in its creation was immediately inspired by God, and not extracted from the mass of heaven and earth." We, on the other hand, do not wonder why Bacon expects to "obtain from philosophy" the knowledge that the "substance" of the rational soul is "extractable" from matter in a way not dissimilar to its "irrational" counterpart (DA 4.3, I: 604–5).[20] Adding to this that the ancient sources tell us that Prometheus made man from water and earth,[21] and considering that he only admixed particles from non-human and therefore nonrational and nonprovidential creatures, we are led to the conclusion that the rational or rationalized creature would have "wandered without purpose" like the nonrational creatures, had Bacon's Prometheus not "breathed" reasonable providence into the "seat" of human providence, notwithstanding the fact that it seems "harsh and incredible to raise and educe reason from irrational principles." But "abundance and excellence of virtue resides in mixture and composition" (DSV XXVI "Prometheus," VI: 671), as it is "mixed" bodies that are receptive to the admixture of virtue.[22]

Prometheus "stealthily ascended to heaven" with a bundle of fennel stalks in his hands. He "kindled them at the chariot of the sun, brought fire to the earth, and shared it with mankind." Jupiter brought "many and grave charges" against the titan, among them the charge that he had "attempted to vitiate Pallas." Prometheus was put in irons and "condemned to perpetual torture." For he was dragged to Mount Caucasus and bound to a column, where an eagle gnawed and consumed his liver by day, which grew again at night. But "it is said that eventually this punishment came to an end." For Hercules sailed across the ocean in a cup that had been "given to him by the Sun," and when he arrived at Mount Caucasus he "liberated Prometheus and shot the eagle with his arrows."

Bacon takes the titan's heroic liberation by the hands of the demigod who was hated by the wife of the god who hated the maker of man to mean that Prometheus was liberated by "fortitude." He emphasizes that it is worth noticing that the virtue of fortitude was not "natural" to Prometheus, but "adventitious," and that it "came by help from without," as it is "not a thing that innate and natural fortitude can attain to." For it came from the sun, that is, from "wisdom, which is like the sun."

Our author goes on to explain that it is added "for the consolation and confirmation of men's minds" that the mighty hero sailed in a cup or pitcher, "lest men should mistrust the narrowness and frailty of their own nature too much, as though it were altogether incapable of this kind of fortitude, on which Seneca divined well when he said that '*It is true greatness to have in one, the frailty of a man, and the security of a God.*'" Bacon ends by saying that the voyage of Hercules, sailing in a pitcher to liberate Prometheus, "seems to present an image of God the Word hastening in the frail vessel of the flesh to redeem the human race" (DSV XXVI "Prometheus," VI: 671, 675–6).

But since the sun signifies wisdom and Pallas is the goddess of wisdom, we begin by inferring that Prometheus kindling his fennel stalks at the chariot of the sun is identical to Prometheus vitiating Pallas. Yet an attempt at obtaining "impersonal" wisdom can only be a "vitiation" of the personation of wisdom if wisdom has first been personified. It must therefore have been because Jupiter's nonfabulous successor had subjected wisdom to his will that Prometheus's attempt upon the personation of wisdom came to constitute a vitiation of wisdom personified. Bacon draws our attention to this point by mentioning the attempted vitiation of Pallas as a "new crime" Jupiter charged Prometheus with in addition to the old crimes, without, however, discussing the substance of this new crime (DSV XXVI "Prometheus," VI: 670). Moreover, it is only these changed circumstances for "Promethean activities" that can explain why the ancient sources nowhere mention a vitiation of the goddess of wisdom by the hands of Prometheus.[23]

But although "the works of wisdom surpass the works of fortitude in dignity and power," as Bacon points out in his fable on philosophy (DSV XI "Orpheus," VI: 647), times that are adverse to wisdom require the virtue of fortitude for the liberation of wisdom from its main adversary.[24] For the "Vertue of *Adversity*, is Fortitude," and "*Adversity* is the Blessing of the New" Testament, since "*Adversity* doth best discover Vertue" (E V "Of Adversitie," 18–9, VI: 386). It may not be out of context to mention here that Bacon's preoccupation with rediscovering virtue goes some way toward explaining why we nowhere find him raising the question "What is virtue"; a question with which we believe him to have been as familiar as were those philosophers who did raise it as a consequence of their philosophical approach of political life, but which we believe him to have suppressed for reasons having to do with his political approach of philosophy. There are some who consider the explicit treatment of the question of virtue indicative of an author's credentials as a political philosopher. For, so they argue, one of the best ways to prevent oneself from taking the possibility and necessity of philosophy for granted is by taking the question of virtue as one's guiding question. But although to a certain extent this argument is correct, we have reason to wonder whether the explicit and almost necessarily protreptical treatment of virtue does not stimulate precisely one's potentially virtuous readers to take for granted the possibility and necessity of virtue itself, to say nothing of the fact that inflated virtue tends to become a receptacle for inflated meanings.

Leaving these considerations with those interested in the problem of virtue, we return to the times of Bacon, where we find that because God the Word had come in the "frail vessel of the flesh," the virtue of adversity had to be personated by a fortitudinous god-man like Hercules, who was himself a son of the highest god and a woman. The heathen Seneca therefore "divined well," albeit unwittingly, that fortitude would come to consist in having "in one, the frailty of a man, and the security of a God." But his speech was "much too high for a Heathen" (E V "Of Adversitie," 18, VI: 386), because he did not yet know the God whose protection would have to be secured in order for "frailed" man to become resilient[25]; in other words, because he did not yet know that the advent of God the Word would call for "adventitious" fortitude. For "Man, when he resteth and assureth himselfe, upon divine Protection, and Favour, gathereth a Force and Faith; which Humane Nature, in it selfe, could not obtaine." Therefore, as Bacon goes on to argue in adverse times, which "bow Mens Mindes to *Religion*," atheism is hateful inasmuch as it "depriveth humane Nature, of the Meanes, to exalt it selfe, above Humane Frailty" (E XVI "Of Atheisme," 53, VI: 415).

That times of adversity indeed require the "adventitious" protection of religion for the liberation from adversity is confirmed by our author in his *Essay* on the "blessing" of adversity. For whereas in the fable on the mythical fire-bringer Bacon had "interdicted" himself "all licence" in explicating his comparison between Hercules and God the Word, "lest perchance" he should "bring strange fire to the altar of the Lord" (DSV XXVI "Prometheus," VI: 676),[26] in the *Essay* on the virtue of adversity he has become "fortitudinous" enough to say that Hercules, sailing "*the length of the great ocean, in an Earthen Pot or Pitcher*," lively describes "Christian resolution;

that saileth, in the fraile Barke of the Flesh, thorow the Waves of the World" (E V "Of Adversitie," 18, VI: 386). By adding the adjective "earthen" to the "pot or pitcher" Hercules sailed in, Bacon moreover places Hercules in the earthen vessel of St. Paul, in which the believers "carry the Word" in order to show that the "all-surpassing power" is of God (2 Cor. 4: 7).[27] He thus "consoles and confirms" human nature by implying that it is capable of the virtue of liberating fortitude, provided that it protects itself with the "earthen pot or pitcher" the "modern Hercules" was compelled to sail in.

But although in the narrative of the fable he discussed Prometheus's attempt at vitiating Pallas before mentioning his liberation by the virtue of Hercules, in the interpretive part of the fable Bacon discusses what he now calls Prometheus's attempt on the "chastity of Minerva" after having treated the liberation of the father of Deucalion, the "instaurer of the human race" whom we have met at the beginning of the present chapter. Bacon explicitly says that he deliberately passed by this point because he did not want to "interrupt what is interconnected," adding that the attempt on the chastity of Minerva, who signifies "nature" in the only other fable in which the goddess of wisdom is called Minerva,[28] seems to consist in nothing else than "men attempting to bring divine wisdom under the dominion of sense and reason" (DSV XXVI "Prometheus," VI: 675). We suggest the following interconnection. The liberation of Prometheus or human nature by means of "adventitious" fortitude presupposes the "inadventitious" wisdom or knowledge of nature that liberates human nature from non-natural or supernatural bondage[29]—a causality Bacon visualizes by saying that Herculean fortitude "comes from" and therefore presupposes the sun, which signifies wisdom, and which rises in the east.

In an earlier fable, treating the "reasonable and prudent conduct of war," the philosopher had already discussed wisdom's "warfare strategy" for these early stages of what one might call the "modern titanomachy." Perseus, a "man from the east," was sent by Pallas to behead Medusa, who was the "only one of the three Gorgons that was mortal." Having cut off Medusa's head, Perseus "connected the severed head to the shield of Pallas," since, Bacon explains, the "excellence" of such a safeguard is "beyond comparison," considering that the shield of Pallas signifies the "providence that provides security like a shield" (DSV VII "Perseus," VI: 641 ff.; DA 2.13, I: 530). Seeing that Perseus also received "secrecy of counsels" from Orcus (DSV VII "Perseus," VI: 642; I: 533), we may infer from Bacon's silence on the connection between the two fables on wisdom at war that after Perseus had brought the embellished shield back to its owner, the goddess of wisdom connected it to the pitcher of Hercules, so that the latter would be protected by God the Word, the "only one of the three Gods that was also mortal."[30]

But in order "not to interrupt what is interconnected," we go on to observe that after having liberated human nature, Hercules passed over his arms to Prometheus unbound, in order for the latter to be able to make an "attempt" at unchastening nature chastened by Christian chastity.[31] Yet although wisdom attacks from the east, the war over the hand of Minerva must be fought globally, since "*East* and *West* have no certaine Points of Heaven: And no more have the *Warres*, either from the *East*, or *West*, any Certainty of Observation" (E LVIII "Of Vicissitude of Things," 174–5, VI: 515). It seems to be primarily for this reason that Mercury, who represents the "Divine Word" in the fable on nature, and who stole the art of speech from the "God of Harmony,"[32] gave Perseus "wings for his ankles, and not for his shoulders." For, Bacon explains, "celerity is required not so much in the first attacks as in the ones that follow." And "no error is more common in war than that the prosecutions and subsidiary attacks do not correspond to the alacrity of the beginnings" (DSV VII "Perseus," VI: 642), considering that the "*Kingdome* of *Heaven* is compared, not to any great Kernell or Nut, but to a *Graine* of *Mustard*-seed; which is one of the least Graines, but hath in it a Propertie and Spirit, hastily to get up and spread" (E XXIX "Of the True Greatness of Kingdomes and Estates," 90, VI: 445). When "Things are once come to the Execution, there is" therefore "no *Secrecy* comparable to *Celerity*" (E XXI "Of Delayes," 69, VI: 428), although celerity itself is a "great part of secrecy" (DA 2.13, I: 533). When we call to mind that the mother of the Sirens, who provided Orpheus with the light but earthly doctrines that countered the heavy ethereal doctrines of the religious Sirens, went on foot but wanted to restitute the "wings" to her daughters,[33] we are led to the conclusion that the war between earth and heaven over nature will primarily be a war of propaganda. In Bacon's time, this war was still in its early stages, as the philosopher himself indicates by saying that the fable on the state of man "carries with it many true and grave speculations both on the surface and underneath." For, he explains, there are some things in it that "were observed long ago," whereas other things "have never been touched at all" (DSV XXVI "Prometheus," VI: 670). And indeed, nobody had as yet dared to "touch" the chaste Minerva, because no "instaured Prometheus" had as yet dared to ascend to heaven to "bring strange fire to the altar of the Lord" by bringing Minerva or divinized wisdom back "under the dominion of sense and reason."[34] But "if Miracles, be the Command over Nature, they appear most in Adversity" (E V "Of Adversitie," 18, VI: 386).[35]

Justification

It was because he desired to "benefit and protect" his work that Prometheus brought fire to the earth and shared it with mankind. For, Bacon explains, it was because man in his origins seems to be a "naked and defenceless thing, slow to help himself, and full of wants," that Prometheus "with haste" applied himself to the "invention of fire," which "in almost all human necessities provides relief and help." But men were "not at all grateful for this memorable benefit." For they "brought before Jupiter an accusation against Prometheus and his invention." This act of theirs, Bacon comments, was "not taken as justice may seem to have required," for the accusation was "very agreeable to Jupiter and the other gods." In fact, they were "so delighted that they not only granted man the use of fire, but also bestowed upon him the gift of eternal youth." Bacon emphasizes that this part of the parable is "remarkable," for, he wonders, "Why should men's ingratitude towards their author be approved and rewarded?" (DSV XXVI "Prometheus," VI: 668 ff.).

But apparently the ancient sources did not consider this part of the parable "remarkable," for nowhere do they mention the question of justice in the context of Jupiter's response to men's "accusation" of their "author" and benefactor.[36] We therefore suspect that by not explicitly answering the question as to the justice of the gods, Bacon wants to arouse thoughtful man's indignation over his supposed author and benefactor unjustly condemning his true benefactor, without whose invention he would still be "naked," "defenceless," helpless, and dependent.[37] This suspicion is sustained by the fact that whereas the ancient Prometheus stole the fire, Bacon's Prometheus "invented" the fire, which implies that the gods who "granted" man the use of fire in order to show their gratitude for man's "ingratitude" toward his benefactor, stole the fire from Prometheus in order to make man grateful to and therefore dependent on them instead of on the inventor who wanted man to depend wholly on himself.[38]

But it is only by adding that the gods also bestowed on man the "gift of eternal youth" that Bacon makes his reversal of the story of the Fall explicit. For whereas the biblical account of the genealogy of evil teaches that Adam lost eternal youth because he had eaten of the tree of the knowledge of good and evil, according to the Baconian account of the genealogy of human dependence, man gained eternal youth because he had accused a titan of having eaten of the tree of the knowledge of good and evil.[39]

On the basis of these conclusions, Bacon elsewhere stirs up "benefacted" man's resentment at God by stimulating him to unite the separated observations that "Being is a gift granted by God," but that "Being separated from Well-Being is as a curse," and that "the greater Being is the greater curse" (DA 8.2, I: 772, 790)—a unification that culminates in the compound observation that the "greater Being" who by His curse (Gen. 3: 17 ff.) "separated" man's "being" from his "well-being" is a "greater curse" for man's well-being than the curse by means of which He cursed man's well-being.

In the *Essay* on judicature, our author takes it one step further by corroborating the injustice of God's curse from the point of view of God's own Law, which states that the land was given in possession to man by the Lord (Deut. 19: 14; 27: 17): "*Cursed* (saith the Law) *is hee that removeth the Land-Marke.*" But "it is the Unjust *Judge*, that is the Capitall Remover of Land-markes, when he Defineth amisse of Lands and Propertie" (E LVI "Of Judicature," 166, VI: 507), as Bacon glosses in order to have us draw the conclusion that the "greater Being" cursed Himself when He unjustly "redefined" the "land-marks" by removing man from Eden, thereby separating his "being" from his "well-being."

Concealed as an exercise in appeasing anger, the *Essay* immediately following the *Essay* that discusses God's injustice carries anger to its logical conclusion by putting it to work. It is because "*Contempt* is that which putteth an Edge upon *Anger*" that ingenious men pick out "Circumstances of *Contempt*," and "sever, as much as may be, the Construction of the Injury, from the Point of *Contempt*." And it must have been for the same reason that Bacon exposed Jupiter's unjust and therefore "contemptible" response to man's accusation of his benefactor.

In the central section of the *Essay*, Bacon counsels that the "best remedy" for the anger he raised in order to remedy it is "to win Time; And to make a Mans Selfe beleeve, that the Opportunity of his Revenge is not yet come: But that he foresees a Time for it"[40] (E LVII "Of Anger," 170–1, VI: 510 ff.). Yet when the opportunity has come, "let a man take heed, the *Revenge*[41] be such, as there is no law to punish: Else, a Mans Enemy, is still before hand." For "the most Tolerable Sort of *Revenge*, is for those wrongs which there is no Law to remedy" (E IIII "Of Revenge," 16–7, VI: 384–5). But eradicating the "Law" requires more than a lifetime. Or a project.

"For Time is the greatest *Innovatour*: And if Time, of course, alter Things to the worse, and Wisedome, and Counsell shall not alter them to the better, what shall be the End?" (E XXIIII "Of Innovations," 75, VI: 433). In other words, as the course of time had come to be directed by Christianity, time's own course could no longer be relied on, because time as if by its own course would only "alter things to the worse." It therefore seems to be primarily because time forces an "innovation" of its own course by means of "wisdom and counsel" that Bacon says that all innovations are the "Births of Time" (E XXIIII "Of Innovations," 75, VI: 433),[42] which is the "author of authors" (NO 1.84, I: 191), and which in turbulent times gives birth to a "masculine" author like Moses, whose birth Bacon alludes to by means of the title of the first version of his *Instauratio Magna: Temporis Partus Masculus sive Instauratio Magna Imperii Humani in Universum* (1603).[43]

"From these and all long Errors of the Way, In which our wandring Predecessors went, And like th' Old Hebrews many years did stray, In Desarts but of small Extent, Bacon, like Moses, led us forth at last, The barren Wilderness he past, Did on the very border stand, Of the blest promis'd Land, And from the Mountain's Top of his exalted Wit, Saw it himself, and shew'd us it."

Had Bacon still lived at the time the poet Abraham Cowley wrote this eulogy on the leader of modern man's "exodus" as part of his ode to the *Royal Society of London for the Improvement of Natural Knowledge* (1660),[44] he would have partly approved of being compared to the man Machiavelli called an "armed prophet."[45] For although the "head" of the "new orders" had observed that among all men praised, the "heads and orderers of religion" are praised most,[46] Bacon knew that it was only on the basis of these new orders that he was able to put the inventor at the head of them, as will become clear from what follows. When we therefore find Bacon saying that "among human actions the introduction of noble inventions appears to hold by far the first place," since "divine honors" are attributed to the "inventors of things," considering that "man is a god to man on account of the arts" (NO 1.129, I: 221–2), we find our author approving the inventor's dependence on "Mosaical arms." In fact, we even find the English inventor almost explicitly praising the arts of his Hebrew predecessor in the highest possible terms. For Bacon argues that the "invention of the means of discovering all things" would be the "truly masculine birth of time" (CV 16, III: 610), since it would be like the birth of Moses, who introduced religion or the means of discovering the Law, which made him "a god to man."[47] And it seems to be because God is a god to man on account of His sovereignty that Bacon argues in his more straightforwardly political *Essays* that the man who is a god to man deserves the highest degree of "*Soveraigne Honour*" on account of his founding of an empire (E LV "Of Honour and Reputation," 164, VI: 505–6).[48]

But the parallel with God's sovereign emperor can be pursued even further, albeit in a somewhat different direction. For as remains the case with

Moses, one is inclined to confuse Bacon with his arms. Although this confusion is deliberate inasmuch as it arises from hatred against greatness that does not consist in diminishing greatness, it is deliberately contributed to by Bacon himself. For it was Bacon himself who said that his New Organon "leaves little to the acuity and strength of wits," and that it "almost levels wits and intellects" (NO 1.61, I: 172). But it was because he was a student of the solitary Italian writer who had counseled that in order for an empire to last long it must be ordered in such a way that it remains "in the care of many," although it can only be ordered well from the beginning by "one man alone," that Bacon hardly ever says that it was a solitary enterprise he engaged in (IMProemium, I: 122; cf. NO 1.113, I: 210).[49]

But before "engaging" in Bacon's enterprise, we should address the question why Bacon himself engaged in it. Since this question runs parallel to the question of the good, we suspect to find an answer in the context of Bacon's discussion of the "exemplar of good." But since it is only at the end of his substantial discussion of the doctrines of the "communicative" and the "individual" good that we find our author discussing the reasons for engaging in his enterprise, we realize that his enterprise is too extraordinary to be subsumed under any "exemplar."

Bacon points out that the doctrines of the communicative and the individual good both treat the good not only "simply," but also "comparatively." To the "comparative treatment of the good" belongs the "weighing of duties between person and person, case and case, private and public, and present and future." By way of example, Bacon refers to what he calls the "severe and atrocious proceeding" of Lucius Junius Brutus against his own sons. He says that by the majority this proceeding was "extolled to heaven," but that "someone else" said of it: "Unhappy man, however posterity reports the deeds" (DA 7.2, I: 730). Bacon wrongly suggests a contradiction between these two points of view, though, because what that "someone else" actually said was that however posterity reports the deeds of the unhappy Brutus, the love of Brutus for his country will prevail.[50] This leads us to suspect that what Bacon is actually suggesting is that the private happiness of Brutus was outweighed by his public duty.[51] We find this suspicion confirmed when we consider that it was primarily the man Brutus was praised by for having "kissed the earth" who "extols" the deeds of the father of Roman liberty "to heaven." For not only does Machiavelli counsel that in order to keep a liberated state free no remedy is "more powerful, more valid, more secure, and more necessary" than to kill the sons of Brutus, who had been persuaded by the Tarquins to conspire against their liberated fatherland, he also remarks that a "man of quality" cannot choose to "abstain" even if he chooses it "truly and without ambition," because he is not believed, and is therefore not allowed to "abstain."[52]

But besides this perennial fusion of public duty and private safety, Christianity's fusion of the public and the private provided the philosopher with a more contemporary reason for not "abstaining." For the reverse side of the Christian fusion of public and private was a partial fusion of the communicative good and the private good of the philosopher, which compelled philosophy to become philanthropic, although philosophy may also have had less compelling reasons for pursuing its own good through philanthropy. For is it not through philanthropy that the philosopher's self-sufficiency is most severely put to the test? And is it not through philanthropy that the malleability of human nature is most effectively measured in times characterized by the divisive power of conscience? Finally, is it not through philanthropy that the philosopher is best able to separate his philanthropy from his benevolence, that is, from the love that sets or keeps his pen in motion and that is similar to philanthropy inasmuch as it loves at a distance, but different inasmuch as it does not love anticipatorily?

By calling his *Instauratio Magna* "our work" (IMPr. 1, I: 125), Bacon testifies to his understanding of the fusion between the good of mankind and the good of himself, although by separating the "thing he had taken in hand" from "himself" (IMProemium, I: 122) he not only alludes to the partial character of this fusion, but also indicates that he should not be confused with his work, notwithstanding the fact that to the extent that the work is also a work of understanding, his good fuses with the good of his work. In other words, although "in the course of every single age the greatest wits have been forced out of their course" (IMPr. 1, I: 127), the fact that the Christian age demands an "innovation" of the course of time itself allows Bacon to "stay on his own course" inasmuch as time is the "mother of truth" (NO 1.84, I: 191).

In any event, as Christianity had successfully "conspired" against the world by convincing the world that its ways had changed, philosophy's forces were too limited to engage in open war, as the great Florentine conspirer had stimulated his philosophic reader to consider. We for our part are stimulated to consider whether the justness of the modern philosophers' estimation of both Christianity's anti-philosophical and its historical forces is not indicated by the fact that three and a half centuries after the hatching of philosophy's secret conspiracy against Christianity a philosopher of rephrased *eros* made an open antichristian conspiracy part of his political philosophy.

But we should redescend from Modernity's destructive heights to its founding moments. In light of the limitedness of its forces, philosophy had to "conspire" against the sons of God instead of "killing" them in order to avoid the fate of Brutus, who was killed on the battlefield by the hands of Tarquin the Proud because he had allowed the latter to conspire with the Etruscans.[53] Mindful of our writer's credentials as a reader, we are not surprised to find that as Machiavelli's magisterial chapter on conspiracies follows his discussion of the virtues and vices of Lucius Junius Brutus, Bacon discusses the conspiracy of Marcus Brutus, Gaius Cassius, and others against the tyrant Caesar immediately after having discussed the case of Lucius Junius Brutus (DA 7.2, I: 730–1). Summarizing the strategy for philosophy's "tyrannicidal" conspiracy in terms used by its "head and orderer," we may say that the modern philosopher must "kiss the earth" like an instaured Brutus, that is, he must become consciously "ambitious" for his "fatherland" while simulating lack of ambition. For he will thus be less observed and have more occasion for "freeing" his fatherland when the opportunity is given to him.[54]

But although "Outward Accidents," such as "Opportunitie," "conduce much to *Fortune*," there "be secret and Hidden Vertues, that bring Forth *Fortune*" (E XL "Of Fortune," 122–3, VI: 472), as Bacon counsels in imitation of his virtuous predecessor.[55] Bacon emphasizes that "Publique *Revenges* are, for the most part, Fortunate," but he warns that revenge is a "Kinde of Wilde Justice" (E IIII "Of Revenge," 16–7, VI: 384–5). Seeing that the means to justice by means of "public revenge" are often unjust, it seems to be primarily in order to bring forth a "fortunate opportunity" for committing the "capital" of capital sins that Bacon goes on to discuss the case whether justice may be deviated from "for the well-being of one's country or for some great future good of that kind."[56] For he says that with regard to this case Jason of Thessaly used to say that "some things must be done unjustly, so that many things may be done justly," but that "there was a ready reply: 'you have the author of present justice; you do not have the guarantee of future justice'" (DA 7.2, I: 731). This reply to the partly successful Thessalian general is nowhere to be found in Plutarch, however, who was Bacon's source,[57] which leads us to suspect that Bacon "replies" to Jason that in order to guarantee "future justice" it is necessary for the "author" of "present justice" to conceal the injustice that is necessary for establishing present justice by means of "wild justice." This suspicion becomes more plausible when we consider that the philosopher goes on to reply to his reply that men "are to pursue what is just for the present" (DA 7.2, I: 731).[58] For although it is "the duty of the judge to regard both the deed and the circumstance of the deed," as Bacon quotes the only heathen author in the center of his *Essay* on judging according to justice (E LVI "Of Judicature," 167, VI: 508),[59] the unjust judge's judges cannot be expected to judge "wild justice" just, since it is contrary to the Law, whose interpreters they are.

We find our suspicion confirmed when we look more closely into Bacon's discussion of rhetoric. For it is in the context of that discussion that our author says that it was the "greatest injustice" in Plato to place rhetoric among the "voluptuary arts," likening it to cookery (DA 6.3, I: 672). But after arguing that medicine is the "bodily counterpart" of justice, and that cookery is to medicine what rhetoric is to justice, what Plato actually says is that cookery and rhetoric are not arts at all, because they lack a "reasoned account of things."[60] Since Bacon preludes his partly just criticism of the Platonic Socrates with the argument that the end of rhetoric is to "fill the imagination with simulacra to assist reason" (DA 6.3, I: 671), we are tempted to conclude that Bacon considers it the task of the "medicinal" art of rhetoric to "assist" reason's injustice by "filling the imagination" with a "simulacrum" of justice, and that judging rhetoric or reasonable injustice unjust by judging it "voluptuary" and therefore unreasonable is considered by him to be the "greatest injustice."[61]

Our rhetorician's second example supports this conclusion. For after having said that there is "no man who does not speak more honestly than he thinks or acts," Bacon says that it was "most soundly" noticed by Thucydides that Cleon was wont to be reproached for "always defending the bad side," as a result of which he was "ever inveighing against eloquence and grace of discourse, knowing well that no one can speak fair of sordid and indignant courses" (DA 6.3, I: 672). Thucydides tells us, however, that Diodotus criticized Cleon for having employed rhetoric exactly in order to speak well in a bad cause.[62] But Thucydides himself was too wise to call Cleon's advocacy of the killing of the Mytilenians bad, although Cleon may have acted unjustly by "speaking less honestly than he thought."[63]

Theologization

Let us turn once more to Bacon's fabulous account of the foundations of modern justice, according to which man lost the "gift of eternal youth" to a serpent in exchange for some life-preserving water from the fountain the serpent had been assigned to guard. "Embittered" toward Jupiter, Prometheus thereupon tried to "deceive" the ruler of gods and mortals with his sacrifice. By way of "vengeance," Jupiter decided to "afflict the human race." He ordered Vulcan to make the "lovely and beautiful Pandora." When she had been made, "each of the gods bestowed a gift upon her." They placed in her hands an elegant vase, which contained "all evils and hardships." Hope subsided into the bottom of the vase though. Pandora first betook herself with her vase to Prometheus, to see if the latter would take and open it, which the forward-thinking philanthropist "cautiously and astutely rejected." She then went to Prometheus's brother Epimetheus, who opened the vase "unhesitatingly." When he saw all kinds of evils flying out, he could only keep hope inside, which lay at the bottom of the vase (DSV XXVI "Prometheus," VI: 669–70).

Following the divine narrative, we can hardly overlook the similarity between Prometheus and Cain, the self-sufficient tiller of the ground who brought some of the ground's fruits as an offering to the Lord.[64] This personal similarity is corroborated by a sequential parallel. For as Cain only brought his offering to the Lord in the course of time (Gen. 4: 3), Bacon points out that it is only after having described the state of man with respect to the "arts and intellectual matters" that the parable "passes to religion," since the "worship of divine things followed upon the cultivation of the arts," although it also followed on man having lost what had been given to him by the divinity that gained man's worship as a gift for having regained his gift, as Bacon stimulates us to infer from the sequence of his narrative (VI: 673).[65]

But whereas the founder of the first earthly city (Gen. 4: 17)[66] was cursed from the earth for having responded unjustly to God's injustice, Prometheus's unjust response to Jupiter's most recent injustice of having had a serpent "bargain back" his gift was repaid by the evils of Pandora.[67] Calling to mind that Pandora signifies "ethereal" or "Dionysiac" pleasure,[68] we are inclined to define Pandoric evil as the evil of bargaining back the good. Opening Pandora's vase proves our inclination correct, since Bacon observes that with Jupiter's vengeance the parable "turns to morals" (VI: 674).

But although the Pandoric evils "deluged" mankind, hope remained at the bottom of the vase that had been opened by the father of Pyrrha, the only woman to survive the "universal deluge." Bacon indicates, however, that it was precisely on account of hope that the son of Prometheus and his wife survived the punishment of Zeus, since he observes that the improvident Epimetheus "revolved around many empty hopes because of his ignorance of things" (VI: 674). This seems to suggest that the prudent and forward-thinking brother of "Afterthought" would have released hope in order to counter the evils that had got the chance to flourish because his brother had put the lid on Pandora's vase prematurely.[69] That Bacon actually had his Prometheus release hope in order to counter "Pandoric discontentments" is confirmed in one of the *Essays*, where we read that the "Part of *Epimetheus*, mought well become *Prometheus*, in the case of Discontentments," since the "Politique and Artificiall Nourishing, and Entertaining of *Hopes* . . . is one of the best Antidotes, against the Poyson of *Discontentments*" (E XV "Of Seditions and Troubles," 48, VI: 411).

But Bacon notes that although by his caution a member of the "school of Prometheus," that is, a member of the school of the "initiates," is able to "remove and drive off many evils and misfortunes," he "deprives himself of many pleasures," and, what is far worse, "torments himself, and wears himself out with worries, solicitude, and inward fears" (DSV XXVI "Prometheus," VI: 674). The implication of this lack of ataraxia on the part of the modern scientist seems to be that he is in need of the "song" of Orpheus, which is pleasurable to him because it anticipates his glory by singing of immortality.[70] In anticipation of the further development of our argument, we may thus conclude that Bacon proposes a threefold "antidotal" programme that answers to the three theological virtues, faith, hope, and charity: progressive faith in the "arts of Cain,"[71] "Promethean" hope, and a practice of charity stimulated and as it were sanctioned by the song of Orpheus.

In order to overcome the old faith, the new faith must build on foundations that cannot be shaken by faith because they answer to the foundations of faith. It is a "Secret, both in Nature, and State; That it is safer to change Many Things, then one" (E XXX "Of Regiment of Health," 100, VI: 453), as Bacon expands Machiavelli's "architectural" counsel to the province of nature.[72] It is thus with a view to laying unshakable foundations that Bacon emphasizes that the "foundations of the sciences must be dug deeper and made firmer," and that the "beginnings of inquiry must be assumed further back than men have done heretofore."[73] For the "common logic accepts many things on faith" (IMDO 11, I: 137), since it proved susceptible to faith, as we may deduce from the fate of the syllogism. And it is with a view to answering to the foundations of faith that our author observes that, since the "human mind is beset by the vainest of idols," that is, by false gods (Lev. 19: 4), there "remains but one way to safety and sanity, namely to begin the entire work of the mind anew, and from the very beginning not to permit the mind to take its own course, but to guide it continually, and to accomplish its work as if by machinery" (NO Pr. 2, I: 152). In other words, the "whole way from the very first perception of the senses must be secured by a certain method" (IMPr. 2, I: 129), which attempts to root out the "adventitious idols" (cf. 1 Cor. 10: 18–22), and to prevent the "innate idols" from materializing themselves "adventitiously" (cf. IMDO 14–5, I: 139). This method is Bacon's *Novum Organum*,[74] a new logic of "induction," which is supplied with "matter" by a natural history that serves as the foundation for the "building" of philosophy, as it "supplies suckling philosophy with its first food" (IMDO 16, 20, I: 140–1).[75]

One of Bacon's more profound reasons for ascribing "maternity" to natural history seems to have been that, at least provisionally, Baconian method presupposed what it was meant to procure. For the intellect must first have been "wholly liberated from" and "purged of" the idols before it can "enter the Kingdom of Man," which is "founded on the sciences," as Bacon points out at the end of his substantial discussion of the idols (NO 1.68, I: 179).[76] By supplying method with "matter" fit for procuring the expulsion of the "adventitious idols," natural history thus in a way procures what method presupposes (cf. NO 1.130, I: 223), which explains why Bacon says that his "ways are entered" by means of natural history (IMDO 16, I: 140).[77]

But all this having been said, two things should be said in response to those who say that Bacon said something different from what he said or implied. For first of all, although historians of egalitarianism have branded Bacon an egalitarian because he stresses that his method "almost levels men's wits" and "leaves little to their excellence" (NO 1.122, I: 217), in actual fact Bacon's methodical egalitarianism ensued from his decision to go back to "common principles," which implies that it ultimately ensued from an essentially inegalitarian principle. The reason why Bacon emphasizes that his part as a guide is "of mean authority" (NO Pr. 3, I: 153) is thus that he considered "mean authority" to be the only means to counter authority. For Bacon thought that the only way to counter authority in science was to "enter the several provinces of the sciences with a greater authority than that conferred by their own principles" (IMDO 11, I: 137), an authority, however, that derived its "greater authority" from the very fact that it did not derive its authority from authority, although Bacon was its author.[78] Moreover, and this brings us to our second and final point, although it may be true that we live in an era in which it has become increasingly fashionable to mask lacking foundations with the foundation of method, because for all we know, we only know what we know by method, it was one of the very founders of modern scientific method who emphasized that the "way of reasoning" is open to one man alone (NO 1.113, I: 210).

Bacon could hardly have summarized his project of raising and raising faith in the "arts of Cain" more aptly than by giving it its title: *Instauratio Magna*. For inasmuch as the project attempts to "restore the commerce between Mind and Things to its former condition" (IMProemium, I: 121), the Instauration is a restoration (IMDO 18, I: 141; NO Pr. 6, I: 154); inasmuch as achieving this former condition is conditional on Bacon's mind having submitted to and therefore "commerced" with things (NO 1.113, I: 210), the Instauration is a repetition; and inasmuch as the "commerce between Mind and Things" can only be restored to its former condition by "trying everything anew," and by undertaking a "wholesale Instauration of the sciences, the arts, and all human learning, raised on the proper foundations" (IMProemium, I: 121; NO 1.31, I: 162), the Instauration is a renewal by means of a "regeneration of the sciences" (IMEpD, I: 124; NO 1.97, I: 202).[79]

But whereas the chroniclers of the Old Testament describe the restoration of the first temple of Solomon as an "instauration" (2 Kings 12: 8; 22: 5–6; 2 Chron. 24: 4, 12; 34: 8, 10), Bacon's Instauration must destroy the proverbial temple dedicated to the worship of idols or false gods (cf. NO 1.115, I: 211) in order to restore the "commerce between Mind and Things" by "laying the foundation in the human intellect for a holy temple after the example of the world" (NO 1.120, I: 214).[80] For after the Babylonians had destroyed the first temple (586 BCE), a new temple was built (516 BCE), which existed for nearly six ages before it was destroyed by the Romans in 70 AD, along with the city of Jerusalem. In allusion to an apocalyptic event of this nature and proportion, some of the prophets of the Old Testament describe the rebuilding of Zion on the old wastes as an "instauration" (Ez. 36: 10, Isa. 61: 4).[81] But before the second temple was physically ruined, it had been supplanted by the body of Christ (John 2: 19–21), which implies that from the death of Christ onward, the temple is to be found in those in whom Christ "dwells through the Spirit" (1 Cor. 3: 16; Hebr. 9: 11–2).[82] Consequently, an "instauration" in Christian times presupposes either the completion or the destruction of the "spiritual" temple. The apostle Paul testifies to this conclusion by using the word "instauration" to describe the bringing together of all things in Christ in the fullness of time (Eph. 1: 10). The philosopher Bacon testifies to the same conclusion by adding the following paraphrase of the prophet Daniel to the frontispiece of the *Instauratio Magna*: "Multi pertransibunt & augebitur scientia" (I: 119). For whereas Daniel was told to keep the prophecy on the end time secret until the end time (Dan. 12: 4), Bacon says that it is "safe now after the event" to interpret the prophecy (VT, III: 221), although he also says that it "pertains to the further progression and augmentation of the sciences" (DA 2.10, I: 514; NO 1.93, I: 200). It can thus be said that it was only after Bacon had destroyed the "spiritual" temple in thought that man could safely start working toward its destruction by building Bacon's "holy" temple with the New Organon for building and destroying temples.[83]

Faith is the "evidence of things not seen" (Hebr. 11: 1). But although its object is not seen, faith admonishes the faithful to demonstrate their faith by works, as Bacon reminds us (NO 1.73, I: 183). It was therefore not because of benevolence but because of faith that Bacon had to provide visible and as it were anticipatory evidence of the benevolence of the arts of Cain, despite their lack of inherent benevolence.[84] For it was only by creating faith in the benevolence of the arts of Cain that Bacon could answer to faith in the benevolence of faith. It is therefore not because of truth but because of benevolence that Bacon argues that philosophy, like faith, is to be "judged by its fruits" (NO 1.73, I: 183).[85] For wisdom has always relied on being "justified by her children," which in modern times are her works of benevolence (AL 1.103, 54, III: 319; cf. Matt. 11: 19 with Luke 7: 35).

Faith is the "substance of things hoped for" (Hebr. 11: 1). But one can only hope for what one believes possible. It is therefore because hope presupposes faith that Bacon opens the section on hope in *Novum Organum* with the consideration that "by far the greatest obstacle" to the advancement of the sciences consists in "men despairing and supposing things impossible" (NO 1.92, I: 198).[86] But the "substance of things hoped for" is not seen. It is therefore in order to "make future and remote things appear as present" (DA 6.3, I: 673) that Bacon says that those things are to be held possible which "may be done in succession of ages, though not within the houre-glasse of one mans life" (AL 2.16, III: 329), although it is within the confines of one life span that the possible is to be held plausible, which can only be achieved if contributing to a remote work is perceived as the highest moral duty. But "the untaught does not receive the words of knowledge, unless he is first told what transpires in his heart" (IMDO 2, I: 134; cf. Prov. 18: 2). It is therefore in order to make the faithful receptive to the arts of Cain that Bacon presents the "restitution and instauration of corruptible things, and, which to a lesser degree is the same thing, the conservation of bodies in their current state, and the retardation of dissolution and putrefaction" as the "noblest work of natural philosophy" (DSV XI "Orpheus," VI: 647–8).[87]

For faith hopes for the resubstantialization of the body since it hopes for the substantialization of its substance, which is immortality. Bacon could therefore only create faith in the arts of Cain by directing a considerable part of them to the work of substantializing faith's hope. But whereas faith works toward the resurrection of the body, the arts of Cain must work toward the body's regeneration. For not only is it through faith of a progressive and therefore self-reinforcing nature that the believer is most effectively kept in motion, seeing that the progression of the work of hope through works of faith substantiates the believer's hope on every rung of the ladder toward its ultimate substantialization; it may also have been the degeneration of the body that originally substantiated man's hope for the resurrection of the body, as Bacon argues in the fifteenth fable of *De Sapientia Veterum*.

Aurora begged Jupiter that Tithonus, with whom she was in love, might never die. But "with a woman's carelessness" she forgot to add to her petition that Tithonus might not be burdened with old age. "Pitying his fate," Jupiter at last changed Tithonus into a cicada. Bacon points out that the fable seems to contain an "ingenious adumbration and description of pleasure." For although man "was deserted by the acts of pleasure, desire and appetite did not die." He therefore "finds delight in recollecting the pleasures of his first state, like a cicada, whose vigour is only in his voice" (DSV XV "Tithonus," VI: 653). Bacon, on the other hand, would probably have found delight in hearing us conclude from his fifteenth fable that it was because God pitied immortal man for losing the "pleasures" of Edenic youth as a result of bodily degeneration that He made man mortal and provided him with the "ethereal" pleasure of regenerating his first state "in speech" as an "adumbration" of the resurrection of his body.

But as long as God is considered to be the author of man's finiteness, faith in the arts of Cain can only be finite. It seems to be for this reason that Bacon emphasizes elsewhere that the prolongation of life is the "noblest" part of medicine, since "delaying and retroverting the course of nature" would make the physician the "dispenser and administrator of the greatest of earthly gifts" (DA 4.2, I: 598–9), and therefore "a god to man."[88] And it seems to be for the same reason that our author published a *Historia Vitae et Mortis* (1623) as part of a more comprehensive "Natural and Experimental History," and that he published it earlier than he had initially planned, considering the "exceptional utility of a work in which the slightest loss of time should be considered precious," especially, as we may notice, if this loss would be caused by the death of the founder of the work. For, Bacon wonders, "how can someone turn the ways of nature who does not know them?"

Because Bacon defines hope as "the most useful of all affections, provided that it feeds the imagination with the prospect of good," the conclusion seems justified that the "exceptional utility" of a natural history of life and death consists in the fact that it holds out to man the prospect of the regeneration of his body, which is good to the extent that the prospect of the resurrection of the body is evil; and "what can be gradually repaired without destroying the original whole is potentially eternal," as Bacon speculates.[89] In anticipation of the materialization of this prospect, the "initiates" are set in motion by their desire for glory and the vulgar by the visibleness of their morality. For it is in consideration of man's mortality that Bacon elsewhere remarks that the "care for civil things is properly and orderly represented as subsequent to the present failure to restore the dead body to life."[90] It is also in view of the mortality of the human condition that Bacon emphasizes that only our works are not distorted by the injuries of time, since "their works follow them" (Rev. 14: 13), as it was said (DA 7.2, I: 722–3). For it is only by means of works that "*Revenge* triumphs over *Death*" (E II "Of Death," 10, VI: 380; cf. Rev. 20: 6 ff.).

As faith works by love, and the works of love are visible, love substantializes faith's hope by adumbrating it, which makes charity the greatest of theological virtues (1 Cor. 13: 1–2, 13), and potentially the greatest of Baconian virtues.[91] For the goodness of Christian goodness consists in the fact that its subservience to faith makes it potentially subservient to Baconian faith, especially because man has a natural inclination toward goodness or the "*love* of others" (E X "Of Love," 33, VI: 398), an inclination that underlies any form of its materialization. And it was of the utmost importance that Christian goodness would in fact become subservient to Baconian faith. For although "Mens Thoughts are much according to their Inclination," their "Deeds are after as they have beene *Accustomed*" (E XXXIX "Of Custome and Education," 120, VI: 470), and the custom of Christian goodness had "altered and subdued" nature.[92]

But considering the fact that custom and habit rule in morals, because the moral virtues are in the human mind "by habit" and not "by nature" (DA 7.3, I: 737; DA 7.1, I: 714), it was only by bending "*Nature* as a Wand, to a Contrary Extreme, whereby to set it right" (E XXXVIII "Of Nature in Men," 119, VI: 469) that Bacon could as it were disincline man's accustomed inclination toward Christian goodness, and re-incline it toward goodness.[93] Yet contraries meet, and in morals habituation necessarily implies dehabituation. The "Habit" of "*Goodnesse*" (E XIII "Of Goodnesse and Goodnesse of Nature," 39, VI: 403) that Bacon had to establish in order to counter Christian goodness thus had to be established by means of the "varying and interchanging of contraries," "[b]ut with an Inclination to the more benigne Extreme," as our author makes a contemporary of Christ say, whom he describes as a "Wise Man" (E XXX "Of Regiment of Health," 101, VI: 453).[94]

It is Bacon's success in securing the transformation from Christian goodness into goodness by concealing the provisional parasitical dependence of goodness on its Christian counterpart that goes a long way toward explaining why scholars have almost unanimously confused charity and philanthropy or "the affecting of the Weale of Men" (E XIII "Of Goodnesse and Goodnesse of Nature," 38, VI: 403).[95] This success may not be wholly unrelated to the fact that at the end of the *Essay* on goodness and Christian goodness Bacon completes the aforementioned transformation with what one philosopher calls a "peculiar mixture of boldness and caution"[96]: caution, because one would almost overlook the fact that the transformation has already been completed when one reads that "If a Man be Gracious, and Curteous to Strangers, it shewes, he is a Citizen of the World; And that his Heart, is no Island, cut off from other Lands; but a Continent, that joynes to them"; and boldness, because by recommending "*St. Pauls* Perfection," who "would wish to be an *Anathema* from *Christ*, for the Salvation of his Brethren" (E XIII "Of Goodnesse and Goodnesse of Nature," 40–1, VI: 405; Rom. 9: 3), Bacon recommends "perfecting" Paul by reversing Paul. For with a boldness bold enough to be a form of caution he recommends man's real "anathemization," for the "liberation of his brothers."[97]

Subsequent to his substantial discussion of the "good," Bacon lays the groundwork for his "habit of goodness" by proposing what he calls a "culture of the mind," whose aim it is to "most effectively reduce the mind to virtue," and to "place it in a state approximating perfection." Such a culture consists in man "proposing to himself and electing right and virtuous ends of his life and actions, of such a kind as to a certain degree may be within his ability to attain." For, Bacon explains, "if the ends of man's actions are honest and good, and his mind fixedly and constantly pursues and obtains them, it follows that the human mind immediately carries itself to all virtues at once" (DA 7.3, I: 740–1).

It must have been primarily these considerations, in addition to the fact that "A Mans *Nature* is best perceived in . . . a new Case or Experiment, for there Custom leaveth him" (E XXXVIII "Of Nature in Men," 120, VI: 470), and the fact that the "Force of *Custome* Copulate, and Conjoyned, and Collegiate, is far Greater" than the "Force of *Custome* Simple and Separate" (E XXXIX "Of Custome and Education," 120, VI: 471), that caused Bacon to propose his project as a project of goodness by proposing the "reliefe of Mans estate" as the "last or furthest end of knowledge" (AL 1.48, 31–2, III: 294).[98] For not only could our author thus employ a natural inclination in order to correct an unnatural aberration; he could also prevent man's "re-inclined" inclination from aberrating again, seeing that by grounding the habit of collective goodness on man's natural inclination to goodness, as it "naturally" spreads "it selfe, towards many" (E X "Of Love," 33, VI: 398), he made philanthropy pleasurable and therefore forceful.[99] Moreover, by making a subservient affection subservient to faith in his project, Bacon prevented it from subserving faith, while at the same time responding to theologians who claimed that it was from "lust of knowledge" that man fell.

It is therefore only on the presupposition of charity's subservience to Baconian faith that we can explain why Bacon had the confidence to say that charity is the only virtue that "admits of no excess" (DA 7.3, I: 742; IMPr. 5, I: 132). And it was only by establishing a ministerial virtue that could safely be practised "excessively" that Bacon could assure himself of the durability of his project, the probability of which was further enhanced by the fact that faith in the arts of Cain was faith of a progressive nature. For not only did this imply that by performing works of charity man could in a way obtain the object of his faith at any rung of the ladder toward perfection, seeing that in a way every rung was perfect in itself, if only because the perfection of the human soul consisted in working toward the perfection of the human body by means of moral virtue; it also explains why Bacon says that the reason why charity is called the "bond of perfection" (Col. 3: 14) is that it "comprehends and binds all virtues together" (DA 7.3, I: 742).[100]

Yet all the previous considerations on the virtue of charity notwithstanding, it should not be forgotten that charity is a "correctiue" of knowledge, seeing that without it knowledge "hath in it some Nature of venome or malignitie, and some effects of that venome which is ventositie or swelling" (AL 1.6, 7, III: 266), that is, knowledge has somewhat of the "venom" of the serpent that manipulated man into the "malignity" of judging knowledge good. But knowledge could only be made "soueraigne" by being sprinkled with the "correctiue spice" of charity. For *"Knowledge bloweth up, but Charitie buildeth up"* (AL 1.6, 7, III; cf. 1 Cor. 8: 1).[101] The "building" of knowledge, that is.

Working toward his proposal of a "culture of the mind," Bacon emphasizes that "young men are no fit auditors of politics until they have been thoroughly imbued with religion and knowledge of morals and duties," lest their "judgments should become so corrupted and depraved that they should come to the opinion that there are no true and solid moral differences of things, but that all things are to be measured by utility and success" (DA 7.3, I: 739). Throughout the seventh book of *De Augmentis,* Bacon repeatedly states that he treats duties not as they are part of "civil society," but "insofar as they pertain to the framing and predisposing of separate minds to the bonds of society" (DA 7.2, I: 730). For "as in architecture it is one thing to frame the posts, the beams, and the other parts of the building, and another thing to fasten and join them together, so the doctrine of the conjugation of men in society differs from the doctrine that teaches them to conform and be well-disposed to its commodities" (DA 7.2, I: 726–7).

We wonder, however, why a man who explicitly calls himself an "architect" (DA 7.1, I: 714–5) should "fasten and join" together the parts of the "building" of society before having "framed" them. Our confusion increases when we find our author saying at the beginning of the eighth book, which treats "civil science," that it is the "order of things" that caused him to treat duties before treating the "art of empire"(DA 8.1, I: 745). For doesn't this amount to putting the cart before the horse?

Having pointed out that he will teach the "art of silence" by his own example, Bacon refers to one of Cicero's letters to Atticus, which he summarizes by saying that after having related a discourse with someone on the different positions with regard to a certain subject, Cicero writes that he was "silent." But when we take the trouble of looking into the letter ourselves, we find that it is only after having written that he was silent that Cicero discourses with his cousin Quintus on the different positions with regard to the subject concerned.[102] Bacon goes on to paraphrase a remark made by Pindar: "Sometimes silence effects more than speech" (DA 8.1, I: 745–6). But the Englishman is silent about the fact that Pindar prefaces this remark with the counsel not to tell the "whole truth with an unveiled face."[103] When we combine these two examples, we are led to the conclusion that Bacon will not only not be silent on the "art of empire," but that he has already spoken on the "art of silence," albeit in a manner appropriate to its subject matter. For the Kingdom of Man had to be modelled on the Kingdom of God, as we are stimulated to deduce from its "architect's" observation that "the causes and beginnings of the greatest things are from God" (NO 1.92–3, I: 199–200).[104]

But seeing that the Kingdom of God "comes not with observation," since it is to be found in the believers (Luke 17: 20–1), the Kingdom of Man had to come in quietly, and "without noise and clamour, before men were to suppose or perceive that it had already been set in motion" (NO 1.93, I: 200).[105] In order to smoothly establish the Kingdom of Man, Bacon thus had to conjugate men by means of the "bond" of charity. For in response to Christianity's fusion of the public and the private or the external and the internal, it was only by binding men internally with the bond of charity that the philosopher could prevent them from experiencing the bonds of the Kingdom of Man as bondage, seeing that bonds working "commodities" are themselves commodious. The fact that the "good" offered by the "art of empire" is the "protection from injuries" (DA 8.1, I: 747) in a way visualizes this connection between charity and the art of empire. For it was by "fastening and joining together" its parts with the bond of charity that our "imperial architect" prevented the frame of the Kingdom of Man from eroding and therefore from causing "injuries." But Bacon had to be silent to the ears of the many generations of "young men" that were to govern the Kingdom of Man, lest they would hear him saying that the "moral differences of things" are less "solid" than they thought, and that "all things are to be measured by utility and success."[106]

Expectation

In the Proemium to the *Instauratio Magna,* Bacon points out that because he was uncertain when his considerations might "occur in the mind" of someone else, he decided to publish them as soon as he was able to complete them. He emphasizes that his haste was "not caused by ambition, but by solicitude," in the sense that in case the "human fate should befall him, there would remain some plan and design of what his mind had grasped" (IMProemium, I: 122).

The fact that Bacon himself repeatedly stresses that the separate parts of his Instauration are either wanting or unfinished, added to the fact that a number of extant works, especially those pertaining to the third part of the Instauration, which concerns the "phenomena of the universe" or "natural and experimental history," unequivocally show signs of haste, has not only contributed significantly to legitimizing the egalitarian fashion of equating the "architect" with the "workman," but has also brought scholars to the unanimous conclusion that Bacon's *Instauratio Magna* is essentially unfinished. This conclusion is corroborated by the fact that with regard to the sixth and final part of the Instauration, to which "the rest is subservient and ministrant," and which "discloses and sets forth that philosophy which is drawn from and established on a legitimate, chaste, and grave kind of inquiry," Bacon himself says that "completing" it and "bringing it to an end" is a thing both "above" his "powers" and "beyond" his "hope." For, he explains, although he made the "beginning," the "outcome" will depend on the "fortune of the human race" (IMDO 29, I: 144).

But when we consider Bacon's counsel that there is "no greater Wisedome, then well to time the Beginnings, and Onsets of Things" (E XXI "Of Delayes," 68, VI: 427), we begin to suspect that the unfinished character of the *Instauratio Magna* may somehow be deliberate. This suspicion gains plausibility when we also take into consideration that Augustine, whom Bacon imitatingly opposed on more than one occasion, divides the history of the world into six parts or ages, of which the sixth and final part remains as yet unfinished. The Church Father argues that the first period runs from Adam to Noah, the second from Noah to Abraham, the third from Abraham to David, the fourth from David to the Babylonian captivity, the fifth from the Babylonian captivity to the advent of Jesus Christ, and the sixth and final period, which is now in progress, from the advent of Jesus Christ to His second coming at the end of times. Because man was created in the image of God on the sixth day, and the Son of God was made the Son of Man in the sixth age, it is in the sixth age of the world that man is "renewed by the gospel," or so we learn from Augustine of Hippo.[107]

But the sixth age of the world began under the reign of Augustus Caesar, who is abundantly praised in Bacon's works, although the effectual fruit of his reign was the "germination" of the "mustard seed" of Christianity. We have to go back to a relatively unknown comparative characterization of Augustus, whom Bacon ranks among the "*Liberatores, or Salvatores*," and his uncle Julius, who is described by Bacon as one of the "*Conditores Imperiorum*; *Founders of States*, and *Common-Wealths*" (E LV "Of Honour and Reputation," 164, VI: 506), in order to find Bacon criticizing Augustus in a matter doing justice to the great dissimulator's actual lack of foresight.[108] For although Bacon points out that Augustus, being "sober and mindful of mortality, seemed [visus] to have delineated and balanced his ends in an admirable order," considering that after having "put himself at the head of things" he "thought of applying himself to some work in order to imprint the image and virtue of his reign on future ages," the English philosopher does not mention any work effected by Augustus, which can only be explained by the fact that the only effectual work of Augustus was the effectuation of the invisible reign of Christ.[109] Since Bacon explicitly says that Augustus was inferior to his uncle Julius in "strength of mind," we suspect that in order to be an effective founder of the kingdom of liberated men, the "instaured Augustus" must adopt the "greatness of mind" from Julius Caesar, who, as Bacon observes, would have perpetuated his reign had he not been "too forgiving towards his enemies."[110] And when we call to mind that the second temple, which was replaced by the body of Christ, existed for only nearly six ages, we find confirmed what we already knew, namely that Bacon himself is the "instaured Augustus," who "renews" the sixth age by means of his Instauration.[111]

But contrary to Augustus, Bacon possessed the "architectural" foresight that in certain ages causes great minds to "knowingly and willingly throw aside the dignity of their wits and names" by becoming "workmen" (DA 7.1, I: 714–5) in order to lay the groundwork for their indirect reign in future ages. For being himself a great mind, the far-sighted Bacon knew that the coincidence of great minds compelled him to provide secondary minds with a plan to work upon; stronger even, that he himself had to make a beginning of the work, considering that a "grain of mustard seed," although itself a "very small thing," is a "pledge of that which is to follow,"[112] and pledges breed followers, as the etiology of Christianity teaches. Similar to Augustus, however, Bacon was "sober and mindful of mortality." He therefore did not indulge in unreasonable hope or overestimations of his own powers, knowing that the durability of the indirect reign of one man is dependent on fortune, although the "*Fortune*, in being the First in an *Invention* . . . doth cause sometimes a wonderfull Overgrowth in *Riches*" (E XXXIIII "Of Riches," 111, VI: 462), and the "Proceeding upon somewhat conceived in Writing, doth for the most part facilitate *Dispatch*" (E XXV "Of Dispatch," 78, VI: 435). Bacon's project can thus be said to have been essentially unfinished to the extent that its outcome was dependent on future generations.[113] But inasmuch as Bacon possessed a greatness of mind similar to that of Julius Caesar, who "referred everything to himself" and was himself the "most credible centre of all his actions,"[114] the project was both unfinished and finished; to the extent that it was a project for the ages, it was unfinished; but insofar as it was a project for Bacon himself, it was finished when Bacon referred it back to himself and began his "seventh day."[115]

We recall that man brought before Jupiter an "accusation" against Prometheus and his invention. Bacon comments that man's accusation "of his own nature and the arts" proceeds from a "most excellent state of mind," and that it "results in good," whereas the contrary is "envied by the gods," and is "unfortunate" (DSV XXVI "Prometheus," VI: 672). It is therefore in order to govern the long-term fortune of his Instauration that our far-sighted author says that the Instauration is to be considered "more as a matter of felicity than of anyone's eminent capacity" (NO 1.78, I: 186). For "Wise Men, to decline the Envie of their owne vertues, use to ascribe them to Providence and *Fortune*" (E XL "Of Fortune," 124, VI: 473).[116] And it is in order to advance the future "felicity" of his Instauration that at the beginning of the work the cunning philosopher pours forth the "most humble and ardent prayers" to God the Father, God the Word, and God the Spirit, entreating them to "vouchsafe through [his] hands" the "endowment of the human family with new mercies" (IMPr. 4, I: 131). For great men act in accordance with the observations of great men, and a great man once observed that it is or is considered "Greatnesse in a Man, to be the Care, of the Higher Powers" (E XL "Of Fortune, 124, VI: 473).[117] It is, moreover, in order to subtly attribute to the "divine mercy and goodness" the possible absence of good that will possibly be attributable to the Instauration that Bacon says that if there is "any good" in his Instauration, it should be "attributed to the infinite divine mercy and goodness" (IMEpD, I: 123).[118] For as those who "extraordinarily extol human nature or the arts" are "less revered in comparison with the divine nature, with the perfection of which they almost compare themselves" (DSV XXVI "Prometheus," VI: 672), those who do not "extol human nature or the arts" will be more "revered" than the "divine nature," should "perfection" remain forthcoming, which explains why Bacon observes that the "wiser Sort of great Persons, bring in ever upon the Stage, some Body, upon whom to derive the *Envie*, that would come upon themselves" (E IX "Of Envy," 30, VI: 396).[119] Finally, it is in order to confuse humility with calculation that Bacon says that "any progress in the matter" is to be traced back to the "true and legitimate humiliation of the human spirit" (IMPr. 3, I: 130). For although the humiliation of the human spirit before nature is a form of humbleness, it is also the humility of the "instaured Brutus," who by his humility "made less of things held to be great" (NO 1.97, I: 202).[120]

At the end of his *Essay* on lies, and poetic lies Bacon says that the "Wickednesse of Falshood, and Breach of Faith, cannot possibly be so highly expressed, as in that it shall be the last Peale, to call the Judgements of God, upon the Generations of Men, It being foretold, that when Christ commeth, *He shall not finde Faith upon the Earth*" (E I "Of Truth," 9, VI: 379). But in truth, not only did Bacon consider it the "highest possible expression" of the "wickedness of falsehood" to suggest that the "breach of faith" will be the foreboding of God's final judgment, he also had the confidence to foretell what Christ Himself had only conjectured with regard to His second coming (Luke 18: 7–8), which is foretold to be His coming in judgment (Rom. 2: 16; 1 Cor. 4: 5; 2 Cor. 5: 10).[121] We ascribe Bacon's confidence to his faith in Baconian faith, especially since he himself says that Christ's conjecture is "meant of the Nature of Times" (E XX "Of Counsell," 66, VI: 425), that is, times of Baconian faith.[122]

In his final *Essay*, Bacon comes full circle by replacing by Baconian faith the poetic lies that in the first *Essay* had replaced faith; in other words, by uniting the subject matter of truth with what one might call the *verità effetuale* of the subject matter of his *Essays*.[123] It is in this final *Essay*, which treats the "vicissitude" of sects and religions, that Bacon makes his faith in Baconian faith somewhat more explicit by commenting Machiavelli's observation "that the *Jealousie* of *Sects*, doth much extinguish the Memory of Things."[124] For although the Florentine "anamnesiologist" was right in traducing Pope Gregory the Great (590–604) for having done "what in him lay, to extinguish all Heathen Antiquities," Bacon does "not finde, that those Zeales, doe any great Effects, nor last long: As it appeared in the Succession of *Sabinian*, who did revive the former Antiquities" (E LVIII "Of Vicissitude of Things," 172, VI: 513). Considering that Machiavelli concluded from his considerations on the vicissitude of the memory of things that sects vary two or three times every five or six thousand years,¹²⁵ we may conclude from Bacon's considerations on Sabinian's revival of the memory of things that according to the Baconian timetable, the effective life span of the Christian sect can be shortened by as many years as can be effected by the restoration of the memory of things.

That this need not necessarily be a considerable number of years already appears from the fact that in the central sections of his final *Essay* our author speaks about the causes of new sects. If a "*New Sect* have not two Properties," it "will not spread. The one is, the Supplanting, or the opposing, of Authority established," and the other is "the Giving License to Pleasures, and a Voluptuous Life." Since Bacon's "Orphic antidote," which for all its mad excesses was a work of moderation because its principle was health, possesses these two properties, and the central "Manner of Plantation of *New* Sects," which is plantation by the "*Eloquence and Wisedome* of Speech and *Perswasion*," is moreover the manner of Bacon's sect, Bacon seems to be foretelling that his own sect will spread.¹²⁶

But although the "True *Religion* is *built upon the Rocke*," whereas "The Rest are tost upon the Waves of Time" (E LVIII "Of Vicissitude of Things," 173, VI: 514), a true religion is a Baconian oxymoron.¹²⁷ In accordance with this truth, Bacon builds not on "rocks," but on the foundation that is "as a cornerstone of both theology and philosophy, with regard to which they almost agree as to what ought to be sought first." For theology declares: "Seek first the Kingdom of God," and philosophy says: "Seek first the good of the mind" (DA 8.2, I: 791). But although the Kingdom of Man was to provide men with the "goods" promised by the Kingdom of God (Matt. 6: 25–34), and philosophers with the latitude to seek the "good of the mind," its twofold foundation, which carried the seeds of the future confusion of voluptuousness and pleasure,¹²⁸ already laid the foundation for its degeneration, if only because one of the main causes of atheism is "*Learned Times*, specially with Peace, and Prosperity" (E XVI "Of Atheisme," 53, VI: 414).

At the end of his final *Essay*, Bacon himself foretells the end of the Kingdom of Man in a way reminiscent of his foretelling of the postponed end of the "Orphic order" at the end of his final fable.¹²⁹ For our chronicler describes the cycle of the state of learning or the Kingdom of Man in the following way: "In the *Youth* of a *State*, *Armes* doe flourish: In the *Middle Age* of a *State*, *Learning*; And then both of them together for a time: In the

Declining Age of a *State, Mechanical Arts* and *Merchandize*."[130] And indeed, in the autumn of the state of learning learned men armed with learning armed the unlearned with learning and inaugurated learning's "old Age," when it "waxeth Dry and Exhaust" (E LVIII "Of Vicissitude of Things," 176, VI: 517). But by that time another Prometheus would send a "purging deluge" upon mankind.[131]

CHAPTER 5

The Perfective Son

> But will God indeed dwell on the earth? behold, the heaven and heaven of heavens cannot contain thee; how much less this house that I have builded?
>
> —1 Kings 8: 27

In the thirtieth paragraph of the Distributio Operis of the *Instauratio Magna*, immediately after saying that completing the sixth part of the work is a thing both "above" his "powers" and "beyond" his "hope," Bacon entreats God to "benignly grant" him to "write an apocalypse and true vision of the Creator's footprints and seals upon the creatures" (IMDO 30, I: 145).

When we consider that Augustine, who has become our contrastive point of reference for Bacon's use of numbers, argues that Noah's ark is thirty cubits high (Gen. 6: 15) because Christ, "who is our height," gave His sanction to the doctrine of the Gospel in His thirtieth year, and that, since Noah was six hundred years old when he entered the ark (Gen. 7: 6), the sixth age of the world is occupied with the "construction of the Church by the preaching of the gospel";[1] when we observe that the two ships on the frontispiece of the *Instauratio Magna* are returning from their journey beyond the pillars of Hercules (I: 119); and when we notice that the thirtieth paragraph is the central paragraph of the fable whose action takes place beyond the pillars of Hercules; when we consider, observe, and notice these three points, we begin to suspect that in *New Atlantis* Bacon paints a picture of the sixth part of the Instauration in motion, at the very height of the renewed sixth age.[2]

But the prospering of the Instauration presupposes Baconian faith. It therefore presupposes an obliviating "apocalypse," as Baconian faith presupposes the oblivion of faith; an apocalypse, moreover, which had to assume the form of a deluge, not only because of its divine predecessor, or because a deluge is one of the "great Winding-sheets, that burie all Things in Oblivion," but also because the coming of the Son of Man will be like the coming of the Flood (Matt. 24: 37–42). In other words, since the "*Philology*" of the vicissitude of sects and religions is "a Circle of Tales" (E LVIII "Of Vicissitude of Things," 172, 176, VI: 512, 517),[3] Bacon first had to come full circle before he could instaure the "circle of sects" and set the sixth part of his Instauration in motion, albeit only in speech. In short, it was only on the presupposition of what one might call a "diluvian forestallment" of John's Apocalypse, which consists of twenty-two chapters, that the fabulist Bacon could write a "true vision of the Creator's footprints and seals upon the creatures" by rewriting the Old Testament, whose Septuagint version consists of twenty-two books, and whose "Blessing" is "*Prosperity*" (E V "Of Adversitie," 18, VI: 386).[4]

It is this rivalry among authors that may explain why the pronoun "I" occurs twenty-two times alongside the twenty-two references to God in *New Atlantis*. But all the unrivaled blessings of Bacon's "prosperous people" notwithstanding, prosperity will not be "without many Feares and Distastes" (E V "Of Adversitie," 19, VI: 386).

New Atlantis was published posthumously (1627) by Rawley, as an appendix to *Sylva Sylvarum*.[5] But although we do not have reason to assume that Rawley is lying when in the preface to *New Atlantis* he says that Bacon "designed" the work "for this place," we do have reason to believe that the preface itself, as well as two highly important editorial remarks, do not contain Bacon's own editorial seal of approval. For whereas Rawley explicitly says that the epistle "to the reader" that he prefixed to *Sylva Sylvarum* is the "same that should have been prefixed to this book if his lordship had lived,"[6] the epistle that he prefixed to *New Atlantis* contains no such comment.

But even if we had not been blessed with Rawley's honesty, the fact that the other posthumously published and seemingly unfinished work, *An Advertisement touching a Holy War* (1629), is prefaced with an extensive epistle stemming from Bacon's own pen would have justified the conclusion that Bacon wanted *New Atlantis* to go unprefaced. And even if the title of the work had not suggested that the subtitle, "A Work Unfinished," and the final words, "The Rest was not Perfected," originate from Rawley at least to the extent that they do not reflect the imperfect or unfinished character of Plato's *Critias* or *Atlanticus*,[7] some of Bacon's own remarks in *De Augmentis* would have presented us with this suggestion.

It is none other than William Rawley who puts us on the track of the possibility that *De Augmentis* contains explicit or implicit references to *New Atlantis*, as it is he who tells us that *New Atlantis* was written earlier than or at the same time as *De Augmentis*.[8] But it is Bacon himself who confirms this possibility by announcing that with regard to the "art of empire" he will be "very brief," which is a "thing next to silence" (DA 8.1, I: 746). For in the context of his explicit discussion of the art of empire our author says that what he said earlier was that with regard to the art of empire he "imposed silence" on himself, before saying that if his free time will "in the future be in travail with anything concerning politics, its offspring will perhaps be either abortive or posthumous." He continues with a substantial discussion of parts of the art of empire, emphasizing, however, that the parts he discusses "do not pertain to the arcana of empire" (DA 8.3, I: 792). Having become familiar with Bacon's "art of transmission," we have become confident enough to conclude from these remarks that in *De Augmentis* Bacon touches the "arcana of empire" only "very briefly," as he confines himself to saying that with regard to politics he confined himself to becoming an "impregnator."[9]

In fact, provided that we have the patience to wait until we get to the very last paragraph of *De Augmentis*, we are even led to the place where the sower of seeds sowed the seeds of politics. For it is at the very end of his most theoretical work that Bacon says that the response of Themistocles comes to his mind, who reproached an ambassador from a small town for having used great words: "Friend, your words require [desiderant] a city" (DA 9.1, I: 837). But by having his mind deceive him, Bacon stimulates us to read in Plutarch that it was actually Lysander who responded to a man from Megara that the latter's words lacked a city to support them.[10] We are thus led to the provisional conclusion that in a manner not dissimilar to its Greek exemplar the city required for words in motion is the subject matter of *New Atlantis*.[11]

Disclosure

It is by means of the accidental that in the very first paragraph of *New Atlantis* Bacon transmits the message that in his fable nothing is accidental. For he goes out of his way to draw our attention to the fact that it was the capriciousness of the winds that caused the crew ("We") to find themselves "in the midst of the greatest wilderness of waters in the world, without victual," after having been tossed to and fro for twelve months on their voyage from Peru toward China and Japan by the Pacific Ocean. But since "Salomon's House" possessed "engines for multiplying and enforcing of winds" (NA 24, III: 157), it can hardly be overlooked that the crew was manipulated into the situation in which they "gave" themselves for "lost men, and prepared for death."[12] Yet they lifted up their "hearts to God above, who *showeth his wonders in the deep*; beseeching him of his mercy, that as in the beginning he discovered the face of the deep, and brought forth dry land, so he would now discover land to" them, so that they "might not perish." But whereas the God of the crew *showed* His wonders in the deep, the God of the Bible had raised a stormy wind exactly in response to sailors *having seen* His wonders in the deep (Ps. 107: 23 ff.), which reminds us of God's reasons for sending the Flood (Gen. 6: 5–7). That it was indeed Bacon's intention to put us in mind of the Flood is proved by the fact that by saying that in the beginning God "discovered" the face of the deep, Bacon alludes to the ambiguity of the first verses of Genesis, presumably in order to suggest that God did not create the deep (Gen. 1: 1–4). This would lead to the further suggestion that the deep in which the crew found themselves had not been caused by God; a suggestion that would underscore the intention of Bacon the "creator" to go back beyond God the Creator in order to renew the Old Testament.

It "came to pass" that the next evening the crew saw "as it were thick clouds," which put them "in some hope of land." God had come to Moses on Mount Sinai in a "thick cloud" so that the people might hear Him speak to Moses, considering that no man shall see God's face and live (Exod. 19: 9, 20–1; 33: 20). The crew, however, "plainly discern[ed]" land at the dawn of the next day. And whereas God had brought the sailors to their desired haven in response to their cry (Ps. 107: 29–30), the crew entered into a "good haven" without knowing if this was the haven they desired when they had beseeched God to "discover" land to them.

But when they "offered to land," straightway they were warned off by signs. A small boat with "about eight persons in it" approached the ship, subtly reminding us of the fact that only eight people survived the Flood (Gen. 6: 18), and that the drama of Bacon's fable unfolds in a "post-diluvian" world. One of them came aboard the ship, "without any distrust at all," a clear confirmation of our suspicion that the ship had been directed toward its haven, albeit unwittingly. He gave the "foremost man" a scroll with the following text: "'Land ye not, none of you; and provide to be gone from this coast within sixteen days, except you have further time given you.'" The scroll was signed with a "stamp of cherubins' wings, not spread, but hanging downwards, and by them a cross" (NA 1, III: 129–30).

Earlier in the paragraph Bacon had remarked that the wind that had driven the sailors to the island had a "point east." After Adam and Eve had eaten of the tree of the knowledge of good and evil, God placed Cherubim at the "east" of the Garden of Eden to guard the way to the tree of life (Gen. 3: 24). This subtle geographical hint thus not only suggests that the island is somehow comparable to Eden, but also that the crew is as yet denied access to paradise. And to the extent that the Cherubim's wings "hanging downwards" intimates that God's glory, which is intrinsic to paradise, is not yet present, Bacon's hint is also in conformity with the prophet Ezekiel's prophecy that the Cherubim will only lift up their wings when God's glory shall stand over them (Ez. 10: 18–9; 11: 22).[13]

The crew was asked if they were Christians, which they confirmed, "fearing the less, because of the cross" they had seen "in the subscription." This unchristian fear of answering the decisive question of Christianity gives us reason to suspect that the crew is weak in faith.[14] In fact, in the sentences immediately following their "confession" it becomes clear that the crew's weak or absent faith was the very reason for their being in the haven of an unknown island. For in order for them to "have licence to come to land," they were asked to "swear" by the "merits of the Saviour" that they had not "shed blood lawfully nor unlawfully within forty days past" (NA 2, III: 131). However, God's Law forbids killing as such, without distinguishing between older and more recent killings (Exod. 20: 13; 21: 12–4). This observation, taken together with the tributatory connotation of the number 40 (cf. Gen. 7: 4; Num. 14: 32–4; Matt. 4: 2; Mark. 1: 13; Luke 4: 2), leads us to suspect that by means of a forty-day trial of lawfulness the lawlessness of the crew was put to trial. But whereas by taking the oath the crew merely demonstrated the conditional nature of their lawfulness, the intrinsic indifference toward the shedding of blood on the part of the insulars shows a first glimpse of the bloody conditions for "bloodlessness" as such.

The next morning, six crewmembers were taken to the "Stranger's House." On the way to their accommodation, people gathered on both sides of the street, and as the crewmembers passed by they "put their arms a little abroad" as a way of welcoming the strangers, thereby reminding us of the Cherubim. Being asked about the number of their "sick," the crewmembers told the officer who accompanied them that seventeen of the fifty-one crewmembers were sick, which gives us reason to believe that superstition was the nature of the "sickness" of the sick.[15]

The officer left, and when an hour later he came back he showed the crewmembers their accommodation. Having been alerted to the significance of the decisions made during the officer's absence, we come across some peculiarities when reading what the officer showed the crewmembers upon his return. For not only did the gallery that had been instituted as an infirmary for sick persons consist of forty cells, although only seventeen crewmembers were sick, there were also only ten chambers that had been prepared in case the sick should regain their health. This suggests not only that some of the sick could turn out to be incurable; the possibility that on the island of bodily health some of the healthy might grow sick also alludes to the fact that superstition has more than one face.

After men and goods had been removed from the ship and the company had settled down somewhat, the narrator ("I") "thought good" to call the company together in order to inform them on their situation, of which he had a superior understanding, as appears from his words. In harmony with the fact that "poetic history" is history "at will" (DA 2.13, I: 517–8), he told his crewmembers that they were "beyond both the old world and the new," and that with regard to their "deliverance past" and the "dangers present and to come," "every man" should "reform his own ways"—a first explicit indication of a natural hierarchy among the crewmembers. This hierarchy was as it were sealed by the fact that the company thanked the narrator "with one voice" for his "good admonition," a fact that reminds us of the "one voice" the people of Israel answered Moses with after he had told them all the words of the Lord (Exod. 24: 3).

The sick were healing so fast that they "thought themselves cast into some divine pool of healing" (NA 3, III: 134–5). But contrary to the sick people at the pool Bethzatha, at the "Sheep Gate," the sick crewmembers were not healed by Jesus (John 5: 2–9; cf. NA 25, 28, III: 157–8).

On the morning after a three-day quarantine, a new man came to the crew, who also "put his arms abroad" when coming in, once again reminding us of the Cherubim. He introduced himself as "by office" governor of the House of Strangers, and "by vocation" a Christian priest. He told the crew that there were some things he "may tell" them, which he thought they would "not be unwilling to hear," thus implying that there were some secrets he was not allowed to tell. He went on to say that the crew had been given licence by the state to stay on land "for the space of six weeks" or longer, although they were to stay within a mile and a half of the walls of the city, unless they would be given "especial leave."

The crew expressed their thanks in the following words: "It seemed to us that we had before us a picture of our salvation in heaven; for we that were awhile since in the jaws of death, were now brought into a place where we found nothing but consolations."[16] Having emphasized that they would obey the "commandment," although it was impossible that their hearts would not be "inflamed to tread further upon this happy and holy ground,"[17] they concluded their thanksgiving by saying that their "tongues should first cleave to the roofs" of their "mouths ere" they "should forget either this reverend person or this whole nation in" their "prayers." The crew's words call to mind the following verses of Psalm 137: "If I forget thee, O Jerusalem, let my right hand forget her cunning; If I do not remember thee, let my tongue cleave to the roof of my mouth ... if I do not set Jerusalem above my highest joy" (Ps. 137: 5–6). But by using "forget" (fifth verse) instead of "not remember" (sixth verse), Bacon cunningly makes the crew say that they were about to forget Jerusalem in favor of their "highest joy."

After the governor-priest had left, the crewmembers said among themselves that they "were come into a land of angels" (NA 4, III: 135–6), a saying that by now we know is to be taken literally.[18]

The next morning the governor-priest came again. There were ten crewmembers present, "the rest were of the meaner sort, or else gone abroad." By this time, the crew had spent six days on the island. A few paragraphs further, however, the narrator says that after "six or seven days were spent" he had "fallen into strait acquaintance" with a Jewish merchant (NA 17, III: 151). We are thus led to suspect that it was none other than the narrator who went abroad, presumably because he had been given "especial leave" on account of his superior understanding, which would imply that the narrations of the following paragraphs will be either not or not wholly true.[19]

The governor-priest told the remaining crewmembers that in the language of the insulars the island on which they found themselves was called "Bensalem," a name underscoring the Old Testamentic nature of Bacon's fable, since "Bensalem" is a combination of the Hebrew words "בֵּן" and "שָׁלוֹם," meaning "perfect son" or "son of peace." It was David who admonished his son Solomon to serve the God of his father with a "perfect heart" (1 Chron. 28: 9; 29: 19). But at the end of his life, Solomon's heart was turned after other gods, whereupon God told him that He would tear the kingdom from him, but that He would give one tribe to his son, for the sake of David and Jerusalem (1 Kings 11: 4–6, 11–4). Although this etymological excursion causes us to find that Bensalem is or mediates a more perfect son of David than Solomon, the perfection of this son will not consist in a pious imitation of Jerusalem, as is subtly indicated by the fact that Bacon uses the Greek "Hierusalem" ("Ἱεροσόλυμα," NA 17, III: 151).

But let us not anticipate events. The governor-priest pointed out that by means of their "solitary situation," the "laws of secrecy" that they had for their "travellers," and their "rare admission of strangers," the Bensalemites knew "well most part of the habitable world," and were themselves "unknown." Therefore, "because he that knoweth least is fittest to ask questions," the crew was allowed to ask questions. They responded that there was "no worldly thing on earth more worthy to be known than the state of that happy land." But since the "land was so remote, and so divided by vast and unknown seas, from the land where our Saviour walked on earth," they above all desired to know who was the "apostle of that nation, and how it was converted to the faith."

Twenty years after the "ascension of our Saviour," so the governor-priest began his relation of Bensalem's conversion, the people of "Renfusa," a "city upon the eastern cost" of the island, saw a "great pillar of light" rising from the sea, with a "large cross of light" on top of it. The Renfusans gathered in boats in order to "go nearer to this marvellous sight." But when the boats had come within sixty yards of the pillar, they "found themselves all bound, and could go no further." The boats stood "as in a theatre, beholding this light as an heavenly sign."[20] It "so fell out" that "one of the wise men of the society of Salomon's House," which was the "very eye" of "this kingdom," was in one of the boats. After "having awhile attentively and devoutly viewed and contemplated" the pillar and the cross, the wise man "fell down upon his face," "raised himself upon his knees," lifted "his hands to heaven," and in the form of a prayer "acknowledge[d] and testif[ied] before this people" that the "thing" they saw before their eyes was God's "Finger and a true Miracle" (NA 7, III: 137).

Having finished his prayer, the wise man "presently found the boat he was in moveable and unbound; whereas all the rest remained still fast." He rowed toward the pillar, but before he had come "near it, the pillar and cross of light brake up, and cast itself abroad, as it were, into a firmament of many stars; which also vanished soon after." There was nothing left to be seen "but a small ark or chest of cedar, dry and not wet with water, though it swam." A "small green branch of palm" grew "in the fore-end of it." The cedar and the palmbranch subtly allude to Solomon's temple (1 Kings 6: 9 ff.), or to its instaured version after the Apocalypse (Ez. 41: 18). The ark itself clearly refers to the ark carrying the testimony (Exod. 25: 10 ff.), which was placed in Solomon's temple (1 Kings 8). But the existence of Salomon's House suggests that the floating ark came from the instaured version of Solomon's temple after the Baconian "apocalypse."

The ark contained a "Book and a Letter." The book contained "all the canonical books of the Old and New Testament," the "Apocalypse itself," and "some other books of the New Testament which were not at that time written" (NA 8, III: 138), which in any case implies that it contained the Acts and the Epistles of Paul. The letter was written by "Bartholomew," an apostle of Jesus Christ, who had been "warned by an angel that appeared" to him in a vision of glory that he "should commit this ark to the floods of the sea" (NA 9, III: 138). In the book as well as in the letter, "a great miracle" was "wrought," to "conform to that of the Apostles in the original Gift of Tongues," so that everyone could read both book and letter "as if they had been written in his own language." And thus, as the governor-priest went on to say, "this land was saved from infidelity . . . by an ark, through the apostolical and miraculous evangelism of St. Bartholomew" (NA 10, III: 138–9).

It seems superfluous to mention that by employing the apostle who had brought the gospel of Matthew to India, Bacon employs the apostle whose name was to be associated with the massacre that started on the eve of the festival devoted to his name (August 24, 1572). We therefore confine ourselves to saying that the fact that the insulars are angels, added to the fact that Bacon follows pseudo-Dionysius the Areopagite in calling the Cherubim "Angels of light" or knowledge (AL 1.52, 33, III: 296),[21] and the fact that Salomon's House sends "Merchants of Light" into the world for the purpose of obtaining "knowledge of the affairs and state" of other countries, especially of the "sciences, arts, manufactures, and inventions of all the world" (NA 15, 46, III: 146–7, 164)—that these facts taken together justify the conclusion that a far-sighted angel or merchant of light warned the apostle Bartholomew that he was to become the unwitting name giver of a massacre inspired by what at that time was the world's most recent invention. Bartholomew thereupon decided to lend his name and apostolate to an attempt at preventing similar faith-based acts of violence from occurring on the island of Bensalem.

The merchant's foresight draws our attention to the fundamental question as to whether historical Christianity was somehow inherent in the beginnings of Christianity in history. But as this is a subject matter of earlier chapters, we go on to formulate our suspicion that in order to obtain politico-religious authority among the Renfusans, the wise men of Salomon's House staged a miracle that elected them as the sole interpreters of miracles, allowing them to once and for all put an end to the miraculous, which explains why they decided to deceive the senses, although they "hate all impostures and lies" (NA 44, III: 164).[22] They were capable of doing so because they had the ability to imitate "meteors," "thunders," and "lightnings," to make "demonstrations of all lights and radiations; and of all colours," and to prepare "engines and instruments for all sorts of motions" (NA 26, 38, 42, III: 158, 161 ff.). What is more, the fact that Salomon's House was also able to imitate "all articulate sounds and letters" (NA 40, III: 162–3) explains not only why the gospel of Bartholomew was written in tongues, but also why it contained books that had as yet not been written. But for the sake of their successful conversion all these deceptions of the senses had to be withheld from the vast majority of the crew, which had landed on the east coast of the island, as they had been withheld from the sheepish inhabitants of Ren-fusa (ῥήν-φύσις), a city on the east coast of the island.

Immediately after having finished his relation of the island's conversion to the faith, the governor "paused, and a messenger came, and called him from" the company (NA 10, III: 139). But should the coming of the messenger itself not have sufficed for conveying his message, the fact that the governor never resumed the thread of his account of Bensalemite faith almost explicitly calls our attention to the fact that a sheepish faith that speaks in bleating tongues cannot but be substanceless.

The next day the governor came again, offering the crew to spend time with them, if they "held his company and conference agreeable." By responding that they thought an hour spent with their host was worth years of their "former" lives, the crew testified to the weakness of their faith. For they demonstrated that seven days on the island of Bensalem had sufficed for them to almost forget Jerusalem. But it is only after having been given a credible account of the Christian island's political isolation that its happiness became credible enough for the crew to "forget all that was dear" in their own countries (NA 16, III: 147), and to convert to the Bensalemite faith.

In accordance with their desire for a political as well as a religious conversion story, the crew wondered why the "happy island" of Bensalem was "known to few," although it "knew most of the nations of the world"; more especially, why the Bensalemites "should have knowledge of the languages, books, [and] affairs" of those that lay "such a distance from them." For it seemed to the crew "a condition and propriety of divine powers and beings, to be hidden and unseen to others, and yet to have others open and as in a light to them."

But since without knowing it the crew had already answered their own question, we have reason to suspect that the governor's answer served some other purpose besides answering the crew's question. The governor himself in a way confirms our suspicion by emphasizing that he had to "reserve some particulars," because with a view to the laws of secrecy it was not "lawful" for him to "reveal" them (NA 11, III: 139–40), thus implying that we have to reveal these particulars by ourselves.

Stimulating the understanding of those capable of understanding, the governor asked the crew to "understand" what "perhaps" they would "scarce think credible," namely that about three thousand years ago (1388 BCE) the "navigation of the world" was "greater than at this day" (1612).[23] At that time, "and an age after, or more," the inhabitants of the "great Atlantis," which the Europeans "call America," flourished. But although the insulars had "large knowledge thereof," there was "sparing memory" of "all this" with the crew (NA 12, III: 140–1). Although, as the governor went on to say, the narration and description of Atlantis by Plato, whom he called "a great man with you," was "all poetical and fabulous," "so much" was "true," that Atlantis, as well as Peru ("Coya") and Mexico ("Tyrambel"), were "mighty and proud kingdoms"; so mighty, that "they both made two great expeditions; they of Tyrambel through the Atlantic to the Mediterranean Sea; and they of Coya through the South Sea upon this our island." With regard to the expedition of the Tyrambelans, Plato "(as it seemeth) had some relation from the Egyptian priest whom he citeth. For assuredly such a thing there was. But whether it were the ancient Athenians that had the glory of the repulse and resistance of those forces," the governor could "say nothing." The voyage of the Coyans "upon" Bensalem would not have had a "better fortune, if they had not met with enemies of greater clemency." For the king of "this island," whose name was "Altabin," and who was a "wise man and a great warrior, knowing well both his own strength and that of his enemies," dismissed the Coyans in safety, after having contented himself with their oath that they "should no more bear arms against him." But the "Divine Revenge overtook not long after those proud enterprises. For within less than the space of one hundred years, the great Atlantis was utterly lost and destroyed: not by an earthquake," as the governor took Plato to have said, "for that tract is little subject to earthquakes," but by a "particular deluge or inundation" (NA 14, III: 141–2).[24]

Having been stimulated to understand what is "scarcely credible," we are beginning to understand that, despite appearances to the contrary, the governor wants us to understand that in a way the "great Atlantis" still exists, and that it is not to be confused with America. For not only does the governor contradict himself by saying that Atlantis may still have flourished "more" than an age after the flourishing of navigation, whereas a few lines further down he says that it was "within the space of one hundred years" that the great Atlantis was lost; he also mentions Atlantis and two actual American countries together before almost silently dropping Atlantis and equating Tyrambel with Plato's Atlantis, and before misrelating both Plato and Acosta. For what the Egyptian priest had actually told Solon was that a deluge had destroyed Atlantis, and what can actually be read in Acosta is that earthquakes are very common in America.²⁵ Moreover, although the governor suggests a certain causal relationship between the "Divine Revenge" and the destruction of Atlantis, he does not mention any "proud" Atlantan "enterprise" that could have provoked the divine revenge. We are inclined to think that by means of these obvious blunders and contradictions the governor, who consistently speaks of the "great" instead of the "old" Atlantis, transmits the message that the great but "poetical and fabulous" Atlantis of the "great" Plato will remain "virtual" until it becomes "actual" by becoming "new." This line of thought would enable us to solve the mystery of the title of the fable, provided that we take the obvious analogies between Bensalem and Plato's Egypt into consideration, and that we do not give in to the scholarly custom of equating the island of Bensalem with the New Atlantis.²⁶ For like Egypt, Bensalem was not visited with deluges; like Egypt, Bensalem had chronicles dating far back; and like Egypt, Bensalem distinguished between history and myths or fables, although it did employ the "poetical and fabulous" for the intervention in history. What is more, as the Egyptian priest had criticized the Greeks' lack of memory, the governor criticized the memory of the European crew, thus alluding to a second analogy in addition to the first, namely an analogy between Greece and Europe.²⁷

But when we pursue this analogy somewhat further, we come across the decisive difference between the Athenians and the Europeans. For whereas Solon had already given laws to Athens when he traveled to Egypt,²⁸ the narrator was given "leave to publish" the "relation of the true state of Salomon's House" at the end of *New Atlantis* (NA 19, 59, III: 156, 166). This chronological difference seems to suggest that whereas the Egyptian priest had transmitted Athens' warlike past to Solon, the island of Bensalem will transmit to a "modern Solon" an exhortation to a warlike future.

We find this suggestion confirmed when we have a closer look at the tension between the name and the deeds of Altabin. For the governor takes great trouble to cast doubts on the question whether the Coyans actually arrived at Bensalem, and it was the Atlantis that the Europeans "call America" that "lay nearest to" Bensalem (NA 14, III: 143). We are thus led to the conclusion that the "enemies of greater clemency" were the Atlantans. This conclusion is not only supported by a careful reading of the narrative, but will also be confirmed in the next paragraph, where we find that "clemency" was not a Bensalemite virtue.²⁹ But although his name suggests otherwise, the king of Atlantis was not "twice [bini] lofty [alta]." For not only is clemency hardly the virtue of a great warrior; the clement Altabin was also not wise since he did not know the true strength of his enemies, as appears from the fact that he allowed the "Divine Revenge" to destroy his island. We are therefore led to the further conclusion that by criticizing the first king of Atlantis, "Atla-*singulus*",[30] as it were, the governor implicitly counsels the future king of Atlantis to improve upon his predecessor by becoming an "Atla-*binus*," that is, by becoming a wise warrior for wisdom, in order for his kingdom to be set in motion.[31] For it is only by disallowing the "divine" its "revenge" that Plato's "great Atlantis" can be actualized by being improved on, seeing that the *Atlanticus* ends with the question whether Zeus punished Atlantis after it had lost its share in the divine. But a non-divine Atlantis in motion is necessarily a *New Atlantis*.[32]

The "modern Solon" must be a lawgiver of the Mosaic type, as Bacon stimulates us to infer from the fact that the "great Atlantis" flourished simultaneous to the Exodus (approximately 1400 BCE). But contrary to his predecessor, the "Mosaic Solon" must lead his people out of the mountains instead of the desert. By describing the mountain people who survived the inundation as "simple and savage people," Bacon leads his lawgiver to Machiavelli, who teaches that, since those who are saved from inundations are all "coarse mountain men" who have no knowledge of antiquity, the lawgiver can conceal his knowledge of antiquity and "pervert it in his mode."[33]

That in fact he has good reason for doing so, the lawgiver can learn by following Bacon's second lead. For it was by making an obliviating flood the beginning of political life that the Athenian stranger in Plato's *Nomoi* illustrated his teaching that in order for political life to flourish it is necessary to obliviate the natural or political terror inherent in all political beginnings.[34]

Bacon illustrates this twofold teaching by having the governor provide the crew with an account of the lawgiver of Bensalem, and by having him present this account in a manner that is in accordance with its underlying teaching, as is indicated by the governor's remark that his account would only "draw nearer to giv[ing]" the crew satisfaction to their "principal question," which was the question as to the "state" of Bensalem (NA 14, III: 144).

About nineteen hundred years ago (288 BCE) a king reigned in Bensalem, whose "memory of all others" the Bensalemites "most adore[d]," and whom they esteemed as their lawgiver. He had a "*large heart*, inscrutable for good." His name was Solamona, a subtle allusion to Machiavelli's *uno solo* ("sola"–"μονάς"), spelled backwards *anomalos*, a political anomaly, who was "wholly bent to make this kingdom and people happy" though.

Taking the self-sufficiency of the island into consideration, and "recalling into his memory the happy and flourishing estate" of the land, "so as it might be a thousand ways altered to the worse, but scarce any way to the better," nothing "wanted to" Solamona's "noble and heroical intentions, but only (as far as human foresight might reach) to give perpetuity to that which was in his time so happily established." Fearing "novelties, and commixture of manners," because the "entrance of strangers . . . at that time (though it was after the calamity of America) was frequent," he ordained Bensalem's "interdicts and prohibitions" touching strangers (NA 15, III: 144).

But in the previous paragraph the governor had said that after the inundation the Bensalemites had lost their "traffic with the Americans, with whom of all others" they had "most commerce," adding that "in the ages following . . . navigation did every where greatly decay," and that "specially far voyages . . . were altogether left and omitted" (NA 14, III: 143). These contradictions amount to the suggestion that the strangers who until the time of Solamona had been allowed entrance to Bensalem were the offspring of the "simple and savage" people who had survived the deluge. This implies that the happiness of the self-sufficient Bensalemites mirrored the forgetful simplicity of their guests, but that in the ten ages since the inundation the arts, and especially the art of warfare, had redeveloped up to the point where a "commixture of manners" had come to pose as great a corruptive threat to the simple happiness of the Bensalemites as it had done eleven ages ago, at the height of navigation. It thus explains why the Bensalemites would have been enemies "of lesser clemency" than the Atlantans, had the Coyans arrived at Bensalem.[35] It may actually have been precisely his memory of the fragility of Bensalemite happiness that made Solamona desire to become responsible for its long duration.[36]

But leaving our thoughts about the long-term motives of Bensalem's lawgiver aside, we find him "[d]esiring to join humanity and policy together; and thinking it against humanity to detain strangers . . . against their wills, and against policy that they should return and discover their knowledge of this estate," before we read that he ordained that "of the strangers that should be permitted to land, as many . . . might depart as would; but as many as would stay should have very good conditions and means to live from the state." He "saw so far" in this respect that the Bensalemites had "memory not of one ship that ever returned; and but of thirteen persons that chose to return" in their "bottoms. What those few that returned may have reported abroad," the governor did not know (NA 15, III: 144–5). But the fact that he had said earlier that Solamona considered it "against policy" that strangers should return, added to the inauspicious connotation of the number thirteen and the indifference of the Bensalemites toward shedding blood, allows of the conclusion that the "bottoms" of the Bensalemites sent the thirteen departees to the bottom of the sea. This conclusion permits the further conclusion that a policy of murder was the precondition for the humanity of the Bensalemites, inasmuch as their humanity could have been threatened by the lives of those murdered.[37]

The weight of Solamona's memory was balanced by his lack of foresight, though, as is intimated by the nominal and epithetical allusions to King Solomon (cf. Prov. 25: 2; 1 Kings 4: 29), and as is confirmed by the fact that, of all people, the lawgiver of the people of Bensalem is not called "wise." For as the imperfect Solomon was unable to prevent his kingdom from being torn from him, Solamona's foresight was imperfect to the extent that it prevented him from foreseeing the coming of Christianity, one of these "thousand ways" in which the happiness of the island could "alter to the worse."[38]

The seventeenth paragraph is the pivoting point of *New Atlantis*.[39] For it is the paragraph in which the narrator reappears and the crew as a whole disappears[40]; it is also the paragraph in which the author of the action of the fable is introduced; and finally, it is the paragraph that brings the first and the second half of the fable together, in an almost marital way, as we shall see.

The narrator points out that what he relates in the seventeenth paragraph transpired after "six or seven days" had been spent on the island. We have already seen that after the narrator had spent six days on the island, he went "abroad." The fact that up to this point he had been almost overly precise in reporting the number of days and hours spent on the island makes us suspect that by being imprecise he suggests that the conversation related in the seventeenth paragraph took place on a voluntary basis, since it took place on the seventh day (cf. Exod. 20: 8–11). We find this suspicion confirmed when we consider the significance of the fact that it was a "merchant" whom the narrator had "fallen into strait acquaintance" with on his seventh day. For by means of a subtle numerical hint Bacon compares the missions of the "Merchants of Light" to the "observation missions" treated in the twelfth book of Plato's *Nomoi*.[41] And it is in the context of discussing these observation missions that the Athenian stranger points out that, in order to "place the noble on a firmer footing" and to "correct what is lacking," "divine human beings," who do not by nature grow more frequently in cities with good than in cities without good laws, must always be "sought and tracked down" by a city with good laws, and must always be allowed to "go to the doors of the wise uninvited."[42] Without himself knowing it, the Athenian stranger has thus not only confirmed our conclusion that it was on account of his wisdom that the narrator had been given "especial leave" to go abroad to visit a merchant whom he describes as a "wise man"; he has also provided us with the beginning of an answer to the question as to why the ship carrying the narrator had been directed to Bensalem. For apparently the narrator had been "tracked down" by the merchant on one of the latter's observation missions as someone who could provide Bensalem with something the island was "lacking."43

But in order not to run ahead of things, we should return to the author of the action of Bacon's fable, whose true importance is underscored not only by the fact that he is the only contemporary Bensalemite who is mentioned by name, but also by the paradoxical fact that he is the only contemporary Bensalemite whose garments are not described. The name of this Jewish merchant was "Joabin," which is the plural form of Joab, a ruthless, but highly important commander in David's army, who often acted contrary to the knowledge or will of his master, but who was indispensable for the success of David's reign (2 Sam. 3: 22–7; 18: 5–17; 19: 5–8). Since Joabin, who was not only "wise," but also "learned, and of great policy," was "excellently seen in the laws and customs" of Bensalem, the narrator desired to know from him "what laws and customs" the Bensalemites had "concerning marriage," whether they "kept marriage well," and whether "they were tied to one wife." Joabin thereupon stimulated the narrator to "understand" that there is "not under the heavens so chaste a nation as this of Bensalem; nor so free from pollution or foulness. It is the virgin of the world," and its "Spirit of Chastity" is like a "fair beautiful Cherubin." He went on to contrast the virginity of the Bensalemites with the pollutedness of the Europeans, upon whom the Bensalemites look with "detestation" because they permit "stews," "dissolute houses," and "courtesans." The Bensalemites say that the Europeans have "put marriage out of office: for," they explain, "marriage is ordained a remedy for unlawful concupiscence, and natural concupiscence seemeth as a spur to marriage. But when men have at hand a more agreeable remedy to their corrupt will, marriage is almost expulsed." They hear the Europeans "defend these things, as done to avoid greater evils; as advoutries, deflouring of virgins, unnatural lust, and the like. But they say this is a preposterous wisdom, and they call it *Lot's offer*, who to save his guests from abusing, offered his daughters." They say "farther that there is little gained in this; for that the same vices and appetites do still remain and abound; unlawful lust being like a furnace, that if you stop the flames altogether, it will quench; but if you give it any vent, it will rage."

Joabin continued by answering the narrator's first and third questions, for he said that the Bensalemites have "many wise and excellent laws touching marriage," that they "allow no polygamy," and that they have ordained that "none do intermarry or contract than until a month be passed from their first interview." He read in a book on a "Feigned Commonwealth," written by a man whom he calls "one of your men," and whom we identify as either Plato or Thomas More, that the "married couple are permitted, before they contract, to see one another naked."[44] The Bensalemites dislike this, however, because they "think it a scorn to give a refusal after so familiar knowledge." But considering the "many hidden defects in men and women's bodies," they have a "more civil way." For near every town they have a couple of pools, which they call *"Adam and Eve's pools,"* where it is permitted to "one of the friends of the man" and "one of the friends of the woman" to see them "severally bathe naked" (NA 17, III: 151 ff.).

We can only begin to understand Joabin's obscure account of Bensalemite marital laws when we begin by considering that it was Joabin's biblical predecessor who once performed the office of "marriage counselor." For it was Joab who was assigned by David to kill Uriah the Hittite, so that David could marry Uriah's wife, the beautiful Bathsheba, who conceived David's child after David had seen her naked (2 Sam. 11: 2–27). After their first son had died because of David's sin, Bathsheba bore David a second son, whose name was Solomon (2 Sam. 12). At the end of David's life, Joab conspired with Adonijah, another son of David's, who wanted to become king instead of Solomon. But through the intervention of Nathan and Bathsheba, it was Solomon who was crowned king, and by whose order Joab was killed (2 Sam. 3: 4; 1 Kings 1: 5–37; 2: 28–34).

When we call to mind that Bensalem mediates a more "perfect son" than Solomon, and that God gave one tribe to Solomon's son for the sake of David, whom Ezekiel prophesied to be the king and shepherd in the instaured kingdom (Ez. 37: 24), we begin to suspect that the "instaured Joab," a "friend" of Bensalem, is attempting to arrange a "marriage" between an "instaured Bathsheba," whom we identify as the "virginal" Bensalem, and an "instaured David," whom we identify as the "polluted" Europe, from which a more perfect "son" than Solomon will be born.[45] Apart from the obvious similarities between the "concupiscence" of Europe and that of David, and between the "virginal chastity" of Bensalem and that of the once chaste Bathsheba, this suspicion becomes plausible when we consider that it is the marriage counselor himself who prophesies the perfection of the fruit of his counseling. For the narrator compares himself to the "widow of Sarepta," who said to "Elias; that he was come to bring to memory our sins." Earlier in the paragraph he said that the Bensalemite Jews call Joabin "the *Eliah* of the *Messiah*," adding that Joabin "would ever acknowledge that Christ was born of a Virgin," that he "would tell how God made him ruler of the Seraphims which guard his throne," and that he was "desirous" to "have it believed" that "when the Messiah should come, and sit in his throne at Hierusalem, the king of Bensalem should sit at his feet, whereas other kings should keep a great distance" (NA 17, III: 151).

Bacon could hardly have signified the perfection of the son of the "virgin of the world" more compellingly than by comparing his midwife to the prophet who before the day of the Son of the Virgin will come to "restore all things" (Mal. 4: 5; Matt. 17: 12; Mark 9: 12).[46] But since no prophet is accepted in his own country (Luke 4: 24–6),[47] the merchant or angel of light, who rules the "Angels of loue" or charity (cf. AL 1.52, 33, III: 296; Isa 6: 1–2), had to bring the narrator to Bensalem in order to remind him of the fact that the sick son of the widow of Zarephath would have died if it had not been for Eliah's intervention (1 Kings 17); in other words, that the "virgin of the world" might bring forth a sick or even stillborn "son," should the narrator decide not to support Joabin in arranging a "marriage" between Europe and Bensalem.

It is this acute necessity of having the virgin "deflowered" by an "instaured David" that provides us with an explanation for the admonishing undertone of Joabin's account. We therefore believe that by saying that "natural concupiscence" is a "spur to marriage," and that marriage is the "remedy" for "unlawful concupiscence," the "restorer of all things" exhorts the instaured David to remedy his natural but unlawful desire for the instaured Bathsheba by marrying her. And by saying that the "wisdom" of the Europeans is "preposterous," the learned Jew seems to suggest that Lot should have prevented the two angels from being "abused" by "marrying" them instead of offering his daughters to the men of Sodom, the "furnace" of whose "lust" was thus given "vent" to "rage" (Gen. 19: 1–11); in other words, he seems to suggest that the angelic Bensalem can only be prevented from being abused by becoming the "wife" of a powerful Europe that is ruled indirectly by the narrator, whom we identify as Francis Bacon the poet.

From what quarter the wind of "abuse" is blowing becomes clear when we allow Joabin to explain the implications of the chaste Bensalemites giving "reverence and obedience" to the "order of nature" (NA 16, III: 148). For should one have overlooked the possible implications of his explicit remark that the Bensalemites "intermarry," Joabin almost explicitly draws our attention to the incestuous nature of Bensalemite intermarriage by saying that he is "desirous" to "have it believed" that the people of Bensalem "were of the generations of Abraham, by another son, whom they call Nachoran" (NA 17, III: 151). For Nahor was the brother of Abraham and Haran. He married Milcah, the daughter of Haran. Among the children of Nahor and

Milcah was Bethuel, the father of Rebekkah, who married Isaac, the son of Abraham. Rebekkah begot Jacob, who had a son with Rachel, the daughter of Laban, who was the brother of Rebekkah (Gen. 11: 22–8; 22: 20–4; 24: 15 ff.; 25: 20–6; 30: 22–4). In other words, it is on account of the very fact that the Bensalemites "revere and obey" the order of nature that they are unchaste, since nature is "incestuous" and therefore "unchaste"—an observation, we mention in passing, which is in accordance with the fact that Bathsheba is primarily remembered for her unchastity.[48] And it is the naturalness of Bensalemite unchastity that directs us to the similarity of the "virgin of the world" to Minerva or nature "unchastened" in speech.[49] This similarity provides us with the answer to our question what direction the danger of abuse comes from. For it is primarily "Christian chastity" that lies in wait for an opportunity to rechasten the newly unchastened virgin. The acuteness of this threat is as it were historically and geographically confirmed by the fact that in and in the wake of the age of discovery the new world was inundated by missionary Christianity.

Only a "marriage of convenience" between wisdom and power could prevent the unchastened from being rechastened. This explains not only what it was that Bensalem was "lacking" when Joabin decided to "track down" the narrator, but also why the narrator was "given leave" to publish the "relation of the true state of Salomon's House" for the "good of other nations," that is, for the good of the city of "Hierusalem," which is the city that is good for wisdom, because it is the city required for words in motion: the New Atlantis. But it was not until Thomas Hobbes "married" wisdom and power by means of his writings that words were set in motion; a historical fact that on a nonhistorical level explains not only why Rawley says that Bacon only "thought" to "have composed" a "frame of Laws, or of the best state or mould of a commonwealth" (NA Pr., III: 127) in the fable of the New Atlantis, but also why apart from Hercules we nowhere find a "*Mortall God*" in Bacon's other fables.

Be that as it may, since Bensalem was "God's bosom, a land unknown" (NA 19, 59, III: 156, 166), her only begotten "son" had to be declared publicly (cf. John 1: 18), which the narrator did on his twenty-second day, as this was the day after he had been given the "relation of the true state of Salomon's House," for which he had been chosen by his fellows since of the natural authority that he had displayed earlier (cf. NA 18 with NA 3, III: 155, 134–5). That it is only the narrator, a "friend" of Europe, who on his last day on the island sees Bensalem "naked" by seeing Salomon's House in addition to the city of Renfusa, is visualized by the original frontispiece to *New Atlantis*, on which we see Chronos guiding a naked woman out of a cave, elucidated by the words "Tempore Patet Occulta Veritas": "Time Discloses Concealed Truth."[50] But despite the fact that the truth had to be reconcealed for the likes of the narrator's crewmembers in Europe, the publication of a picture of a scantily dressed woman was a precondition for the effectuation and durability of the "marriage" between power and wisdom. This was not the task of Joabin, however, who was "commanded away" when he was about to answer the central question of the narrator, which was the question whether the Bensalemites "kept marriage well" (NA 17–8, III: 154).

Knowing no sin, the philosopher knows no shame when finding himself standing on the edge of "Eve's pool," observing the "naked woman" (cf. Gen. 2: 25). But despite or because of his shamelessness, the narrator was able not only to see the "defects" of the "woman's body," but also to transmit them in a way honouring Bensalem's laws of secrecy. *New Atlantis* is a testament to the "learnedness," "policy," and "wisdom" of Joabin, because the work testifies to the rightness of the merchant's decision to "track down" Francis Bacon on account of the latter's credentials as a master of the art of transmission. For by adding an account of the preconditions for Salomon's House to what he had been "given leave" to publish, Bacon delineated to his lawgiver the road toward the New Atlantis, while at the same time transmitting to his equals an account of the price that was to be paid for words in motion, that is, for the inconvenient "marriage" between wisdom and power for the sake of the protection of wisdom by power.[51]

Sheepish citizens, concealed old murders, and harmless new superstitions were only the foreboding of new and greater evils, however. And it is the grimness of this prospect that may provide us with an explanation for the far-sighted Bacon's admonition that the "Vertue of *Prosperitie*, is Temperance" (E V "Of Adversitie," 18, VI: 386), seeing that prosperity is the blessing of the Testament that allows of historical progress and therefore of intemperance.

But to him who has eyes to see, the master of the "art of silence" also speaks about the fragilities of those parts of the "woman's body" that he was permitted to make public, to say nothing of their deficiencies, not the least of which consisted in the exclusive practice of *Scientia Activa* or *Philosophia Secunda* (IMDO I: 134) by the initiates of Salomon's House. For the Cherubim with spread wings on the mercy seat of the ark carrying the testimony (Exod. 25: 18–22) lead us to find that the Cherub with "displayed" wings on the top of the Father of Salomon's House's chariot (NA 18, III: 154–5) suggests that the latter was carrying the "true" state of Salomon's House.[52] The displayed wings themselves, on the other hand, allude to the protective function of the angels of light, since after the ark had been brought to Solomon's temple, it was placed under the spread wings of the Cherubim (1 Kings 6: 23–8, 8: 6–8; 2 Chron. 3: 10–3). And since the true "End" of Salomon's House was the "knowledge of Causes, and secret motions of things; and the enlarging of the bounds of Human Empire, to the effecting of all things possible" (NA 20, III: 156), protection of and from the hostility of those pursuing other ends was imperative, as is indicated by the fact that the members of Salomon's House took an "oath of secrecy, for the concealing of those" inventions that they thought "fit to keep secret" and not to "reveal" to the "state" (NA 55, III: 165).[53] But the protection of the Cherubim could not be relied on beyond the world of poetic history, unless we take the narrator's remark that every man was to reform his own ways (NA 3, III: 134) to imply that even Satan had to transform himself into an angel of light (2 Cor. 11: 14).[54]

Obfuscation

Now that the city required for words in motion has been described and Bacon's philosophic successors have received the admonishment to set it in motion, we can hardly think of any other work stemming from Bacon's pen, let alone a work that could add anything to Bacon's teaching. But in fact there is such a work, and in fact it does supplement Bacon's teaching. For it was in 1629 that Rawley published a small dialogue written by his Lordship, in which six complementary characters discuss the possibility and legitimacy of charitable warfare.[55]

That Bacon wanted to see *An Advertisement touching a Holy War* published already appears from the fact that the dialogue is prefaced with an extensive epistle from Bacon's own hand. But that our author may actually have intended to see the work published subsequent to *New Atlantis* does not become plausible until we start considering the tension involved in the following two statements. For whereas in the Epistle Dedicatory to the *Advertisement* itself Bacon says that the dialogue contains an argument "mixt between contemplative and active" (AHW EpD, VII: 15), in a memorial of access (March 1622) he had described the work as a work of his "contemplative" "Pen."[56]

This tension can be resolved in a twofold way, corresponding to the twofold understanding of the relation between action and contemplation. Let us first consider the question of historical sequence, which corresponds to action and contemplation as a "conjunctive two." The dialogue takes place in the house of the namesake of the comic poet Eupolis (445–410 BCE). But it is not merely his name that the host of a conversation on making history shares with the author of comedies. For it can hardly be overlooked that it was in imitation of a comedy (*Demes*) about four men being raised from the dead in order to put right the state of affairs at Athens that four men were invited into the house of Eupolis in order to speak "of the affairs of Christendom" (AHW 4, VII: 18). When we take this analogy to allude to a concern with the good of the city on the part of Eupolis; when we look upon the man whose name means "good city" (Εὐ πόλις) as the dialogic predecessor of the city that is good for wisdom; when we recall that the city that is good for wisdom is founded on war; and when we anticipate Eupolis laying the foundations for a holy war (AHW 18–9, VII: 23–4), we see *An Advertisement touching a Holy War* coming to sight as a work that conjoins action and contemplation by contemplating the need for action before laying a contemplative foundation for action, and advertising action for the sake of contemplation.

Having thus given a historical explanation for the fact that *New Atlantis* is succeeded by a work that was to secure its foundations, we go on to emphasize that it is not merely the question of truth as such that leads us to the question of philosophic substance, which corresponds to action and contemplation as a "disjunctive two." It is especially the unsatisfactory character of partial truth that compels us to look for a more substantial explanation for Bacon's decision to conclude his oeuvre by advertising a "holy war." For besides the fact that the question of war has already been given an exhaustive treatment in Bacon's other works,[57] it is only minds sensible to the art of writing and capable of the art of reading that can discern the true object of the war that is advertised in *An Advertisement touching a Holy War*; minds, in other words, that are often of a noble and magnanimous disposition, and that consider only the whole of mankind a worthy recipient of their virtues; minds, in short, like those of the French *philosophes* who were to carry out Bacon's project in the form of an *Encyclopédie ou Dictionnaire Raisonné des Sciences, des Arts et des Métiers*. It is minds of this order that are able not only to see through some of Bacon's lies, but also to justify these lies by referring them to the common good as they understand it; minds of this order also, that have some understanding for the fact that the Antichrist came or had to come in the name of Christianity, and that Bacon used others for purposes of his own, which could only have been the purposes of mankind; minds of this order, in sum, that go a long way toward disclosing Bacon's ultimate intention. It is, to drive our point home, minds of the caliber of a Diderot or a d'Alembert that will find the ultimate moral legitimation of Bacon's immorality and therefore of their own project of universal enlightenment in *An Advertisement touching a Holy War*, the dialogue that culminates in the self-incrimination of a Christian zealot by means of considerations of philanthropy and the common good. It must therefore, to state the conclusion to which we are driven in the form of an analogy that is more than an analogy, have been in conscious imitation of Machiavelli's well-known *Esortazione a liberare la Italia da' barbari* that Bacon's *Advertisement touching a Holy War* placed the philosophic teaching in the service of the communicative good. It was, however, only by giving his eccentric work a touch of comedy of the sort to which merely serious men are immune that Bacon could leave ajar the door that leads to the true center of gravity of his thought.

But what is most grave can best be hidden behind the overwhelming face of the grave, as already appears from the Epistle Dedicatory to the only one of Bacon's writings that goes "into the temple," and is dedicated to a bishop.[58] Bacon goes out of his way to convince most of his readers of the fact that after his conviction for bribery (1621) he had no "ambition of rising again," before saying that the reason why he chose an argument both "mixt of religious and civil considerations" and "mixt between contemplative and active" was that there may be an "*Exoriere aliquis*" (AHW EpD, VII: 11 ff., 15). It was Dido who spoke the words: "May some avenger rise from our bones," before she committed suicide.[59] By omitting the second part of Dido's speech, Bacon not only appeals to the moral indignation of some of his more careful readers, he also assures them that he himself was ultimately driven by considerations of justice, a conviction that he moreover corroborates by suggesting that it was primarily the fact that his examples Demosthenes, Cicero, and Seneca died a "violent death" after having been "restored with great glory" that withheld him from actually rising again (AHW EpD, VII: 11 ff.). The philosopher thus conveys to his followers the message that they are to carry out his legacy by means of a "holy war." And Bacon cannot have disparaged the glory that was to befall him through posthumous victory in war, as is in a way demonstrated by the fact that it was primarily the eulogies of his followers that were to guarantee the ongoing visibility of his public persona. But the Englishman's posthumous attempt to achieve glory should not be taken to justify the conclusion that he valued his legacy or the least "durable part" of his memory for reasons other than securing his most careful readers indirect access to the common good.

Four persons "of eminent quality, but of several dispositions" met in Paris, in the house of Eupolis, in the year 1621.[60] Eupolis himself was also present. While they were "set in conference," Pollio "came in to them from Court," and said "after his witty and pleasant manner" that the four "were able to make a good World," because they were "as differing as the four Elements," although they were "friends." As for Eupolis," Pollio added, "because he is temperate and without passion, he may be the Fifth Essence" (AHW 1, VII: 17).

Pollio's first intervention makes explicit what the description of the setting of the dialogue had already suggested, namely that both Pollio himself and Eupolis are to be separated from the rest of the company. Pollio's own importance, which will be demonstrated *ad oculos* as the course of the dialogue's argument causes him to emerge as the author of its action, is already indicated by the fact that he is the first speaker, although he is the last person mentioned of the six "persons that speak." The prominence of Eupolis is further underscored by the fact that he is presented as the "sublunary version" of Aristotle's fifth element, which is higher than and prior to the other four elements.[61] He is characterized as a "Politique," a clear allusion to the moderates who during the Wars of Religion (1562–1598) had propagated a policy of subordinating religion to civil authority. It is thus at the very beginning of the dialogue that we are invited to infer from the character of Eupolis that the argument "mixt of religious and civil considerations" is to be looked on as an argument for subjecting religious to civil considerations.

But although by his "inviting" presence Eupolis already proved himself highly active, his concealing passivity was in need of being balanced by the concealed activity of Pollio, whose namesake Gaius Asinius Pollio (76 BCE–4 AD) was the author of a history of the civil wars of his time, in addition to being Roman consul, and a patron of poets. For the characterization of Pollio as a "Courtier" stimulates us not only to detect a first sign of lightness in an otherwise weighty atmosphere, but also to read in Castiglione that the nonchalance ("sprezzatura") of a courtier conceals design.[62] And the polishing courtier had good reason for concealing his design, as is already suggested by his name. For the man who bore on his shoulders the task of "liberating the world from its perpetual fear" had to "walk on fires that lurk beneath deceptive ashes," seeing that he was the author of the history of an unfinished war.[63]

In the next three paragraphs a dialogue in miniature takes place between the Greek politique and the Roman courtier, which as it were seals their alliance by suggesting a pre-existent alliance. For by telling Pollio that he both "professes" and "practises referring all things to himself," Eupolis implicitly asked the courtier to ask him about those who practice referring all things to themselves without professing it, so that he could answer that they are "the less hardy, and the more dangerous" (AHW 2–4, VII: 17–8).

This rehearsed dialogue allows the two allies not only to draw our attention to the danger of those who refer all things to themselves while professing to refer all things to God, but also to make us "wakeful" to the possibility that in Pollio's case it might be the other way around; in other words, that contrary to what he professes, the seemingly self-centered courtier does not refer all things to himself.

Eupolis went on to inform Pollio publicly on the reason for his coming, namely that before his coming the five men "were speaking of the affairs of Christendom at this day" (AHW 4, VII: 18). Feigning courtier-like indifference, Pollio responded that their "discourses had need content" his "ears very well, to make them intreat" his "eyes to keep open," although he promised them to "keep watch the best" he can, provided that they gave him "leave to awake" them, should he think their discourses "do but sleep" (AHW 5, VII: 18).

Eupolis thanked Pollio for doing the company the greatest possible "favor" by being wakeful, and he went on to tell his ally that the reason why the latter had to come in when he had come in was that before he had come in, Martius had begun a speech that "seemed to be the trumpet of a War." Having warned Martius that his auditory was "not a little amended by the presence of Pollio" (AHW 6, VII: 18), Eupolis gave the floor to the "Militar Man," who recommenced his speech by pointing out that he had observed that during the past fifty years "Christian princes and potentates" had been "wanting to the propagation of the Faith by their arms." In fact, the man named after the Roman god of war remembered only three "noble and memorable actions upon the infidels, wherein the Christian hath been the invader."

But the third example exhibited a defect in the memory of the propagator of armed faith (AHW 7, VII: 18–9). For it was not Clement VIII (1592–1605), the Pope who was responsible for the burning of the philosopher Giordano Bruno (1600), who had admonished the incursions of the Polish King Sigismund III in Russia (1605–1618), but Pope Paul V (1605–1621). Should we have been too drowsy to notice this blunder, the wakeful courtier draws our attention to Martius's selective memory by reminding the war-trumpeter of the "extirpation of the Moors of Valentia" (1609–1614), which belonged to the less noble and memorable actions of Christian warfare (AHW 8, VII: 19), as Eupolis was to make explicit a few moments later by saying that it "sorted not aptly with actions of war, being upon subjects, and without resistance" (AHW 11, VII: 20).

But by being "a little at a stop" at Pollio's question, Martius gave the first indication of his moldability, considering that a true believer would not have wanted words to defend even or precisely what defies reasonable defense. Yet besides molding Martius, Pollio's second intervention also served a second purpose. For it enabled the manipulative courtier to demonstrate the nature and therefore the internal dividedness of Christianity at war by drawing two warlike Christian warriors into the discussion on Christian warfare.

It is at this point in the dialogue that its comic features first come into play in an almost ostentatious way. For in the next two paragraphs we become witnesses to some small-scale verbal warfare between a "Protestant Zelant" and a "Romish Catholic Zelant," whose names as it were anticipated their zealotry, seeing that Gamaliel, whose name means "God's reward," was a teacher of Paul (Acts 22: 3), and Zebedaeus, meaning "my gift is God," was the father of the apostles James and John, whom Christ surnamed the "sons of thunder" (Mark 3: 17; cf. Luke 9: 51–6).[64] Gamaliel never approved of the expulsion of approximately three hundred thousand Catholic converts of Muslim origin from Spain, primarily because "God was not well pleased with that deed," as appeared from certain providential signs. His Roman Catholic counterpart, on the other hand, considered "that great action" to be as "Christ's fan" (AHW 9, 10, VII: 20), which is the instrument for purging floors or countries (Matt. 3: 12).

Having witnessed the intricacy of trumpeting a war for the propagation of the Christian faith, the trumpeter of wars made a decisive concession by saying that if he "should speak only as a natural man," he "should persuade the same thing," albeit this time for reasons of "secular greatness and terrene honor" (AHW 12, VII: 20). But by being able to speak as a "natural man," the malleable Martius proved himself a natural man and thus the potential instrument of a natural man. This was already indicated by the fact that the man named after the father of the founder of Rome is the only one of the six interlocutors without a historical predecessor. And it was foretold by the philosopher who relates that the god of war was tamed by Venus, that is, by the forces of peace.[65]

Pollio's central intervention, which is also the center of his three interruptions of Martius, constitutes both the numerical and the substantial center of the dialogue. For it is the thirteenth and central paragraph that causes the movement of Martius's soul to reach its metanoic height. By asking Martius to remember that "wild and savage people are like beasts and birds, which are *ferae naturae*, the property of which passeth with the possession, and goeth to the occupant," whereas "of civil people, it is not so" (AHW 13, VII: 21), Pollio not only forced the "Militar Man" to ask himself whether the difference between civil and savage people was as immutable as his questioner had made it appear to be, he also stimulated the warlover to consider the consequences of a negative answer for the justification of war.

Martius's response at the beginning of his central appearance in the dialogue proved the courtier a successful molder of warlike souls. For by saying that he knew "no such difference amongst reasonable souls, but that whatsoever is in order to the greatest and most general good of people may justify the action, be the people more or less civil" (AHW 14, VII: 21), the instrument of war said without saying or knowing that he had already become the instrument of the reasonable man who was to encourage him to advertise a war against the decivilized enslavers of people.[66]

Application of his own newly discovered criteria for justified warfare thereupon almost naturally carried the propagator of war to the propagation of a war against the Turks. For the Turks were a "barbarous" and "cruel tyranny," a "heap of vassals and slaves"; "in a word, a very reproach of human society" (AHW 14, VII: 22). But even if Martius had not made the Turks the object of his war, the fact that Bacon himself elsewhere and in his own name seems to justify a "holy war against the Turk" would have misled us into thinking that the second half of *An Advertisement touching a Holy War* advertises a holy war against the Turks. We are talking about the same Bacon, however, who not only "desecrated" a holy war against the Turks by saying that it is "not for cause of religion, but upon a just fear" that "Christian princes and states have always a sufficient ground of invasive war" against the Turks, but who also emphasized that "offensive wars for religion are seldom to be approved, or never, unless they have some mixture of civil titles"[67]; the same Bacon, moreover, who in what one might call an inner-Christian context only used the notion of a "holy war" to describe what he considered to be the well-deserved response to the authors of the doctrine of papally and therefore religiously sanctioned regicide.[68] Should one nevertheless remain unconvinced of Bacon's unholy use of the holy, one should not remain unmoved by the observation that it was the subject of coloring one's true intentions "with the pretence of war against the Turk" that made our great dissimulator think of what the "wise" Guicciardini had said of Ferdinand the Catholic (1479–1516), namely that the latter "*did always mask and veil his appetites with a demonstration of a devout and holy intention to the advancement of the Church and the public good.*"[69]

Returning to our text, we find that it was by means of his fourth intervention that the courtier prepared the substitution of zealous Christianity for the Turks; a substitution that was to be consummated in as subtle a way as it had been in the *Essay* on pious cruelty.[70] In response to Martius's secular justification of a war against the infidels, Pollio reminded his soldier of the fact that the Turks are not infidels, before as it were visualizing the justice of a war against another group of non-infidels. For by remarking that the Turks "acknowledge the Father, creator of heaven and earth, being the first person in the Trinity" (AHW 15, VII: 22–3), Pollio provoked a self-incriminating and unwittingly comical response from Zebedaeus, which consummated the object shift that was to govern the rest of the dialogue.

With a "countenance of great reprehension and severity" so remarkable that it caused the author of the dialogue to deviate from his habit of not describing the miens of his characters, Zebedaeus responded to Pollio that they should "take heed" not to fall "at unawares into the heresy of Manuel Comnenus, Emperor of Graecia, who affirmed that Mahomet's God was the true God." For, he explained, this heresy was "not only rejected and condemned by the synod," but also "imputed to the Emperor as extreme madness; being reproached to him . . . by the Bishop of Thessalonica, in those bitter and strange words as are not to be named" (AHW 16, VII: 23). But by completely misrelating the case, the zealot draws our attention to the fact that his own madness was even more extreme than the madness of the maddest actor in the case he misrelated. For in response to Komnenos's (1143–1180) proposition to have the anathemization of the god of Mohammed expunged from the catechetical books, Eustathios, the archbishop of Thessaloniki, did indeed speak the following "bitter and strange words": "My brains would be in my feet and I would be wholly unworthy of this garb . . . were I to regard as true God the pederast who was as brutish as a camel and master and teacher of every abominable act." And in a way not dissimilar to his spiritual successor, the maddened archbishop was "visibly shaken by pious zeal" when he shouted the aforementioned words. But after Komnenos had "charmed" Eustathios with his "reasonableness," the anathema was removed before the synods came to an end without having rejected and condemned Komnenos's heresy.[71]

The preceding collision between Pollio and Zebedaeus had a profound effect on Martius. For now that he had seen that there were "some that are excellent interpreters of the divine law, though in several ways," the instrument of war came to the conclusion that he had "reason to distrust" his "own judgment." He therefore realized that it would be an "error to speak further," until he should "see some sound foundation laid of the lawfulness of the action, by them that are better versed in that argument" (AHW 17, VII: 23).[72]

Having praised the newly acquired "moderation" of the man who had just proved not to be "transported, in an action that warms the blood and is appearing holy," Eupolis went on to do justice to his own name and character by proposing to "make some motion touching the distribution" of the conference into parts (AHW 18, VII: 23); by providing, to put it in Bacon's own words, a "platform" that may "draw on the building" (AHW EpD, VII: 15). In light of the foregoing, he thought it "would not sort amiss if Zebedaeus would be pleased to handle the question, Whether a war for the propagation of the Christian faith, without other cause of hostility, be lawful or no, and in what cases?" Since the second part was to complete the point of "lawfulness taken simply," it was assigned to the complement of Zebedaeus. It was therefore Gamaliel who was to treat the question "whether it be not obligatory to Christian princes and states to design" the aforementioned war. The third part concerned the "comparative"; that is, "it being granted that" the war was "either lawful or binding, yet whether other things" were "not to be preferred before it; as extirpations of heretics, reconcilements of schisms, pursuit of lawful temporal quarrels, and the like; and how far" the "enterprise ought either to wait upon these other matters, or to be mingled with them, or to pass by them and give law to them as inferior unto itself?" Because this was a "great part," and Eusebius had "yet said nothing," Eupolis proposed to the others to "lay it upon" Eusebius "by way of mulct or pain." As Pollio had a "sharp wit of discovery towards what is solid and real and what is specious and airy," Eupolis suspected that the courtier would "esteem all this but impossibilities, and eagles in the clouds." He therefore entreated his ally to "crush" the "argument with his best forces," so that by the "light" they would "take from him" they would "either cast it away if it be found but a bladder, or discharge it of so much as" was "vain and not sperable." Eupolis himself would then "do his best to prove the enterprise possible, and to show how all impediments may be either removed or overcome." Finally, provided that they did "not desert it before," it would be "fit for Martius" to "resume his further discourse, as well for the persuasive, as for the consult touching the means, preparations, and all that may conduce unto the enterprise" (AHW 19, VII: 23–3).

All "allowed the distribution" and "accepted the parts," but because the "day was spent," they agreed to "defer" the conference "to the next morning." Pollio decided to talk out of turn, however, presumably because he could increase the weight and resonance of his last say by having the last say at the end of the first day.[73] After all, it was a wise man who had once counseled that "in Causes of weight" the "Matter" should be "propounded one day, and not spoken to, till the next day" (E XX "Of Counsel," 67, VI: 426).

By saying that he was "of opinion, that except you could bray Christendom in a mortar, and mould it into a new paste, there is no possibility of a Holy War," the courtier almost assumed his largely unassuming role as author of the action of the dialogue. For what his observation implied was that the "braying" of Christianity constituted the most acute object of war even for those interlocutors who sincerely desired a holy war. It was therefore by means of his fifth and final intervention that Pollio set the "Great World" of the five elements in motion and provided the platform of the good city with its foundation, after the "good city" himself had provided a foundation for "greatness" by making it the common good of the differing elements of a "good World" (cf. AHW 1–2, VII: 17).

But immediately after having disclosed the "solid and real," the courtier went on to intimate that under the current circumstances the thought of "braying" Christianity was to be esteemed "specious and airy." For by saying that he "was ever of the opinion that the Philosophers's Stone, and a Holy War, were but the *rendez-vous* of cracked brains, that wore their feather in their head instead of their hat," Pollio implied that the braying of Christianity was currently impossible, since a Christianity having been brayed would tear the "feathers" out of the "heads" of brains behind holy wars, apart from the question whether it would still bring forth such brains, and apart from the question whether brains wearing feathers in their "hats" would still mask their true detachment by transposing these feathers to their heads in order to become advertisers of holy wars.

But the courtier considered a change of the current circumstances currently possible and desirable, as becomes clear from his own qualification of his earlier statement. For Pollio went on to tell his five interlocutors that if they should be "of another mind, especially after" having heard what he "can say," he would be "ready to certify with Hippocrates, that Athens is mad and Democritus is only sober." But what the father of the medicinal arts actually told the people from Abdera, who had accused Democritus of madness, was that only the wise Democritus was capable of making men sober.[74] We suspect Pollio of having misquoted Hippocrates in order to stimulate "Athens," that is, the four men whose predecessors had been raised from the dead by a comic poet, to prove him sober by becoming "mad" for the sake of long-term "sobriety"; more specifically, in order to admonish Martius to supplement his spiritual commander's wisdom with the power or warlike madness that is necessary in order for the building of the Great Instauration to be erected on the platform of the New Atlantis. If proved correct, our suspicion would provide us with an explanation for the fact that the pole of wisdom is compared to the "Philosopher's Stone" of alchemy. For as the good city presupposes a "maddening" alliance between wisdom and the power of warriors for the holy, the "ends or pretences" of alchemy are "noble," although its means are "monstrous" (AL 1.36, 27, III: 289; DA 3.5, I: 574).

Our suspicion is proved correct when we find the courtier saying what he had said he could say. For Pollio went on to tell the company that they should "[t]ake order" that when the "decrepit" Pope Gregory XV (1621–1623) would be "dead, there be chosen a Pope of fresh years, between fifty and three-score; and see that he take the name of Urban" (AHW 20, VII: 25). This description exactly fits Pope Urban VIII, who was fifty-five years old when he was elected on August 6, 1623, two years after a conference touching a holy war had taken place in the house of Eupolis. Considering that what the courtier said was something only a prophet or a fraud would have been able to say, we conclude that the wise Pollio assumed the role of a prophet in order to "sanctify" the unholy for the holy, meanwhile amusing some few of Bacon's readers.[75]

Immediately after Pollio had finished his prophecy, Eupolis returned favors by taking the last say from his ally in order to mask the gravity of the latter's final intervention by making it seem solely light-hearted. For he prayed that the courtier would be a "little more serious in this conference" (AHW 21, VII: 25).[76]

The next morning, Martius once again proved himself a diligent auditor of Pollio. For having seen the preliminary necessity of "braying" Christianity confirmed by prophecy, the means to war resumed and completed his discourse on the means to war by offering himself as the means to Pollio's war. It was fit for Martius to speak out of turn, though, considering that the original enterprise had been deserted for that of Pollio. We need not be surprised that it is in the twenty-second paragraph of *An Advertisement touching a Holy War* that the advertiser of wars advertises Pollio's war by declaring that he often noticed how things that had "showed impossible, by the declaration of the means to effect them, as by a back light, have appeared possible, the way through them being discerned" (AHW 22, VII: 25–6).[77] The author of the dialogue confirms our interpretation by employing the word "advertisement" for the first and only time in the *Advertisement* in order to describe Martius's final speech as a "grave and solid advertisement" (VII: 26). For by advertising Pollio's war, the warlover made the "specious and airy" "grave and solid" and as it were "holy" for himself and many others, although the undirectedness of his desire for war revealed the levity of his gravity to the eyes of some few others.

Now that the war had officially been advertised, it was the Fifth Essence's turn to speak out of turn. For since Martius had anticipated Eupolis's part in the conference, the mover of the parts of the conference had to assign himself a new part in order to keep the "Great World" of the five elements in motion. Considering that the war itself had already been proved possible, this part could only be the treatment of the question as to how far the war was to be pursued. But the Politique answered this question in a political way. For after having said that he "ought to have added or inserted" the question "into the point of lawfulness," he went on to give three possible answers to the question, before referring the question itself to Zebedaeus, who might "fall into it as an incident to the point of lawfulness." The war was to be pursued either to the extent of "displanting and extermination of people," or in order to "enforce a new belief, and to vindicate or punish infidelity," or "only" in order for the "temporal sword" to "open a door for the spiritual sword to enter," so that the latter could do its work by means of "persuasion, instruction, and such means as are proper for souls and consciences" (AHW 23, VII: 26).

Having been alerted to the fact that Eupolis inserted his complete answer into the speech of Zebedaeus, we find that it was by discussing the sixth of the six particular cases into which he had divided the general question of lawfulness that the zealot demonstrated that the war against the likes of himself had already begun. For after having explicitly referred to Eupolis's question, the "good city's" unwitting mouthpiece paraphrased the Politique's first two possible answers, before saying that it was to be considered whether a "Holy War" should be "moderated and limited; lest whilst we remember we are Christians, we forget that others are men" (AHW 25, VII: 27). Combining these hints, we invert the self-incriminating zealot's statement and conclude that the war against the holy will have been pursued too far if the spiritual warriors forget that "persuasion" and "instruction" are the proper means to mould the "souls and consciences" of Christians, who are also men. It was, however, not by forgetting that all Christians are men that Bacon's soldiers were to pursue their holy war too far, but by denying that not all men are potential philosophers.

What follows is the final and by far longest speech of the dialogue; a speech that is composed of so many clauses, discriminations, and distinctions that some few will read it as a satire of scholastic method. At the very beginning of his speech touching the lawfulness of a holy war, the "Romish Catholic Zelant" unwittingly indicated that the lawfulness of the war against the holy was to be measured by his own particular case, because he said that it was "but a loose thing" to "speak of lawfulness without the particular cases." Zebedaeus himself thereupon put his judicious auditors in a position to judge his case, since after having picked up the thread of Martius's case against the Turks he stated that a war "to suppress that empire" would be a "just war" even if one would "set aside the cause of religion" (AHW 24, VII: 28).[78] But this was not the only example of Zebedaeus decisively weakening his case by attempting to strengthen it. For the zealot went on to concede that as the "cause of a war ought to be just, so the justice of that cause ought to be evident; not obscure, not scrupulous," which compelled him also to concede that the "inward intention" was to be left to the "court of heaven."

Having conceded that the "justice of every action consisteth in the merits of the cause, the warrant of the jurisdiction, and the form of the prosecution," Zebedaeus held out to his auditors the prospect of arguing on the basis of the aforementioned criteria that a "war against the Turk is lawful, both by the laws of nature and nations, and by the law divine, which is the perfection of the other two" (AHW 25, VII: 28). He began with the "law of nature," of which he considered "the philosopher Aristotle" to be "no ill interpreter." For, he explained, Aristotle "hath set many men on work with a witty speech of *naturâ dominus*, and *naturâ servus*; affirming expressly, and positively, *that from the very nativity some things are born to rule, and some things to obey.*"[79] He emphasized, however, that this "oracle hath been taken in divers senses." For some have taken it for a "speech of ostentation," whereas others take it for a "speculative platform, that reason and nature would that the best should govern; but not in any wise to create a right." But Zebedaeus himself took it "neither for a brag, nor for a wish; but for a truth, as" Aristotle "limiteth it." For, he explained, Aristotle says that "if there can be found such an inequality between man and man as there is between man and beast or between soul and body, it investeth a right of government,"[80] which seems "rather an impossible case than an untrue sentence." But not only was the "judgment true, and the case possible"; the case also had "a being, both in particular men and nations."

Applying Zebedaeus's argument to his own case, we find that if the inequality between man and zealot should be as great as the inequality between man and beast, the truth of Aristotle's sentence would "invest" the reasonable "platform" of the good city with a natural right to govern bestialized men.

But being a zealot, Zebedaeus was well aware of the problem of consent, which is a problem not only for zealots. He therefore observed that it is "idle" to say that "the more capable, or the better deserver, hath such a right to govern as he may compulsorily bring under the less worthy," because men "will never agree upon it, who is the more worthy." The zealot solved the problem by deciding to argue not "in the comparative," but "in the privative," which led him to the conclusion that if a "heap of people" is "altogether unable or indign to govern," there is a "just cause of war for another nation, that is civil or policed, to subdue them."

Having thus unwittingly argued that wisdom can avoid the problem of consent by avoiding upward comparisons, Zebedaeus went on to contract in deed what he had just expanded in speech. For by saying that he wanted the mistake "banished" that he should have excluded "personal" tyrannies from his case, the zealot made the mistake of not excluding his own case from his case. That Zebedaeus indeed included his own case in his case becomes clear from the conclusion he drew from including personal tyrannies in his case. For whereas visible personal tyrannies know no visible constitutions and laws, the zealot concluded that when the "constitution" and the "fundamental customs and laws" of a state were "against the laws of nature and nations," a war against that state was "lawful."

Zebedaeus argued next that there was no "better ground to declare" that there were "nations or states or societies of people" against whom it was "lawful to make a war without a precedent injury or provocation" than to "look into the original donation of government. Observe it well," he emphasized, "especially the inducement or preface. Saith God: *Let us make man after our own image, and let him have dominion over the fishes of the sea, and the fowls of the air, and the beasts of the land, etc.*"

Observing especially the man who said that we should observe the author of the central of the thirteen biblical quotations that occur in *An Advertisement touching a Holy War*, we observe that it is exactly the men who say that God "donated" them a right to govern that tend to make a preventive war against them lawful. In the next sentence, Zebedaeus himself implicitly confirms this observation by saying that Vitoria extracted a "most true and divine aphorism" from the aforementioned biblical quotation: "*Non fundatur dominium nisi in imagine Dei*," which is the "charter of foundation" that makes it "the more easy to judge of the forfeiture or reseizure. Deface the image, and you devest the right." For in his famous *De Indis* (1537–1538), the Roman Catholic theologian Francisco de Vitoria had actually concluded both from the aforementioned and from many other biblical and non-biblical quotations and paraphrases employed by Zebedaeus that unbelief does not constitute a just title for the deprivation of the right to govern.[81]

On the foundation of Vitoria's aphorism, the zealot concluded that according to the "sound interpreters," the "right of government" ceases when the "image of God, of Natural Reason" is "totally or mostly defaced" (AHW 25, VII: 29–30). But interpreted "soundly" and therefore as it were *ex contrario zelotem*, the zealot's conclusion leads us to the conclusion that governments founded on the "image of God" lose the right to govern when they substantially "deface" the "image" of natural reason.[82] In fact, the zealot himself would have been compelled to draw this conclusion, since he was still arguing on the basis of the "laws of nature and nations," a basis, we notice in passing, which he never officially transcended.[83]

Having unofficially passed from the "laws of nature" to the "laws of nations," Zebedaeus stated that it would be a "great error, and a narrowness or straitness of mind, if any man think that nations have nothing to do with another, except there be either a union in sovereignty or a conjunction in pacts or leagues." For, he explained, there are "other bands of society, and implicit confederations," of which the highest is "the supreme and indissoluble consanguinity and society between men in general." Significantly enough, though, the zealot could only illustrate this general human "consanguinity and society" by having the apostle Paul call "to witness" the "heathen poet" Aratus, who said: "*We are all his generation*" (Acts 17: 22–8).[84]

Almost as if realizing the paucity of Christian witnesses to his argument, Zebedaeus went on to say that especially Christians, unto whom it was "revealed in particularity, that all men came from one lump of earth, and that two singular persons were the parents from whom all the generations of the world are descended, . . . ought to acknowledge that no nations are wholly aliens and strangers the one to the other," and should "not to be less charitable than the person introduced by the comic poet, *Homo sum, humani nihil a me alienum puto.*" It was the old Athenian Chremes who had justified his meddling into the affairs of Menedemus by saying that because he was a man, he held nothing human to be alien from him.[85] But when we call to mind that the courtier who by meddling into the affairs of the "good city" had demonstrated not to "refer all things to himself" was introduced by a man named after a comic poet, we find that Zebedaeus unwittingly exhorted his fellow-Christians to become "charitable" by starting to refer things to men in general, although he once again needed a heathen poet to make his point.

Returning to the "tacit league or confederation" of charitable men, Zebedaeus concluded that it was a league or confederation "against such routs and shoals of people, as have utterly degenerate from the laws of nature; as have in their very body and frame of state a monstrosity; and may be truly accounted common enemies and grievances of mankind; or disgraces and reproaches to human nature." But by adding that this was "not to be measured so much by the principles of jurists, as by *lex charitatis*; *lex proximi*; which includes the Samaritan as well as the Levite," he prepared the climax of his self-incriminating enterprise. For the fact that the Roman Catholic zealot did not include the priest among charitable men (cf. Luke 10: 25–37) was only an adumbration of the ultimate implication of his last words, namely that if to "deny" the "original laws" on which his "opinion" was "grounded" "were almost to be a schismatic in nature" (AHW 25, VII: 35–6), it was Zebedaeus himself who was to be branded an "enemy of mankind," if "measured" by the moral laws he denied by "grounding" his "opinion" on them.

Now that the war for the sake of "braying" Christianity had been proved possible, lawful, and obligatory, it remained only for Eusebius to speak about the "comparative." But although the "moderate Divine" is mentioned among the "Persons that speak," it is only his namesake Bishop Eusebius of Caesarea (263–339 AD) who speaks by being quoted. For in his first appearance Martius refers to the words that "Our Lord" "said from heaven to Constantine, *In hoc signo vince*" (AHW 7, VII: 18), words that mark the conversion of the Roman Emperor to the Christian faith in 312 AD. Eusebius relates that it was "at noon" that Constantine saw the trophy of a cross of light in the heavens, above the sun, bearing the inscription: "Conquer in this sign."[86] But although the warrior who was recruited in the dialogue that began at the "heat of the day" (AHW 5, VII: 18) was to conquer in the sign of the cross, he did not conquer in the spirit of the cross. Having observed that the existence of "schismatics in nature" made the enterprise schismatic by nature (cf. 1 Cor. 12: 25–7), Eusebius concluded that the braying of Christianity was not to result in a "reconcilement of schisms." The man named after the author of a *Historia Ecclesiastica* therefore refused, even "by way of mulct or pain," to contribute to the writing of an anti-ecclesiastical history.[87]

But although the "voice of moderation" decided to remain silent, it was through the silence of moderation that Bacon spoke his last words, which were words of moderation. For since it was by advertising "obedient madness" that the *Advertisement* protected "Athens" from other kinds of madness, it was by silencing a "moderate Divine" that the man who dedicated his advertisement of a holy war to a moderate divine demonstrated his moderation in transmitting matters divine. We find Bacon's moderation in speech confirmed in speech when we read that philosophy is "only sober" in the dialogue that advertises madness in deed. It was therefore by causing the madness of philosophy to be overwhelmed by the madness of philanthropy that our author proved, on the most profound level, the truth of his statement that "the greatest things are indebted to their beginnings" (DA 9.1, I: 837). For by convincing the greater part of thoughtful posterity that he was ultimately driven by the "anti-theological ire" a philosopher's historical narrative once traced back to the surface of Machiavelli's teaching, Bacon both prevented and impeded the access to his thought, the core of which is unbiblical, but also uncharitable.

Epilogue

> Nun gut, wer bist du denn? . . . Ein Teil von jener Kraft, Die stets das Böse will und stets das Gute schafft. . . .
>
> —Goethe: *Faust: Der Tragödie erster Teil*

This book has attempted to trace back the movement of thought that finds its commonly known expression in modern science's claim to philanthropy. At its very outset, however, the enterprise encountered the towering obstacle that the successful transformation of science into modern science had "almost dissolved and undone" the virtue of science, which is indispensable for scientific thought. Fortunately, therefore, it turned out to have been precisely the neutrality that had dictated science's self-instrumentalization that opened the way to its de-instrumentalization and therefore to the rediscovery of its virtue. Science can indeed be "turned to ambiguous uses." To the extent that our study of the scientist Francis Bacon caused us to discover science's benevolence, if not its true philanthropy, adversity thus "discovered" more than just the virtue of fortitude.

But if philosophy is to be judged by its fruits, it is not fruitful philosophic activity but the fruits of philosophy's activity that make philosophy benevolent.[1] And although the fruits of early modern philosophy's activity were to secure the access to fruitful philosophic activity by making philosophy seem to be inherently benevolent, fruits and fruitfulness became inseparable when the active "private good" of the philosophically active became dependent on the "communicative good," and philosophy came to understand itself primarily as a "parte or Member" of the "greater Bodye" of the Kingdom of Man, which was built on a twofold foundation.[2]

We need not wonder that the problem of the twofoldness of the benevolence of philosophy, which runs parallel to the problem of the twofoldness of good and the twofold roots of pleasure, has its strongest presence in the work in which Bacon identifies himself with the "initiates." For it is in the context of his presentation of the sixth part of his *Instauratio Magna*, which "discloses and sets forth" *Philosophia Secunda* (IMDO, I: 134) in contradistinction to *Philosophia Prima* as it indicates what philosophy originally meant and adumbrates philosophy as the mature Bacon understood and lived it[3]—it is in this context that Bacon makes the well-known statement that "nature can only be conquered by being obeyed." Because "ignorance of causes causes the effect to fail," the twin objectives human knowledge and human power "come together in one." That which "in contemplation is as the cause," is "in operation as the rule" (IMDO 29, I: 144; NO 1.3, I: 157). But when the conquest of nature becomes the highest goal of philosophy, *Philosophia Prima* becomes secondary to its primary task, which is "power measurement." Knowledge will only be desired for the sake of power, causes will only be contemplated for the sake of operation, and the natural will only be looked upon as a prefiguration of the man-made. The desire to conquer what was believed to be necessary will culminate in a desire to conquer necessity. But the true power of philosophy is dependent on an awareness of the power of necessity, as it is only on the basis of an awareness of the ultimate dependence of fortune on unconquerable chance that the philosopher can obtain the self-knowledge whose prospect keeps him in motion, even or precisely if his nature turns out to be at odds with the times in which he lives or the times for which he writes.

The almost natural tendency of a philosophy of power to weaken the power of philosophy is illustrated by the very first aphorism of *Novum Organum*, in which Bacon states that man "does and understands only as much as he has observed, in deed or in thought, concerning the order of nature" (NO 1.1, I: 157). For he who desires to act in deed will only observe in thought insofar as thought is conducive to a deed other than the deed of thought. He may dissect nature but he will not understand nature, because the possibility of understanding nature presupposes that the deed of thought is not restrained by the thought of deed.

Bacon indicates his awareness of the decisive difference between knowledge for the sake of knowledge and knowledge for the sake of power by observing that the road to human power and the road to human knowledge "lie very close together [coniunctissimae]," and "are almost [fere] identical" (NO 2.4, I: 229). But his statement that the "principal thing" that "above everything else" the initiates are after is the "contemplation of truth," which is "a thing worthier and loftier than all utility and magnitude of works" (NO 1.124, I: I: 217–8), did not prevent power as the goal of knowledge from overwhelming knowledge as the goal of knowledge. Under the circumstances, this could hardly have been otherwise considering that Bacon eulogized contemplation in his capacity as an initiate in the "science" that was to replace science.

The replacement of knowledge as the goal of knowledge by power as the goal of knowledge becomes somewhat more understandable in light of the preceding replacement of the knowing god of natural theology by the powerful God of the Bible. But the justification of power by philanthropy may have contributed even more decisively to the obliviation of the edifying potential of knowledge understood as a goal in itself. For in conformity with the fact that philanthropy is only a "subservient" virtue to be traced back to a "natural inclination," Bacon's Prometheus only taught man to start regarding himself as the "center of the world." The "reliefe of Mans estate" did not relieve man of the burden of designing his estate. The Kingdom of Man was already erosive before it started to erode. The "providence" that Prometheus breathed into man's rational soul may have been reasonable in light of the rule of unreasonable providence, it was also empty. In fact, it is hardly an exaggeration to say that it sowed the seeds of the rule of the last men in the Kingdom of Man, which is itself an "adventitious idol" (IMDO 14, I: 139; NO 1.44, I: 164).

In light of the fate of Aristotelianism, Bacon was justified in removing final causes and the concept of a Mind from his metaphysical doctrine. But he was compelled to couple the absence of final causes with the nominal presence of the biblical God. After man himself had become the reproducer of causes, it was only a small step to remove God from the fabric of things. For neither man nor the world can be the image of an absconded God. Whereas Bacon himself could still reap the singular benefits of "writing in the new," the philosophically potent recipients of the "new unmixed" were thus almost deprived of an instrument for measuring God and therefore themselves. Truth is not always the "daughter of time."

What accelerated the narrowing of man's contemplative horizon was the fact that measuring a conquered God was increasingly believed to have become obsolete. Philosophy's greatest political victory thus turned out to have been philosophy's greatest political defeat. It is true that Bacon supports his statement that it is the "contemplation of truth" that the "initiates" are after by saying that the initiates are "laying in the human intellect the foundations of a true exemplar of the world, as it is discovered and not as someone's own reason dictates it to be" (NO 1.124, I: 218). But this exemplar left no room for the exemplary, because it left no room for the dictates of reason; it "almost levelled wits and intellects." Salomon's House was indeed the instaured version of Solomon's temple; it was not inhabited by philosophers, as already appears from the false modesty with which the modern scientist declares that he knows almost nothing. For is it not precisely by making wholes out of parts that he satisfies his desire for knowledge of a certain whole? And is it not because his desire for the whole is anerotic that he believes the whole can be a whole without man being a whole first? Finally, is it not because he believes the whole of man to be merely the sum of man's evolved parts that he does not see the whole as a problem?[4]

The humbleness of the modern scientist may have replaced the humility of the believer, it certainly replaced the magnanimity whose "scientification" opens the way to science.[5] It is therefore of the utmost importance to thoroughly consider the significance of the fact that it is in the *Essay* on atheism that Bacon emphasizes that denying a God destroys "Magnanimity, and the Raising of Humane Nature" (E XVI "Of Atheisme," 53, VI: 414). But Orpheus sang the praises of the gods so loud that the twofold nature of his singing could no longer be discerned. His meditations on divine things were drowned by his laudation of the divinities, so that a Savoyard Vicar was needed to disentangle the two by substituting the cult of the gods for the praise of the gods.[6] But the fate of Bacon's Orpheus was irreversible. His sweet doctrines were too voluptuous to become pleasurable; they were too light to become heavy. They almost reversed the Baconian turn.[7]

This is not to say that d'Alembert's eulogy on Bacon is unfounded, although its twofold foundation provided Rousseau with the foundation for his ensuing silence. For the only way for Bacon to liberate science from faith was by replacing faith by faith in modern science. But what d'Alembert believed to be enlightenment, Rousseau knew to be obfuscation. Science that understands itself in essentially adversarial terms becomes its own adversary after having conquered its historical adversary. A mere reference to the fact that the public atheism of our times is founded solely on modern science may suffice as an illustration of the success of adversarial science.[8] We localize the root of this success in Bacon's insufficient articulation of the Antichrist's dependence on the paternity of the Father. For it was only by coming in the name of the Father that he could be received in the name of his father, who was the father of Baconianism. But after having been received by others than his wife, "Juno's suitor" grew together with his suit. Should a son of Baconian times thus decide to embark on a journey from Baconianism toward Bacon, he needs a discerning eye in case he encounters the secularizing agent of one of the founding agents of the movement called Modern Philosophy.

In the context of his discussion of the "Problem of Socrates," Friedrich Nietzsche states that "if one has the need of making a tyrant out of reason, as Socrates did, there must be no small danger that something else makes the tyrant."⁹ But even if it is true that by confining himself to defining or misdefining the Problem of Socrates Nietzsche overlooked pre-Nietzschean responses to the danger of the instrumentalization of reason, which is a danger coeval with political philosophy, his admonition that "[w]e know far from enough about Lord Bacon, the first realist in the great sense of the word, to know all he did, what he wanted, and what he experienced with himself," can be considered a first step toward defining the problem of Bacon.

Nietzsche himself in a way defines Bacon's realism by saying that the "strength to the most powerful reality of vision" presupposes the "most powerful strength to deed, to the outrageousness of deed, to crime."¹⁰ And Bacon did possess the strength that the realism of his vision presupposed, as is proved by the fact that he made his Prometheus found the doctrine of man's humanity on a crime against this very humanity. But it did not take long before the titan who was a god to man because he breathed "providence" into man's rational soul was overwhelmed by the physician who is divinized because he prolongs the life of man's irrational body. And whereas the architect of the human soul still invited the few to question his work of art in a vertical direction, the glory that befalls the workman of the human body only stimulates the many to expand bodily health horizontally, as in a way can be seen from the replacement of the missionary by the doctor without borders. But on a horizontal ladder the achievement of health is followed by boredom concealed as maintenance, indistinctiveness masked by cosmetics, and nihilism in the guise of aestheticism.

The tyranny against humanity puts an overemphasis on man's humanity in a somewhat different light. But the fact of the matter is that a humanity whose fullness is already given with the human body almost as of necessity leads to a flat humanitarianism that does not point beyond itself. And indeed, the "modern web" turned out already to have been cleansed before it had been fully spun, which goes a long way towards explaining why it is so difficult to trace back its threads to their spinners.

Bacon rightly remarks that the word "*Humanitie* (as it is used) is a little too light, to expresse" the virtue of philanthropy (E XIII "Of Goodnesse and Goodnesse of Nature," 39, VI: 403), seeing that what is called man's humanity only lies at the basis of his philanthropy. But it is also a little too heavy. For a workman spurred on by philanthropy became the god of humanity.¹¹ The problem of Bacon: the replacement of the Promethean by the Aesculapian.

"They are happie Men, whose *Natures* sort with their Vocations; Otherwise they may say, *Multùm Incola fuit Anima mea*: when they converse in those Things, they doe not Affect" (E XXXVIII "Of Nature in Men,"

120, VI: 470). As Bacon was not one of those very few philosophers who drew their own lives into what one might call the "eccentric center" of their political philosophy, we must take his statement on happiness as our guide when attempting to "call home his mind" from its "peregrinations" after having followed it on the peregrinations of which his writings give an account. If it is true that the things Bacon "affected" necessarily made him affect political things, and led him to "converse" even with those inimical to the things he affected; if it is true that when Bacon became a political philosopher, his active life and his alliance with political power became political actions of his contemplative life; if it is true that the political philosopher Bacon was always at home in his homelessness because he affected the political nature of the philosophic life that had come to determine the boundaries of his good; if it is true that when Bacon's love necessitated him to be absent from himself, he was able to "circumscribe and collect" himself even among enemies, and be at home wherever he could unify his "vocation" of actualizing his natural potential with his nature[12]; if all this is true, and if it is also true that Bacon's work was finished when Bacon finished his work and became who in a way he always had been, then Bacon must have been a happy man.

John Aubrey tells us that Hobbes told him that Bacon's death had been caused by an experiment. One day Bacon saw snow lying on the ground, and the thought came to his mind that flesh might be preserved in snow, as in salt. He bought a hen and had it eviscerated. As he was stuffing the body of the hen with snow, the snow so chilled him that he fell extremely ill. He was taken to the Earl of Arundel's house, where he was put in a bed that was warmed with a pan. But because it was a damp bed that had not been used for about a year, he caught such a cold that within two or three days he died of suffocation.[13] Divested of its sensationalism, Aubrey's account of Bacon's death contains one of the deepest accounts of Bacon's life that we possess, especially if it is considered in conjunction with the more reliable surmise that the Englishman "died from an overdose of inhaled nitre or opiates," taken in an attempt to prolong his own life.[14] For a life of experimenting will end while experimenting; its cause will cause its end.[15] We are reminded of the death of Pliny the Elder, who was burnt to death while investigating the workings of the erupting Mount Vesuvius.[16] But although with regard to the immediate cause of death this comparison is not misleading, we also hear the echo of the last words of the dying Socrates: "We still owe Aesculapius a cock."[17] For Bacon did not sacrifice to his god.

Notes

Abbreviations

1. *The Advancement of Learning. The Oxford Francis Bacon IV*, ed. Michael Kiernan (Oxford: Clarendon Press, 2000).

2. *Works* VII, 14; *Works* XIV, 436–7. Besides the fact that the *Advancement of Learning* was entered in the Stationers' Register in two separate entries, Book I on August 19, 1605, and Book II on September 19, 1605 (Kiernan, *Advancement of Learning*, lvii–lviii), there is also some inner-textual evidence from which it appears that the two books were written at different times, and that the second book was written with less care than the first book (see *Works* III, 256; *Works* X, 88, 248). In a letter (1605) to the Earl of Salisbury (*Works* X, 253), Bacon confesses that the *Advancement of Learning* was a "work" of his "vacant time, which if it had been more, the work had been better." Considering the fact that later changes and additions were mainly limited to the second book, the bearing of Bacon's confession seems to be mainly limited to the second book. An explanation for our author's haste in publishing *The twoo Bookes* may be found in the fact that he wanted to seize the occasion of the learned King James' accession to the throne (1603).

3. *The Essayes or Counsels, Civill and Morall. The Oxford Francis Bacon XV*, ed. Michael Kiernan (Oxford: Clarendon Press, 2006).

Introduction

1. AL 1.48, 31–2, III: 294–5; DSV XXVI "Prometheus," VI: 668–76; IMPr. 4 and 5, I: 131–2; NO 1.129, I: 221 ff.; DA 7.3, I: 740 ff.; E XIII "Of Goodnesse and Goodnesse of Nature," 38 ff., VI: 403 ff.

2. Antonio Pérez-Ramos, *Francis Bacon's Idea of Science and the Maker's Knowledge Tradition* (Oxford: Clarendon Press, 1988), 29–30.

3. The substantial confusion underlying what seems to be a merely nominal distinction can best be illustrated by the example of a scholar who claims to be criticizing what he calls the "segmentation" of the study of Bacon into "various specialised and technical interests." For although this scholar mentions "science,

philosophy and history" as examples of such fields of specialized and technical study as are "removed from the province of the general reader," he also claims that in all of his works, Bacon indulges in the "typically Renaissance" wish to "improve the amount and quality of knowledge available to mankind, so as to alleviate the miseries of human existence." As general readers, we cannot help but wonder, though, how we are to judge these claims without first becoming historians, philosophers, and scientists. Brian Vickers, *Francis Bacon and Renaissance Prose* (Cambridge: Cambridge UP, 1968), 257 ff.; Vickers, *Francis Bacon* (Harlow: Longman Gr. LTD, 1978), 20; Vickers, ed., *Francis Bacon: A Critical Edition of the Major Works* (Oxford: Oxford UP, 1996), xx, xxviii.

4. Because Paolo Rossi's work "has opened a new phase in Bacon studies," it is "perhaps the most important work on Bacon in [the twentieth] century" (Pérez-Ramos, *Bacon's Idea of Science*, 30, fn. 55).

5. Paolo Rossi, "Baconianism," in *The Dictionary of the History of Ideas: Studies of Selected Pivotal Ideas*, ed. Philip P. Wiener, vol. 1 (New York: Charles Scribner's Sons, 1973–4), 173–8; Rossi, *From Magic to Science*, transl. S. Rabinovitch (Chicago: University of Chicago Press, 1968), xiii, 67; Rossi, *Bacon's Idea of Science*, in *The Cambridge Companion to Bacon*, ed. Markku Peltonen (Cambridge/New York: Cambridge UP, 1996), 37; Rossi, *The Birth of Modern Science*, transl. C. de Nardi Ipsen (Oxford: Blackwell Publishers, 2000), 49.

6. Perez Zagorin, *Francis Bacon* (Princeton: Princeton UP, 1998), x, 24, 27–8, 224.

7. Lisa Jardine and Alan Stewart, *Hostage to Fortune: The Troubled Life of Francis Bacon 1561–1626* (London: Phoenix Giant, 1999), 19. Jardine elsewhere describes Bacon as a "well-educated English gentleman" but "not a professional philosopher" (Lisa Jardine, *Francis Bacon: Discovery and the Art of Discourse* (London/New York: Cambridge UP, 1974), 1, 7).

8. William Rawley, "The Life of the Honourable Author," in *Works* I, 15.

9. Markku Peltonen, "Introduction," in Peltonen, *Cambridge Companion*, 10.

10. Stephen Gaukroger, *Francis Bacon and the Transformation of Early-Modern Philosophy* (Cambridge/New York: Cambridge UP, 2001), 1, 5, 52, 221–2, 226.

11. Alfred North Whitehead, *Science and the Modern World* (New York: Free Press, 1967), 16.

12. Stephen Gaukroger, *The Emergence of a Scientific Culture: Science and the Shaping of Modernity 1210–1685* (Oxford: Clarendon Press, 2006), 3; and: *Bacon and the Transformation of Early-Modern Philosophy*, 91.

13. Charles Webster, *The Great Instauration: Science, Medicine and Reform 1626–1660* (London: Duckworth, 1975), 336; see also 12, 25, 383, 514.

14. Joseph de Maistre, *Examen de la Philosophie de Bacon; où l'on traite différentes questions de Philosophie Rationelle*, vol. 2 (Paris/Lyon: Poussielgue-Rusand, 1836), 267 (my translation). Since the moralist Macaulay makes Bacon's philosophic relevance ultimately dependent on the latter's moral character, his condemnation of Bacon in a way runs parallel to de Maistre's judgment: "Had his life been passed in literary retirement, he would, in all probability, have deserved to be considered,

not only as a great philosopher, but as a worthy and good-natured member of society" (Thomas Babington Macaulay, "Lord Bacon," in *Critical and Historical Essays, contributed to the Edinburgh Review*, vol. 2, 5th ed. (London: Longman et al., 1848), 373; see also 321–2, 335–6, 429).

15. Denis Diderot, *Prospectus*, in *Œuvres Complètes de Diderot*, vol. 13 (Paris: Garnier Frères, 1876), 133–4; Jean le Rond d'Alembert, "Discours Préliminaire," par. 101–3, in *Encyclopédie ou Dictionnaire Raisonnée des Sciences, des Arts et des Métiers*, vol. 1 (Paris, 1751), xxiv–xxv (my translation).

16. Voltaire, *Lettres Philosophiques*, "Douzième Lettre: Sur le Chancelier Bacon," par. 4, 10, 21, in *Œuvres Complètes de Voltaire*, new ed., vol. 22 (Paris, 1879), 116–21 (my translation). David Hume seems to have held a similar view about Bacon, as appears from his statement that if Bacon is considered "merely as an author and philosopher," he is to be looked on as "very estimable," yet as inferior to his contemporary Galileo, since he only "pointed out at a distance the road to true philosophy" (Hume, *The History of England*, vol. 4 (Boston: Little, 1854), 376–7). The chapter on "Baconisme ou Philosophie de Bacon" in the *Encyclopédie* [vol. 2, Paris, 1751, 8 ff.] seems to be mainly a paraphrase of Voltaire's *Lettre*, although it adds that Bacon was one of those "grand architects" who "cannot bring themselves to follow in others' footsteps, and begin by destroying everything, before raising their edifice according to a wholly new design" (my translation).

17. Jean-Jacques Rousseau, *Discours sur les sciences et les arts*, Sec. partie, par. 24 (*Œuvres Complètes de Jean-Jacques Rousseau*, vol. 3, Bibliothèque de la Pléiade (Paris: Gallimard, 1964), 29; my translation).

18. For this reason, Peter Pesic is right in saying that the "intense . . . tides of adoration and revulsion indicate not only how important Bacon was, but also how disturbing he continues to be" (Pesic, *Labyrinth: A Search for the Hidden Meaning of Science* (Cambridge: MIT Press, 2001), 5).

19. Hans Blumenberg, *Die Legitimität der Neuzeit* (Frankfurt am Main: Suhrkamp, 1996), 150–1; cf. also 86, 129, 152 (my translations).

20. Ibid., 86.

21. Ibid., 129, 554. This would make it almost ironical that Blumenberg devised his conception of worldliness in opposition to the secularization-thesis of Karl Löwith. Cf. ibid., 35 ff. with Karl Löwith, *Meaning in History* (Chicago: University of Chicago Press, 1949), 200–3.

22. Ibid., 75–6. It seems to be more than a coincidence that Blumenberg turns out not to have been a very careful reader of Bacon. In trying to weaken philosophy's apparent claim of having inaugurated Modernity as an absolute and timeless beginning in time, the historian introduces Bacon as a thinker to whom history was nothing but a "system of idols, which come to an end without this end being understandable as a consequence of their validity, without their disempowerment being understandable from the intolerability of their reign." Blumenberg summarizes: "The characteristics of self-assertion are concealed in favour of the evidentness of an abiogenesis [Urzeugung]; the crisis disappears in the dark of a past which can only have been a background for the new light" (ibid., 159–60). But

to say nothing of the difficulty of understanding how even a timeless "abiogenesis" in time could free itself from the burden of being and becoming historical, it is Bacon himself who emphasizes that at least three of the four idols can "never be wholly torn out" (IMDO 15, I: 139; DA 5.4, I: 643), and that truth is not to be sought in the "felicity of any period of time" (NO 1.56, I: 170). In justification of his thesis that Bacon concealed the characteristics of self-assertion on which he was dialectically dependent, Blumenberg discusses Bacon's contribution to the justification of worldly curiosity "on the basis of the reframing [Umdisponierung] of biblical paradise into a utopia of the human goal of history." But could this "reframing" have been anything else than a secularization of some sort? Regardless of the answer to this question, Blumenberg's Bacon had to "retain the theological premises in order to derive a new legitimitacy from them." To support this claim, Blumenberg almost solely refers to *Valerius Terminus*, an early (1603), lapidary, and posthumously published (1734) fragment, which contains more explicit and at first sight more radical notions than Bacon's later works, and which for good reasons was not published by the mature Bacon himself. We therefore need not bother that Blumenberg claims to have disclosed some inconsistencies in Bacon that did not bother Bacon himself (ibid., 447–56).

23. See for suggestions along similar lines Robert Pippin, "Blumenberg and the Modernity Problem," *Review of Metaphysics* 40, no. 3 (1987), 537, 542 ff., 554, and Richard Kennington, "Blumenberg and the Legitimacy of the Modern Age," in *The Ambiguous Legacy of the Enlightenment*, ed. W. A. Rusher (Lanham (MD): UP of America, 1995), 22–37.

24. An analogous line of argument was applied to Bacon by Charles Whitney, a scholar who describes Modernity as "discontinuity between tradition and innovation," and who characterizes the works of Bacon, a modern captive of Modernity, as captured by these discontinuities "or contradictions." Whitney encourages us to "look beyond Bacon's self-assurance to the struggle taking place in his work between different discourses and ideologies." In a mood of fashionable postmodern dialectical playfulness he adds that, "beyond the apparent limits" that Bacon sets on the influence of tradition, "there is an intertextuality suggesting both unspoken commitment and unspoken bondage" to his work. According to Whitney, Bacon discontinuously and semiconsciously oscillated between his revolutionary historical self-consciousness and his contribution to the reform of philosophy's "traditional matrix." Fortunately enough, though, Whitney's higher matrix allows him to conclude that the apparent incompleteness of many of Bacon's works providentially "conceals a protective turning away, a conscious or unconscious recognition that elaborating the plan will compromise its freshness and originality, and expose it as a function of knowledge already institutionalized." But regardless of Whitney's admirable qualities as an analyst of Bacon's internal struggles, the results of his analysis conceptually presuppose his higher historical stance. The analysis itself therefore cannot but remain problematic until its author's higher historical stance has become problematic. Charles Whitney, *Francis Bacon and Modernity* (New Haven: Yale UP, 1986), 4, 10 ff., 191.

25. Michael Allen Gillespie, *The Theological Origins of Modernity* (Chicago: University of Chicago Press, 2008), 12.
26. Ibid., 14 ff., 29, 35; cf. also 22, 262, 274.
27. Ibid., 37 ff.
28. See Aristotle, *Politica*, 1253a2–3; but cf. 1253a9.
29. Cf. Plato, *Euthyphro*, 3d6–9, and Leo Strauss, "An Untitled Lecture on Plato's Euthyphron," *Interpretation* 24, no. 1 (1996), 5–23, 7 ff. Cf. also Howard B. White, *Peace among the Willows: The Political Philosophy of Francis Bacon* (The Hague: Martinus Nijhoff, 1968), 11: "[A]ny careful and systematic consideration of needs of society, on Bacon's part, would make Bacon a political philosopher."
30. Cf. White, *Peace among the Willows*, 203: "[T]he highest form of *philanthropia* is that of the philosopher," who "extends his love towards the most people . . ." with Leo Strauss, *On Tyranny: Revised and Expanded Edition*, eds. Victor Gourevitch and Michael S. Roth (New York: Free Press, 1991), 88 ff., 197–8, and John Aubrey, *Francis Bacon*, in *Aubrey's Brief Lives*, ed. Oliver Lawson Dick (Boston: David R. Godine, 1999), 9: "In short, all that were great and good loved and honoured [Bacon]."
31. Cf. Plato, *Symposium*, 204a1–8.
32. Jonathan Swift, *The Battel of the Books*, in Swift, *A Tale of a Tub and Other Works*, ed. Marcus Walsh (Cambridge: Cambridge UP, 2010), par. 6 and 14 (147, 151–2).
33. Cf. Leo Strauss, *Thoughts on Machiavelli* (Chicago: University of Chicago Press, 1958), 231, and Strauss, *Philosophie und Gesetz: Beiträge zum Verständnis Maimunis und seiner Vorläufer*, in Strauss, *Philosophie und Gesetz – Frühe Schriften*, vol. 2 of *Gesammelte Schriften*, ed. Heinrich Meier (Stuttgart: Metzler, 1997), 30–1, fn. 2. Cf. also Heinrich Meier, *Das theologisch-politische Problem: Zum Thema von Leo Strauss* (Stuttgart: Metzler, 2003), 22 ff.

Chapter 1

1. *Works* XIV, 539.
2. Hans-Georg Gadamer, *Wahrheit und Methode: Grundzüge einer philosophischen Hermeneutik*, in *Gesammelte Werke*, vol. 1 (Tübingen: Mohr (Siebeck), 1990), 397 (my translation).
3. Rawley, *The Life of the Honourable Author*, *Works* I, 11.
4. See Gadamer, *Wahrheit und Methode*, 296–312. Cf. Leo Strauss and Hans-Georg Gadamer, "Correspondence concerning Wahrheit und Methode," *Independent Journal of Philosophy* 2 (1978), 5–12, esp. 6 (pt. 1 and 2) and 9 (pt. 1 and 2).
5. Gadamer, *Wahrheit und Methode*, 299.
6. Ibid., 270–6, 281–2, 301, 304. Cf. Friedrich Nietzsche, *Unzeitgemässe Betrachtungen; Zweites Stück: Vom Nutzen und Nachtheil der Historie für das Leben*, 1.1–5, 1.6–12, 5.3, and 9.7.
7. *Works* I, 4.

8. In his epistle to the English version of Jewel's *Apologia*, the Archbishop of Canterbury Matthew Parker emphasizes that by translating the *Apologia* Lady Bacon "expressed an acceptable dutye to the glorye of God," and "deserved well of this Churche of Christe." John Jewel, *An Apologie or Answere in Defence of the Churche of Englande* (London, 1564).

9. Jardine and Stewart, *Hostage to Fortune*, 33.

10. Macaulay, *Lord Bacon*, 296.

11. Richard Bancroft, *A Sermon preached at Paules Crosse*, February 9, 1589 (London, 1589), 51.

12. *Works* VIII, 113.

13. Based on the chorus of Seneca's *Oedipus*: "quidquid excessit modum pendet instabili loco" (909–10).

14. *Works* VIII, 95.

15. Markku Peltonen, a scholar who is interested in the context of Bacon's ideas, considers it "essential to an understanding of Bacon's political writings to recognize that he was preoccupied with the same range of questions as his contemporaries." But by directly comparing essays with letters, letters with fables, fables with dialogues, and dialogues with professional tracts, Peltonen fails to read Bacon's ideas in context. Markku Peltonen, *Classical Humanism and Republicanism in English Political Thought 1570–1640* (Cambridge/New York: Cambridge UP, 1995), 143, 157, 194–8.

16. Tacitus relates the quarrel between Lyons and Vienna. Galba had confiscated the revenues of Lyons, whereas he had conferred many honors on Vienna. The inhabitants of Lyons tried to incite the individual Roman soldiers to destroy Vienna, and they presented a public petition to the end that the army would march to avenge them, and that it would destroy the headquarters of the Gallic war. They brought matters to such a pass that even the ambassadors and the party leaders despaired of quenching the army's anger. But since the Viennese were not ignorant of the danger they found themselves in, they met the approaching army covered with olive branches and headbands. By seizing their weapons, knees, and feet in supplication they succeeded in appeasing the soldiers. Fabius Valens moreover presented each of the soldiers with three hundred sestertii. Thus the antiquity and dignity of the Viennese colony had its weight ["Tum vetustas dignitasque coloniae valuit"]. There was, however, a persistent rumor that Valens himself had been bought with a heavy bribe (*Historiae*, I. 65–6). In Bacon's version, a certain colony asserted the right of asylum before the senate. Since the Emperor and the senate did not listen to the colony's ambassadors impartially, they offered Titus Vinius a good sum of money to support them, which resulted in them obtaining their goal. After misquoting Tacitus ("Tum dignitas et antiquitas coloniae valuit"), Bacon explains that what Tacitus' words meant was that the "arguments which had appeared light before, had gained new weight by the bribe." Bacon goes on to criticize an unnamed critic, whom he only describes as "not one of the meanest," for expunging the word *tum* and replacing it by *tantum*. But the unnamed critic, who turns out to be Justus Lipsius, did not replace *tum* by *tantum* in his edition

of Tacitus' works. He only added the comment: "Fortè, Tatum." Ironically enough, "Tatum" probably was a printing error. Justus Lipsius, *In Cornelii Tacit. Hist. Lib. Notae*, in Cornelius Tacitus, *Annalium et Historiarum Libri qui extant*, edited by Justus Lipsius (Lyon, 1576), 17 (cf. 562, l.19). In later editions, the comment was completely removed (e.g., Tacitus, C. Cornelius. Opera quae Exstant, edited by Justus Lipsius (Antwerp, 1585), 153).

17. Bacon uses the example of Acts 9: 25. The Jews, who had taken counsel to kill Saul, watched the gates day and night to carry out their intention. The disciples took Saul by night and let him down by the wall in a basket (Acts 9: 23–5). Although the Vulgate has: "submittentes in sporta," Bacon criticizes an unnamed priest—whom I have been unable to identify—for apparently having substituted "*Demissus est per portam*" for "*Demissus est per sportam*," because "*sporta*" was a "hard word, and out of his reading."

18. Spedding overlooked the blunder in the *Advancement of Learning*, as appears from the fact that he only comments that "this illustration, which, as reflecting on a Priest might have been offensive to Rome," was replaced by Bacon with another in *De Augmentis* (*Works* III, 414, fn. 2). Wright did notice the blunder, but he concluded that it "leaves no room for the point of the story as Bacon tells it." What Wright failed to notice, however, is that it is only the blunder that makes the point of the story (William Aldis Wright, ed., *Bacon: The Advancement of Learning* (Oxford: Clarendon Press, 1869), 306). Kiernan also noticed the blunder, but what he failed to notice was its intentional character, which would have become obvious had he compared it with its substitute in *De Augmentis*. Instead, he only remarks that "the muffed biblical quotation may have been reason enough to substitute anecdotes" (Kiernan, *Advancement of Learning*, 326–7). With regard to the blunder in *De Augmentis*, Ellis assumes that Bacon had "an imperfect memory of the passage" (*Works* I, 708, fn. 1).

19. The best example is Benjamin Farrington, who translated *Temporis Partus Masculus*, *Cogita et Visa* and *Redargutio Philosophiarum* into English (Farrington, *The Philosophy of Francis Bacon: An Essay on its Development from 1603 to 1609 with New Translations of Fundamental Texts* (Chicago: University of Chicago Press, 1966)). Although Farrington overlooks many of Bacon's misquotations, he comments extensively on Bacon's misquotation of Daniel 12:4 (cf. 132 with RP, III: 584), which he takes to be an "excellent example of Bacon's habit of quoting from memory." He only wonders why Bacon also misquoted Daniel 12:4 on the frontispiece of the *Instauratio Magna* (I: 119), "where one might expect special care." We, on the other hand, do not wonder why Farrington's translations seem not to have been rendered with special care.

20. Cf. AL 1.17, 15, III: 275 with Livy, *Ab Urbe Condita, Praef.*

21. In order to compare the king of Spain with Perseus of Macedon, Bacon quotes Livy (*Ab Urbe Condita*, XLII.18: "quem non iustum modo apparare bellum regio animo, sed per omnia clandestina grassari scelera latrociniorum ac ueneficiorum cernebant"), but he omits a passage ("modo apparare"). It was, however, only by omitting this passage that Bacon could adapt the Livy quotation to the case

in hand, since the Spanish king had "for a time laid aside the prosecution of his attempts against this [England's, TvM] realm by open forces." *A True Report on the Detestable Treason, intended by Dr. Rodrigo Lopez* (1594), *Works* VIII, 275–6.

22. William A. Sessions, "Francis Bacon and the Classics: the Discovery of Discovery," in *Francis Bacon's Legacy of Texts: "The Art of Discovery grows with Discovery,"* ed. William A. Sessions (New York: AMS Press, 1990), 241.

23. *Works* VII, 194.

24. Henry Pott, *The Promus of Formularies and Elegancies by Francis Bacon* (London: Longmans et al., 1883). Cf. 323/334 (nos. 985/1026) with Ovid, *Ars Amatoria*, II. 312 ("nec vultu destrue dicta tuo"), and DA 8.1, I: 747. Bacon replaces "dicta" by "verba," and he uses the misquotation in the context of his discussion of the prudence of conversation. The original context of the Ovid quotation may already have induced Bacon to distort the quotation in his commonplace book, seeing that Ovid added: "Si latet, ars prodest" (*Ars Am.*, II. 313). Interestingly enough, though, the other *Ars Amatoria*-quotations (333–6, nos. 1020–30) are all quoted correctly, but are never used by Bacon.

25. Cf. 354 (no. 1088): ". . . superanda est omnis fortuna ferendo" (idem Virgil, *Aeneid*, V. 710) with AL 2, 147, III: 433: "Vincenda est omnis fortuna ferendo." Having observed that, although fortune itself is "without our command," with regard to the "Conditions" of fortune "our worke" is only "limited and tied," Bacon misquotes Virgil, before emphasizing that when he speaks of the suffering that conquers fortune he speaks of "wise and industrious sufferinge." The misquotation occurs in the context of an argument on the limits within which "Morall Philosophy," while giving "constant attention to the doctrines of Diuinity," may "yeeld of her selfe . . . sound and profitable directions" regarding the "cure of mens Mindes" (AL 2, 146–7, III: 433–4).

26. The first book is entitled: "Notae omnifariae *ex conceptu proprio* exceptae, sparsum ut occurrunt et cursim ad memoria tantum, sive Civilis et Contemplator Cursor." The title of the second book is: "*Contemplationes digniores* sive Notae omnifariae *ex conceptu proprio* per otiu exceptae, et *adhibita deliberatione descriptae*; sive Civilis et Contemplator Sedatus." The notes contained in the second book seem to be of a more fundamental nature, since they were written down not only from a particular conception, but also on the basis of a specific deliberation. That Bacon also took the original context of his quotations and misquotations into consideration appears from what he says about the third book. For the "principall use" of this book is to "receyve such parts and passages of Authors" as he himself "*shall note and underline in the bookes themselves.*" The fourth book is "meerely a como place book." *Comentarius Solutus sive Pandecta, sive Ancilla Memoriae*, in *Works* XI, 59–60 (emphases mine).

27. In the *Advancement of Learning*, Bacon gives evidence of his awareness of the accidentality of prudence by redefining "Methode of Tradition" as "Wisedome of Tradition" at the end of the chapter on the prudence of transmission (AL 2, 122 and 126, III: 403 and 407). Cf. Plato, *Meno*, 88c6–d3 with Strauss and Gadamer, *Correspondence*, 5–6, and Gadamer, *Wahrheit und Methode*, 404.

28. Bacon describes the origins of the intellectual process in the following way: The sense, which is the "door of the intellect," is only struck by individuals; the images of these individuals, in other words, the impressions they make on the sense, fix themselves in memory; they are reflected on and ruminated by the human soul, which "simply reviews them, makes playful imitations of them, or arranges their composition and division." From these three fountains: memory, imagination, and reason, flow the three emanations history, poetry, and philosophy (DA 2.1, I: 494–5).

29. Further on in the chapter, Bacon narrows down or reduces the three objects of parabolic covering to philosophy (DA 2.13, I: 521).

30. After distinguishing between the exoteric and the acroamatic method, Bacon takes the two methods together and refers to them as the "acroamatic or enigmatic" (DA 6.2, I: 664–5).

31. Although it seems to be more than a coincidence that Bacon places his discussion of ciphers in the chapter immediately preceding the chapter in which he discusses exoteric–esoteric writing, it is of great importance for a proper understanding of the nature of exoteric–esoteric writing not to confuse the two. The principal virtues of ciphers are that they are "impossible to decipher" and "do not arouse suspicion." To elude examination completely, one should have two alphabets, one of true letters, and one of nonsignificants, and two epistles, an exterior and an interior epistle. Since the highest degree of a cipher is to make "anything signify anything," thoughts can be communicated by means of "any object that is capable of two differences" (DA 6.1, I: 658 ff.). But although ciphers and exoteric–esoteric writing partly overlap in the case of censorship, they cannot be equated, since the exterior of the cipher does not signify anything, and the interior can only be read by those who are part of the convention that decided on the meaning of the true letters that constitute the interior.

32. The misunderstanding of exoteric–esoteric writing as a relic of medieval occultism has done great damage to the proper understanding of both the subject matter itself and Bacon's art of writing. After linking it to occultism, hermeticism, cabala, alchemy, and astrology, Perez Zagorin describes esotericism as the "belief in the survival of a wisdom known to pagan sages of antiquity that could bring spiritual insight and purification to initiates and endow them with a power over things." It is this undiscerning reading of the meaning of esotericism that goes a long way toward explaining why Zagorin has not been "able to discover" any "hint" that Bacon "concealed in any of his writings an esoteric message . . . directed to a . . . discerning reader." Perez Zagorin, *Ways of Lying: Dissimulation, Persecution, and Conformity in Early Modern Europe* (Cambridge: Harvard UP, 1990), 257 ff., 271 ff.; and Zagorin, *Francis Bacon*, 43–4, 49, 183–4, 264 (fn. 56).

33. John Toland, *Clidophorus: or of the Exoteric and Esoteric Philosophy*, in Toland, *Tetradymus* (London, 1720), 94, par. XIII. See also Toland, *Pantheisticon* (London, 1751), 99: Esotericism is "practiced not by the Antients alone; for to declare the Truth, it is more in Use among the Moderns, although they profess it is less allowed." The entry in the *Encyclopédie*, entitled "Exotérique & Esotérique"

(vol. 6 (Paris, 1756), 273–4), says that exoteric–esoteric writing as it was used by ancient philosophers was regarded only by "la *plûpart* des modernes" as "un plaisir ridicule, fondé sur le mystère, ou comme une petitesse d'esprit qui cherchoit à tromper" (emphasis mine). Lessing states that Leibniz's defence of the belief in eternal damnation was identical to "was alle alte Philosophen in ihrem exoterischen Vortrage zu tun pflegten," in that he made a distinction between an exoteric and an esoteric manner of teaching and writing. The irony in Lessing's addition that "freylich unsere neuesten Philosophen viel zu weise geworden sind" for this "Klugheit" of Leibniz's is too apparent to require further explanation (Gotthold Ephraim Lessing, *Leibniz von den ewigen Strafen*, in Lessing, *Werke 1770–1773*, vol. 7 of *Werke und Briefe in zwölf Bänden*, eds. W. Barner et al (Frankfurt am Main: Deutscher Klassiker Verlag, 2000), 483). That Leibniz was indeed aware of the distinction that was to be repeated by Nietzsche (*Jenseits von Gut und Böse; Zweites Hauptstück: der freie Geist*, Aph. 30) is proved by the fact that he makes Theophilus say that the ancients distinguished "entre la manière d'écrire exotérique, c'est-à-dire populaire, et l'acroamatique, qui est pour ceux qui s'occupent à découvrir la vérité" (Gottfried Wilhelm Leibniz, *Nouveaux Essais sur l'Entendement Humain*, Book 2, ch. 29, par. 12, in *Œuvres Philosophiques de Leibniz*, vol. 1, 2nd ed. (Paris, 1900), 219). In the two central sections (XVI and XVII) of *De Stilo Philosophico Nizolii*, Leibniz extends his discussion of the distinction between the exoteric and the acroamatic to the moderns. This is the same Leibniz who in his laudation on Bacon said: "Merito *Verulamium* magni facimus. Etiam obscuriora ejus habent aliquid in recessu" ("Epistolae mutuae G. G. Leibnitii et. Fr. Guil. Bierlingii," in Leibniz, *Opera Philologica*, vol. 5 of *Opera Omnia*, ed. Ludovicus Dutens (Geneva, 1768), Epistola V, Responsio Leibnitii, 368, V).

34. Being himself a loyal son of the Father, de Maistre was able to observe that "Bacon citant la Bible, on peut être certain . . . qu'il est sur le point de blasphémer" (*Examen de la Philosophie de Bacon*, vol. 2, 121).

35. See Francis Bacon, *De Sapientia Veterum Liber* (London, 1609 and 1617), 125. See also Heidi Studer, *Grapes Ill-Trodden: Francis Bacon and the Wisdom of the Ancients*, unpubl. diss. (University of Toronto, 1992), 270, fn. 564.

36. Moses Maimonides, *The Guide of the Perplexed*, transl. Shlomo Pines (Chicago: University of Chicago Press, 1963), 12.

37. But cf. Niccolò Machiavelli, *Discorsi Sopra la Prima Deca di Tito Livio*, *Proemio*, vol. 1, (Rome: Salerno Ed., 2001), 8: "E benché questa impresa sia *difficile*, nondimanco . . . credo portarlo in modo che ad un altro restarà *breve* cammino a condurlo a loco destinato" (emphases mine).

38. We therefore only qualifiedly agree with Lemmi, according to whom Bacon "accepted Natale Conti as the leading light on the subject" (Charles W. Lemmi, *The Classic Deities in Bacon: A study in Mythological Symbolism* (New York: Octagon Books, 1971), 45).

39. With regard to the conformity in the names and the indications the names provide, Bacon gives the examples of Typhon, Pan, and Nemesis, meaning "swelling," "universe," and "revenge" (DSV Pr., VI: 626). In the subtitles of the

fables themselves, however, Typhon, Pan, and Nemesis are described as "rebellion," "nature," and the "vicissitude of things" (DSV II "Typhon," VI: 630; DSV VI "Pan," VI: 635 and DSV XXII "Nemesis," VI: 662). Apparently, the names in themselves do not suffice to unshadow the things signified.

40. Cf. DA 8.2, I: 769: "Fables were formerly substitutes and supplements for examples, but now that times abound with history, one aims more straightly and alacritously at a mark that is alive," with CV 11, III: 600: "The path to true knowledge through fictions was closed long ago, or at least narrowed."

41. Leo Strauss, *Persecution and the Art of Writing* (Glencoe (IL): Free Press, 1952), 17–8, 21, 24–5, 32–7. Cf. also Alfarabi, *The Philosophy of Plato*, in Alfarabi, *Philosophy of Plato and Aristotle*, rev. ed., transl. Mushin Mahdi (Ithaca: Cornell UP, 2001), 67 (x. 38).

42. Letter (1607–8) to Tobie Matthew, *Works* XI, 9. See also Letter (February 17, 1610) to Tobie Matthew, *Works* XI, 144–5.

43. Letter (1605) to Tobie Matthew, *Works* X, 255–6.

44. Letter (1623) to James, *Works* XIV, 436.

45. Marta Fattori, "Sir Francis Bacon and the Holy Office," *British Journal for the History of Philosophy* 13, no. 1 (2005), 21–49. According to the Dominican Giulio Maria Bianchi, who was one of the consultants, the "author of the book has attempted with remarkable ability to hide the errors scattered throughout his work, and he expressed himself with such an adroit circumspection that it is difficult to criticise his exposition" (44).

46. Jonathan Swift, *A Tritical Essay upon the Faculties of the Mind*, in *The Writings of Jonathan Swift*, eds. R. A. Greenberg and W. B. Piper (New York: Norton, 1973), 422–6.

47. Stanley Fish is right in drawing our attention to the form of the *Essays*, and, more specifically, to what he calls the "*experience* that form provides." He is also right in saying that the *Essays* "are to be read not as a series of encapsulations or expressions, but as a refining process that is being enacted by the reader," a refining process in which the reader transforms the *Essay* "from a vehicle whose form is designed to secure belief into an instrument of inquiry and examination." But by claiming that it is the experience of the form, "rather than the materials out of which it is composed," that is "scientific about the *Essays*," Fish leads us to suspect that he has paid insufficient attention to the matter underlying the form. Stanley E. Fish, "Georgics of the Mind: The Experience of Bacon's Essays," in Fish, *Self-consuming Artifacts: the Experience of Seventeenth-Century Literature* (Berkeley: University of California Press, 1972), 81, 92, 98.

48. William Blake comments on this passage in the following way: "What Bacon calls Lies is Truth itself." William Blake, *Annotations to Bacon*, in *Blake: Complete Writings*, ed. Geoffrey Keynes (London: Oxford UP, 1969), 396–410, 397.

49. See Lucian, *Philopseudes sive Incredulus*, 1–4, 10, 32, 40.

50. Cf. Jean-Jacques Rousseau, *Les rêveries du Promeneur Solitaire*, Quatrième Promenade, par. 5–23, par. 23, 19, and 5 (*Œuvres Complètes de Jean-Jacques Rousseau*, vol. 1, Bibliothèque de la Pléiade (Paris: Gallimard, 1959), 1025, 1030–1):

The justice of the "homme vrai" consists in his "rendre fidèlement à chacun ce qui lui est dû en choses qui sont véritablement siennes." Everything which, "contraire à la vérité, blesse la justice en quelque façon que ce soit, c'est mensonge," or, in other words, "cacher une vérité que l'on doit manifester."

51. Cf. Michel de Montaigne, *Essais*, II.XVIII *Du Démentir*, par. 16 (ed. Pierre Villey, new ed., vol. 2 (Paris: PUF, 1992), 666): "C'est un vilein vice que le mentir et qu'un ancien peint bien honteusement quand il dict que c'est donner tesmoignage de *mespriser* Dieu, et quand et quand de *craindre* les hommes. Il n'est pas possible d'en representer plus richement l'horreur, la vilité et le desreglement. Car que peut on imaginer plus vilain que *d'estrecouart* à l'endroit des hommes et *brave* à l'endroit de Dieu?" (emphases mine).

52. In the penultimate sentence of the final *Essay*, Bacon says that it is "not good, to looke too long, upon" the "turning Wheeles of *Vicissitude*, lest we become Giddy" (E LVIII "Of Vicissitude of Things," 176, VI: 517). But giddiness is good for the "discoursing Wits" who in the second sentence of the first *Essay* are described as men who "delight in Giddinesse" (E I "Of Truth," 7, VI: 377).

53. In *A Promus of Formularies and Elegancies* Bacon noted down: "Quid est veritas?" (Pott, *Promus*, 154 (no. 241)). Cf. the first sentence of the *Essay*: "*What is Truth*; said jesting Pilate; And would not stay for an Answer" (E I "Of Truth," 7, VI: 377) with John 18: 36–8 and 14: 6–7, and Friedrich Nietzsche, *Menschliches Allzumenschliches* II, Abt. 1, Aph. 8, and Nietzsche, *Der Antichrist*, Aph. 8, 46–7.

54. At the beginning of the *Essay*, Bacon misrelates a passage from Tacitus. For whereas the Roman historian speaks of the simulation of Tiberius (*Ann.*, V.1), the English philosopher says that Tacitus said that Livia attributed "*Dissimulation* to *Tiberius.*" Bacon describes the dissimulator as a man who has "that Penetration of Judgment, as he can discerne, what Things are to be laid open, and what to be secretted, and what to be shewed at Halfe lights, and to whom and when," but he dissembles this description in the following sentence by observing that to this man "a Habit of *Dissimulation*, is a Hinderance, and a Poorenesse." He goes on to say that if a man cannot obtain to the aforementioned judgment, "it is left to him, generally, to be Close, and a *Dissembler*. For where a Man cannot choose, or vary in Particulars, there it is good to take the safest and wariest Way in general." But is not being able to choose and to vary in particulars exactly what the judgment of the dissembler consists in? Bacon continues by observing that certainly the "ablest Men, that ever were, have had all an Opennesse, and Francknesse of dealing; And a name of Certainty, and Veracity." At such times, when they thought "the Case indeed, required *Dissimulation*, if then they used it, it came to passe, that the former Opinion, spred abroad of their good Faith, and Clearnesse of dealing, made them almost Invisible." But was not their openness and frankness already an act of dissimulation, considering that it aimed at acquiring a reputation of good faith, which would make them almost invisible? Finally, Bacon states that the greatest disadvantage of simulation and dissimulation is that it "depriveth a Man, of one of the most principall Instruments for Action; which is *Trust* and *Beleefe*" (E VI "Of Simulation and Dissimulation," 20 ff., VI: 387 ff.). But it seems to be not so much

depriving a man of trust and belief that is dissimulation's greatest disadvantage as it is providing him with trust and belief that is its greatest advantage.

55. *Ad Att.*, XVI. 15.

56. Significantly enough, although Bacon states hat envy is the "proper Attribute of the Devill," God is not mentioned in the *Essay* on envy (E IX "Of Envy," 31, VI: 397).

57. Since the meaning of the number twenty-two depends on an interpretation of the contrast between Bacon's political testament and the Old Testament, the reader is referred to chapter 5 for an interpretation of this contrast.

58. Aurelius Augustinus, *De Doctrina Christiana*, II. XXXVIII, II. XVI; *De Civitate Dei*, XI. 30; *Letter* LV. XVII to Januarius; *In Joannis Evangelium Tractatus*, CXXII.

Chapter 2

1. *Works* X, 253.

2. In a letter (1595) to Essex, Bacon comforts himself by emphasizing that with regard to the point of "estate" and "means," he "partly" leans "to Thales' opinion, That a philosopher may be rich if he will" (*Works* VIII, 372). The partial character of Bacon's approval of Thales can be explained by the fact that Thales actually became a rich man (*Diogenes Laertius*, I.26). Bacon repeatedly emphasized that he pursued a legal career primarily because of the inconvenient accident of his bad financial situation. See e.g., Letter (March 21, 1595) to Burghley (*Works* VIII, 358); Letter (1595) to Essex (ibid., 372); Letter (1597) to Burghley (*Works* IX, 49).

3. In his early letter (1592) to Burghley, Bacon already makes it clear that he was never determined to become a political man: "And if your Lordship will not carry me on, I will not do as Anaxagoras did, who reduced himself with contemplation unto voluntary poverty; but this I will do; I will sell the inheritance that I have, and purchase some lease of quick revenue, or some office of gain that shall be executed by deputy, and so give over all care of service, and become some sorry book-maker, or a true pioner in that mine of truth, which (he [Democritus, TvM] said) lay so deep" (*Works* VIII, 109). With regard to the difference in orientation between Anaxagoras and Democritus, cf. *Diogenes Laertius*, II. 6–7 with IX. 38, and IX. 72. See also Letter (1594) to Essex (*Works* VIII, 291).

4. *Works* VIII, 108.

5. In *De Augmentis*, Bacon describes himself as "a man naturally more fit for learning than for anything else, who was dragged into public affairs by some unknown fate and contrary to his genius" (DA 8.3, I: 792). See also Letter (February 28, 1621) to Tobie Matthew (*Works* XIV, 335).

6. *Works* VIII, 109.

7. See Aristotle, *Ethica Nicomachea*, 1176a30–1179a32. Cf. Ronna Burger, *Aristotle's Dialogue with Socrates: On the Nicomachean Ethics* (Chicago: University of Chicago Press, 2008), 198–207.

8. The movement of thought leading up to this observation and conclusion will be traced back in chapters 3 and 4.

9. Besides making the good itself twofold, Bacon also divides the private good into the "active" and the "passive" good, considering that "there is a twofold appetite in creatures: one for multiplying and propagating themselves, and the other for self-preservation and protection." The discussion of the private good intervenes in two substantial discussions of the communicative good (cf. DA 7.1, I: 717–21 with DA 7.2, I: 722–6 and DA 7.2, I: 726–31). The first paragraph of Bacon's discussion of the private good is the central paragraph of a series of thirteen paragraphs, the first of which introduces the twofoldness of good, and the last of which resumes the discussion of the communicative good. Bacon subdivides the passive good into the "perfective" and the "conservative" good. The treatment of the perfective good, which is the highest degree of the passive good, occupies the center and the central paragraph of Bacon's threefold discussion of the private good (DA 7.2, I: 722–6).

10. *Works* III, 518–9.

11. Cf. Xenophon, *Memorabilia*, III. 9.4 with Plato, *Phaedrus*, 249d5–252c2; *Protagoras*, 323b3–8; *Charmides*, 167a1–7; *Laches*, 194d1–195a6.

12. In *De Augmentis*, Bacon points out that "civil science" is conversant about a subject that "of all others is most immersed in matter," and can only "with great difficulty be reduced to axioms." He emphasizes, however, that there are some circumstances to reduce this difficulty. The second and central circumstance is that whereas ethics propounds to itself "imbuing and enlarging the mind with internal goodness," civil science requires only "external goodness," for that "suffices for society." Therefore, Bacon continues, "it not seldom comes to pass that the government is good, but the times are evil, seeing that in sacred history it often occurs that *sed adhuc populus non direxerat cor suum ad Dominum Deum patrum suorum* [2 Chron. 20: 33] is added to the narrative of good and pious kings." Bacon concludes from this that ethics is "more difficult" than civil science (DA 8.1, I: 746). But since the Lord broke off Jehoshaphat's works because the once pious Jehoshaphat had joined with the impious Ahaziah (2 Chron. 17: 3–6; 20: 35 ff.), the true evil of Jehoshaphat's times consisted not so much in the fact that the people had not directed their hearts to the Lord God of their fathers as in the fact that external goodness did not suffice for Jehoshaphat to be a good king in the eyes of the Lord. Bacon therefore seems to suggest that, contrary to external appearances, civil or political science is at least as difficult as ethics, if in fact it did not absorb ethics when ethics was absorbed by piety (cf. Aristotle, *Ethica Eudemia*, 1221a1, and *Ethica Nicomachea*, 1125a19–27).

13. Cf. 1 Kings 4: 29 (Geneva): "And God gave Solomon wisdom and understanding exceeding much, and a large heart, even as the sand that is on the seashore." Although Bacon refers to Solomon's wisdom, understanding, and heart, he puts "heart" in the center and in the central paragraph of the Epistle Dedicatory, which is the only paragraph in which "God" is mentioned (AL 2, 3–4, III: 261–2).

14. See Tacitus, *Ann.*, XIII. 3. That Bacon induces us to consider this contrast becomes plausible once we start paying attention to the following points. First,

Bacon only says that it is with regard to the king's gift of speech that he calls to mind what Tacitus says of Augustus Caesar. He does not say that he equates the eloquence of Augustus with the king's gift of speech. Second, after stimulating us to "note it well," Bacon presents a description of the Neronian type of eloquence: "[S]peech that is vttered with labour and difficultie, or speech that sauoreth of the affectation of art and precepts, or speech that is framed after the imitation of some patterne of eloquence, though neuer so excellent: All this hath somewhat seruile, and holding of the subiect." Third, although Bacon says that his "Maiesties manner of speech is indeed Prince-like, flowing as from a fountaine, & yet streaming & branching it selfe into Natures order, full of facilitie, & felicitie, imitating none & inimitable by any," he leaves it unclear what the characteristics of prince-like speech are, from what fountain it flows, and why it is inimitable by any (AL 1.2, 4, III: 262).

15. Bacon induces us to elaborate on this point by saying: "[L]et a man seriously & diligently reuolue and peruse the succession of the Emperours of Rome, of which *Caesar* the Dictator, who liued some yeeres before Christ, and *Marcus Antoninus* were the best learned: and so descend to the Emperours of *Grecia*, or of the West, and then to the lines of *France, Spaine, England, Scotland*, and the rest, and he shall find this iudgement is truly made" (AL 1.2, 4, III: 263). In the section immediately succeeding the central section on the "attributes and acts of God," Bacon himself instigates this elaboration by observing that the "felicitie of times vnder learned Princes, (to keepe . . . the Lawe of breuitie, by vsing the most eminent and selected examples) doth best appeare in that age, which passed from the death of *Domitianus* the Emperour, vntill the raigne of *Commodus:* comprehending a succession of sixe Princes, all learned or singular fauourers and Aduancers of learning: which age for temporall respects, was the most happie and flourishing, that euer the Romane Empire, (which was then a modele of the world) enioyed" (AL 1.69, 39–40, III: 303). Machiavelli, who himself "writeth as a natural man" (*An Advertisement touching the Controversies of the Church of England*, Works VIII, 80), had induced Bacon to study these examples by advising the prince to "[p]ongasi . . . innanzi uno principe i tempi da Nerva a Marco, e conferiscagli con quegli che erano stati prima e che furono poi; e dipoi elegga in quali volesse essere nato, o a quali volesse essere preposto. Perché in quelli governati da' buoni vedrà . . . ripieno di pace e di giustizia il mondo . . . vedrà ogni quiete e ogni bene . . . vedrà i tempi aurei dove ciascuno può tenere e difendere quella oppinione che vuole. Vedrà infine trionfare il mondo . . . Se considererà dipoi tritamente i tempi degli altri imperadori, gli vedrà atroci per le guerre, discordi per le sedizioni, nella pace e nella guerra crudeli . . . tante guerre civili, tante esterne . . . sopra tutto la virtú essere imputate a peccato capitale. . . . E conoscerà allora benissimo quanti oblighi Roma . . . e il mondo abbia con Cesare" (*Discorsi*, I. 10, 73 ff.; paragraph numbering based on *Discourses on Livy*, eds. Nathan Tarcov and Harvey C. Mansfield (Chicago: University of Chicago Press, 1996).

Returning to Bacon's discussion of the Roman Emperors, we find our author pointing out that he will make "some commemoration" of these princes, wherein, "although the matter will bee vulgar," yet because "it is pertinent to the point in hand, *Neque semper arcum tendit Apollo* and to name them onely were too naked and cursorie" (AL 1.69, 40, III: 303). By thus alluding to Horace, Bacon seems to stimulate us to read the Roman poet prophesying to Licinius that if things are bad now, they will not remain that way, for sometimes Apollo wakes the silent muse with his lyre, and he does not always draw his bow (*Carmina*, X. II.19). We are thus stimulated to look upon Bacon's discussion of the Roman Emperors as somehow serving the medicinal purpose of undrawing the bow of what the Greek god of prophecy represents.

But first things first: Bacon ingeniously and artfully describes the gradual Christianization of the Roman Empire that took place during the period he discusses. With regard to Nerva (Emperor 96–98), he points out that "the last Act of his short raigne left to memorie, was a missiue to his adopted sonne *Traian*, proceeding vpon some inward discontent, at the ingratitude of the times, comprehended in a verse of *Homers*." Bacon goes on to quote this verse, in which Chryses, a priest of Apollo who was unable to ransom his daughter, beseeches Apollo to have his arrows make the Danaans pay for the tears they had caused the priest to shed (*Iliad*, I. 42; AL 1.69, 40, III: 303). Dio Cassius explains how Nerva's discontent could be comprehended in Bacon's quotation of Homer. For the historian relates that Casperius Aelianus incited the soldiers to mutiny against Nerva, and to demand certain persons for execution. Although Nerva resisted Aelianus, he accomplished nothing. He thereupon appointed Trajan emperor (*Historia Romana*, LXVIII. 3–4). Passing over his allusion to the correlation between human impotence and divine justice, we suspect that Bacon takes the failure of Nerva's revenge to signify the decisive enfeeblement of the Roman Empire. For "[w]hen a *Warre-like State* grows *Soft* and *Effeminate*, they may be sure of a *Warre*," as he puts it in his *Essay* on the vicissitude of human things (E LVIII "Of Vicissitude of Things," 175, VI: 516). Trajan (Emperor 98–117), the second Emperor Bacon discusses, receives the following words of praise: "[I]f wee will hearken to the speech of our Sauiour, that sayth, *Hee that receiueth a Prophet in the name of a Prophet, shall have a Prophets reward*, hee deserueth to bee placed amongst the most learned Princes" (cf. Matt. 10: 41). Bacon adds that it was also in "this Princes time" that "the persecutions against the Christians receiued intermission, vppon the certificate of *Plinius secundus* . . . by *Traian* aduanced" (see Pliny, *Epistularum Libri Decem*, X. 96–7). How much Trajan's "vertue and gouernement was admired & renowned, surely no testimonie of graue and faithfull History doth more liuely set forth," than the "legend tale" of Gregorius Magnus (Pope 590–604), "who was noted for the extream enuy he bare towards all Heathen excellencie: and yet he is reported out of the loue and estimation of *Traians* morall vertues, to haue made vnto God, passionate and feruent prayers, for the deliuerie of his soule out of Hell" (see Dante Alighieri, *La Divina Commedia*, Purgatorio X. 73–6, and: Johannes Diaconus, *Sancti Gregorii Magni Vita*, LXXV; AL 1.70, 40, III: 304). In other words, although Trajan did not receive Christianity in the name of Christ, the Holy Father Gregorius Magnus bestowed on him the reward of a Christian. With regard to Hadrian (Emperor 117–138), the third Emperor he discusses, Bacon points out that "[i]t pleased God . . . to vse the curiositie of this Emperour, as an inducement to the peace of his Church in those dayes: for hauing Christ in veneration, not as a God or Sauiour, but as a wonder or noueltie: and hauing his picture in his Gallerie, matched with *Apollonius* (with whom in his vaine imagination, he thought he had some conformitie) yet it serued the turne to allay the bitter hatred of those times against the Christian name: so as the Church had peace during his time" (AL 1.71, 40–1, III: 304–5). Yet the *Scriptores Historiae Augustae* tell us that it was Alexander Severus (Emperor 222–235) who mingled the statues of deified emperors with those of Apollonius, Abraham, Christ, and Orpheus, and with portraits of his ancestors. Severus wanted to build a temple for Christ and give him a place among the gods; something Adrian had already considered doing. But he was prevented from carrying out his plan, since those who examined the sacred victims assured him that if he carried it out all men would become Christians (*Scriptores Historiae Augustae, Alexander Severus*, XXIX, XLIII). The fourth Emperor discussed by Bacon, Antoninus Pius (Emperor 138–161), "likewise approached a degree neerer vnto Christianitie, and became as *Agrippa* sayd vnto *S. Paule, Halfe a Christian*; holding their Religion and Law in good opinion: and not only ceasing persecution, but giuing way to the aduancement of Christians" (Dio Cassius, LXX. 3; AL 1.72, 41–2, III: 305–6). But what King Agrippa actually said in response to Paul's question if he believed the prophets—to which question Paul had added the words: "I know that Thou believest"—was the following: "*Almost* thou persuadest me to become a Christian" (Acts 26: 24–9, emphasis mine). During the time of the last two Emperors Bacon discusses, Lucius Aurelius Verus (co-Emperor 161–169) and Marcus Aurelius Antoninus (Emperor 161–180), "the Church for the most part was in peace." Bacon concludes his discussion of the decline of the Roman Empire by saying that "in this sequence of six Princes, we doe see the blessed effects of Learning in souereigntie, painted forth in the greatest Table of the world" (AL 1.73, 42, III: 306).

In the *Advancement of Learning*, Bacon remarks that the "prouidence of God hath made the distribution: for it hath pleased God to ordaine and illustrate *two* exemplar States of the worlde, for Armes, learning, Morall Vertue, Policie, and Lawes. The STATE of GRECIA, and the STATE of ROME: The Histories whereof occupying the MIDDLE PART of time, haue more ancient to them, Histories which may by one common name, be tearmed the ANTIQVITIES of the WORLD; and *after* them, Histories which may bee likewise called by the name of MODERNE HISTORIE" (emphases mine). Bacon expresses his wish that there be a "perfect Course of Historie . . . for *Rome*, from *Romulus* to *Iustinianus*, who may be truly saide to be *Vltimus Romanorum*" (AL 2, 66–7, III: 335). We notice in passing that the Christian Emperor Justinian (Emperor 527–565) indeed failed to restore the Roman Empire, but that according to Tacitus (*Ann.*, IV. 34) and Suetonius (*Vita Tiberi*, 61), Cremutius Cordus applied the aforementioned epithet to Cassius and Brutus, the assassins of Julius Caesar, the man to whom according to Machiavelli the world owed so many obligations. Bacon's remarks on the perfect history of exemplary states did not find their way into *De Augmentis*.

Returning to our discussion of the representative of divine providence, we conclude that according to our political historiographer it was in the days of Caesar or some years after him that Apollo was enabled to draw and shoot his bow. His arrow went through the time of Marcus Aurelius Antoninus, the time of the so-called Emperors of the Greeks who ruled the Eastern Roman Empire after the division of the Roman Empire (395), and the time of Justinian, who failed to reunite the Roman Empire. It finally arrived in modern times, where Bacon passed it on to the awakened but as yet silent muse, along with the lyre that was to instaure the arrow's motion by being played on in *De Sapientia Veterum*. But let us not anticipate events.

In the eighth book of *De Augmentis,* Bacon presents his consideration on the extension of the boundaries of empire, but he asks us to what purpose he is presenting this consideration, since the Roman Empire is "believed" to have been the last of earthly monarchies (DA 8.3, I: 803).

16. Although Bacon states that Virgil made "a kinde of separation betweene policie and gouernment, and betweene Arts and Sciences, . . . attributing and challenging the one to the Romanes, and leauing & yeelding the other to the Grecians," Virgil does not explicitly mention the Greeks in the verses that Bacon refers to (*Aeneid*, VI. 847–53). It was Meletus, Socrates's accuser on behalf of the poets, who brought against Socrates the accusations that Bacon paraphrases. Anytus was Socrates's accuser on behalf of the craftsmen and the politicians (Plato, *Apology*, 19a9–b8, 23e4–24a2, 24b4–c2; *Euthyphro*, 2b5–3a5).

17. Plutarch, *Life of Marcus Cato,* XXIV–XXV.

18. The zetetic skeptic Montaigne made the same blunder. See *Essais* I. XX *Que Philosopher c'est Apprendre à Mourir*, par. 26.

19. It does not seem to be a coincidence that it is in the context of discussing the arms of learning that Bacon alludes to the ambiguous relationship between learning and arms. He states that "experience doth warrant, that both in persons and in times, there hath been a *meeting, and concurrence* in learning and Armes, flourishing and excelling in the same men, and the same ages," but he restates this statement at the end of the paragraph by saying that "Armes and Learning . . . haue a concurrence *or nere sequence* in times" (AL 1.8, III: 269, emphases mine). Further on in the book, Bacon treats the subject of learning's "power and efficacie in inablement towards martiall and militarie vertue and prowesse" on a more substantial basis. After having illustrated the problem of the "concurrence of militarie vertue and learning" by means of the examples of Alexander the Great and Julius Caesar, Bacon considers it necessary, "in regard of the rarenesse of circumstance," to mention one other person: "*Xenophon* the Philosopher." The rareness of this philosopher consisted in the fact that he "did so sodenly passe, from extreame scorne, to extreame wonder." Bacon refers to Xenophon's (Theopompus') response to Phalinus, who upon the death of Cyrus had delivered to the Greeks, who were besieged and almost unarmed, the message from King Artaxerxes that they were to "deliuer vp their Armes" and to "submit themselues to the Kings mercy" (Xenophon, *Anabasis*, II. 1.8–13). Bacon's Xenophon responded by saying that the Greeks had "*but these two thinges left*," their "*Armes*" and their "*Vertue*," and if they would "*yeeld*" up their

arms," how were they to make use of their virtue? Even though Phalinus scorned Xenophon by saying that he would be *"much abused"* to think that his virtue could *"withstand the Kings power,"* Bacon observes that it was exactly on the basis of the virtue contained in Xenophon's remark that "this young Scholler, or Philosopher," who "neuer had seene the Warres before," and who "neither had any commaund in the Armie," after "all the Captains were murthered in parlye by treason, conducted . . . ten Thousande foote, through the heart of all the Kinges high Countreys from *Babilon* to *Grecia* in safetie, in despight of all the Kings forces," and "to the astonishment of the world" (AL 1.75–95, 43–9, III: 307–14). In light of the fact that the philosopher Xenophon was not the only person to "sodenly passe, from extreame scorne, to extreame wonder," we need not wonder that Bacon presents his discussion of the learned Roman Emperors before his discussion of armed founders in order to draw our attention to the fact that the arms of learning can be turned to ambiguous uses.

20. Rousseau therefore locates the "raison" of this "precepte indifférent par lui-même et inexplicable dans tout autre Systême" in the "intention de donner d'abord aux actions humaines une moralité qu'elles n'eussent de longtems acquise" (*Discours sur l'inégalité*, ed. Heinrich Meier, 5th ed. (Paderborn: Ferdinand Schöningh, 2001), note IX, 320; see also fn. 388 *ad locum*).

21. But cf. Rousseau: "Regimber contre une défense . . . arbitraire, est un penchant naturel, mais qui, loin d'être vicieux en lui-même, est conforme à l'ordre des choses et à la bonne constitution de l'homme . . ." (*Lettre à Monseigneur de Beaumont*, in *Œuvres Complètes de Jean-Jacques Rousseau*, vol. 4, Bibliothèque de la Pléiade (Paris: Gallimard, 1969), 939–40 (fn.)).

22. Cf. Augustine, *De Civitate Dei*, XIV. 12, and *De Natura Boni Contra Manichaeos*, XXXV; Martin Luther, *Vorlesungen über 1. Mose von 1535–45*, in *Martin Luthers Werke. Kritische Gesamtausgabe*, vol. 42 (Weimar: H. Böhlaus Nachfolger, 1911), 71 ff., 122; Johannes Calvin, *Institutio Religionis Christianae*, II. 1.4, and Calvin, *Commentariorum in Quinque Libros Mosis*, in *Opera quae Supersunt Omnia*, eds. G. Baum et al., vol. 23 (Brunswick, 1882), 44–5, 60–1, 67.

23. Cf. Leo Strauss, *Natural Right and History* (Chicago: University of Chicago Press, 1971), 86–92.

24. The Hebrew term for nature is unknown to the Hebrew Bible.

25. In a letter that was originally intended to be prefixed to *Of the Coulers of Good and Evill* (1597), Bacon emphasizes that although he "brought in a new manner of handling" the "argument to make it pleasant and lightsome," he does not "pretend so to have overcome the *nature of the subject*, but that the full understanding and use of it will be somewhat dark, and best pleasing the taste of such wits as are *patient* to stay the digesting and soluting unto themselves of that which is sharp and subtile." In the preface to the *Coulers*, he adds that the different colors "sway the ordinary judgment either of a weak man, or of a *wise man not fully and considerately attending and pondering the matter*" (*Works* VII, 70–1, 77, emphases mine). The *Coulers* thus primarily seems to be a relaxation exercise for wise men, preparing them for a full and considerate inquiry into the dark matter

of the principles or origins of good and evil. In the eighth *Couler*, for example, Bacon observes that what a man brings on himself by his own guilt is the greater evil, and what is externally imposed on him the lesser. For, he explains, the sting of conscience doubles adversity, whereas the consciousness of being free from guilt affords great consolation in calamity (ibid., 86–7; DA 6.3, I: 679–80). In the *Advancement of Learning,* Bacon criticizes those who treated the subject of ethics for not having "stayed a little longer vpon the Enquirye, concerning the Rootes of Good and euill, and the Strings of those Rootes," and for not having "consulted with Nature." He announces, however, that he will endeavour to "resume" and "open" this inquiry "in a more cleare Manner" (AL 2, 136, III: 420), which provides us with one of the more profound explanations for Bacon's seemingly reductionistic statement that his new method also pertains to the science of ethics (NO 1.127, I: 219–20). It is also as part of his endeavor to inquire into the roots of good and evil that Bacon observes in one of his *Essays* that he likes a "*Plantation* in a Pure Soile; that is, where People are not *Displanted*, to the end, to *Plant* in Others. For else," he explains, "it is rather an Extirpation, then a *Plantation*," and it is the "sinfullest Thing in the world, to forsake or destitute a *Plantation*, once in *Forwardnesse*: For besides the Dishonour, it is the Guiltinesse of Bloud, of many Commiserable Persons" (E XXXIII "Of Plantations," 106, 108, VI: 457, 459). Seventeen *Essays* have passed when in the thirty-third *Essay* the noun "Extirpation" occurs for the second and final time in the *Essays.*

26. Cf. Cicero, *De Natura Deorum*, I. 60. Bacon's familiarity with the conversation between the tyrant Hiero and the poet Simonides appears from his reference to it in the *Apophthegms New and Old*, 230 (*Works* VII, 158).

27. Cf. Martin Luther, *Predigt über Johannes* 17: 1–3 (August 15, 1528), in *Predigten*, ed. Emanuel Hirsch, vol. 7 of *Luthers Werke in Auswahl*, ed. Otto Clemen, 3rd ed. (Berlin: de Gruyter, 1962), 217, and Calvin, *Institutio*, I. 5.9, 5.12. Cf. also Leo Strauss, *The City and Man* (Chicago: Rand McNally, 1964), 241, and Heinrich Meier, *Die Lehre Carl Schmitts: Vier Kapitel zur Unterscheidung Politischer Theologie und Politischer Philosophie*, 3rd ed. (Stuttgart: Metzler, 2009), 138 ff. (esp. fn. 53, 55), 299–300.

28. Cf. Plato, *Lysis*, 220d4–7 with *Theaetetus*, 176a5–b1.

29. See Heinrich Meier, *Warum Politische Philosophie?* (Stuttgart: Metzler, 2000), 15–32.

30. Cf. Luther, *Vorlesungen über 1. Mose*, 47.

31. Cf. Leo Strauss, "On the Interpretation of Genesis," *L'Homme. Revue française d'anthropologie* 21, no. 1 (1981), 5–20, 17–8.

32. In order to underscore the distance between man and God, Luther and Calvin emphasize that "homo non ad otium, sed ad laborem, etiam in innocentiae statu, conditus est" (Luther, *Vorlesungen über 1. Mose*, 78), and that "fuerat [Adam, TvM] non ad ignaviam conditus, sed ad aliquid agendum. Ideo praefecerat eum Dominus horto colendo" (Calvin, *Commentariorum in Quinque Libros Mosis*, 73).

33. After stating that "*Envy* ever redoubleth from Speech and Fame," Bacon observes that "*Cains Envy*, was the more vile, and Malignant, towards his brother

Abel; Because, when his Sacrifice was better accepted, there was no Body to looke on" (E IX "Of Envy," 28, VI: 394). But if Cain's envy was more vile and malignant because, besides God, there was nobody to witness his sacrifice, speech and fame would have diminished rather than redoubled his envy. Bacon therefore seems to suggest that Cain's envy would have been less vile and malignant had somebody witnessed that it was justified. The reference to Cain's envy constitutes the twenty-second reference to envy in the *Essay* on envy.

34. Cf. 1 John 3: 12: "Cain . . . slew his brother . . . [b]ecause his own works were evil, and his brother's righteous" with Luther (*Vorlesungen über 1. Mose*, 186): "Si ipsum opus spectes, non poteris Habel praeferre Cain," and Calvin (*Commentariorum in Quinque Libros Mosis*, 87): "Superbia haec omnibus hypocritis est ingenita, ut Deum obsequii specie sibi *obnoxium* habere velint . . . Sola quidem illis obstat *impietas* quominus Deum sibi concilient" (emphases mine).

35. *Ann.*, III. 76 (emphasis mine).

36. Cf. Plato, *Politeia*, 473c11–e5 with *Second Letter*, 314b9–c5 and *Seventh Letter*, 325b6–326b4.

37. Cf. Machiavelli, *Principe*, XV.

38. Cf. Johannes Calvin, *Harmonia ex Evangelistis Tribus* (Geneva, 1582), 269–70.

39. Cf. Seneca, *Ad Lucilium*, VII. 11: "haec . . . ego *non multis*, sed tibi; satis enim magnum alter alteri theatrum sumus" (emphasis mine).

40. Cf. E X "Of Love," 32, VI: 397: "It is a poore Saying of *Epicurus*; *Satis magnum Alter Alteri Theatrum sumus:* As if Man, made for the contemplation of Heaven, and all Noble Objects, should doe nothing, but *kneele before a little Idoll*, and make himselfe subject, though not of the Mouth (as Beasts are) yet of the Eye; which was given him for higher Purposes" (emphasis mine).

41. Plutarch, *Life of Agesilaus*, XIII.

42. Bacon illustrates this point by means of five examples, all of which implicitly counsel the learned to expose their knowledge of the nature of custom by adapting their manners to popular judgment (AL 1.103, 53–4, III: 319). The first example is that of Aesop's cock, who found a gem, but preferred one barleycorn to all the gems of the world (cf. Matt. 13: 31–2; Mark 4: 30 ff.; Luke 13: 18–9). The second example is Midas, who, "being chosen Iudge" between Apollo, the "President of the Muses," and Pan, the "God of the Flockes," "iudged for Plentie." Ovid tells us that Midas expressed his unsolicited judgment in a music contest between Apollo and Pan. Since he was charmed by the song of Pan, he called the judgment of the sacred Mount Timolus in favor of Apollo unjust. For this reason, Apollo punished him by making him wear the ears of an ass (Ovid, *Metam.*, XI. 150–179). The third and central example is Paris, who was offered royal power by Hera, martial prudence by Pallas Athena, and Helen's beauty by Aphrodite (Euripides, *Troiades*, 914–60). According to Bacon, he "iudged for Beautie, and loue against *Wisedomeand Power*" (emphasis mine). The fourth example is Agrippina, Nero's mother, who, when she heard that she would be killed by her son, said: "*Occidat matrem, modo imperet.*" Bacon takes Agrippina to have "preferred Empire

with any condition neuer so detestable," which apparently was an empire in which the most detestable but transitory condition was her own death (cf. Tacitus, *Ann.*, XIV. 9: "occidat, dum imperet"). The fifth and final example is Ulysses, "*Qui vetulam praetulit immortalitati.*" Although Ulysses knew that the goddess and queen Kalypso was immortal, he preferred the mortal Penelope. According to Plutarch's Kalypso, Ulysses was willing to struggle through many new dangers, seemingly to return to a mortal woman, but in fact only to make himself more renowned than he already was, pursuing an empty phantom rather than what is truly good (cf. Homer, *Odyss.*, V. 215–20, and Plutarch, *Gryllus*, 985f–986a).

43. For the reasoning underlying this fourfold capacity of parables, see *supra*, 19 ff.

44. Cf. Machiavelli, *Principe*, Lettera Dedicatoria.

45. Aesop, *Fable 196: The Ageing Lion and the Fox*, in Aesop, *The Complete Fables*, transl. Olivia and Robert Temple (London: Penguin, 1998), 145.

46. Cicero, *Ad Att.*, II. 1.

47. Cicero takes Plato's observation, "tantum contendere in re publica, quantum probare tuis civibus posses; vim neque parenti nec patriae afferre oporteret," to have been the cause of his not engaging in the republic; for he saw that the Athenian people could not be ruled by persuasion, and he looked upon compulsion as arbitrary. Applying Plato's consideration to himself, however, Cicero goes on to state that engaging or not engaging in the republic was not an open question to him (letter to Lentulus, *Ad Fam.*, I. 9).

48. Plato, *Crito*, 51b2–10.

49. Cf. Plato, *Politeia*, 499b1–d5.

50. Plutarch points out that Cato learned the art of speaking in public after having been made priest of Apollo. This art led him to think that political philosophy, like a great city, should maintain for its security a warlike element. Cato did not recite his exercises in the art of speech, and was therefore blamed for his silence. He responded that he would begin to speak as soon as he had that to say which should not be left unsaid. *Life of Cato the Younger*, IV.

51. Aeschylus, *Agamemnon*, 1202–330.

52. According to Apollodorus (*Library*, III. 12.5), the ancient Apollo only wanted to sleep with Cassandra. But it is only by means of marriage that the modern Cassandra can demonstrate her submission to the God who loves His prophets in an anticipatory manner, as Bacon stimulates us to infer from our own retrospective manner of loving them. Knowing, however, that His love could only be reciprocated by His prophets being believed in anticipation of the calamity that they prophesy, it must have been in anticipation of the reciprocation of His love that God bestowed on His prophets the gift of credibility in addition to the gift of divining the truth that He is believed to be, even or precisely in calamity.

53. *Supra*, 25–6, 48 ff.

54. Cf. *supra*, 27–8, 46.

55. See Apollodorus, *Library*, III. 10.2, and: *Homeric Hymn to Hermes*, 463–568. Cf. the interpretation of Mercury's theft in the Praefatio: by means of "friendly

discourses and prudent edicts" the "minds of the subjects can be reconciled as it were *thievingly* [quasi *furtim*]" (DSV Pr., VI: 626, emphasis mine).

56. Hesiod, *Theog.*, 886–900, 924–9.

57. Deleyre therefore rightly translates "philautia" as "Amour propre." Alexandre Deleyre, *Analyse de la Philosophie du Chancelier François Bacon*, vol. 2 (Amsterdam: Artskée & Merkus, 1755), 245. Cf. Rousseau, *Discours sur l'inégalité*, note XV.

58. Cf. Aristotle, *Ethica Nicomachea*, 1178b7–23; *Protreptikos*, B 43 (ed. Düring).

59. The original version has "providentiae divinae," the later version in *De Augmentis* has "sapientiae divinae" (cf. VI: 640 with DA 2.13, I: 529).

60. In the Praefatio Bacon had remarked that detracting from the honour of parables would be "an audacity sounding like profanity," seeing that "religion delights in such veils and shadows" (DSV Pr., VI: 625).

61. Ovid, *Metam.*, XI. 146–93.

62. Apollodorus, *Library*, I. 4.1.

63. The narrative has "proximus," the interpretation has "alter" (cf. VI: 635 with VI: 639).

64. Echo's speech defect was originally a punishment inflicted by jealous Juno, who was angry with Echo for having been detained by the latter's speech, so that Jupiter could indulge in his amours (Ovid, *Metam.*, III. 351–401).

65. Seianus had increased his own power by poisoning Tiberius's son Drusus, and by inducing Tiberius himself to withdraw into Campania. See Tacitus, *Ann.*, IV. 2, 7, 38–41, 57; Dio Cassius, LVIII. 6–11.

66. Cf. Leo Strauss, *Die geistige Lage der Gegenwart*, in Strauss, *Philosophie und Gesetz – Frühe Schriften*, vol. 2 of *Gesammelte Schriften*, ed. Heinrich Meier (Stuttgart: Metzler, 1997), 456.

67. The example of the non-fabulous king Louis XI is the only modern example in *De Sapientia Veterum*. In one of his *Essays*, Bacon remarks that the "closeness" of Louis was his "Tormentour" (E XXVI "Of Frendship," 83, VI: 440).

68. Euripides, *Bacchae*, 25–54.

69. Ibid., 337–43.

70. Cf. DA 2.13, I: 530: to the sleeping Endymion the moon descended "of her own accord [sponte]," for "divine influences sometimes stream spontaneously into the intellect when it is asleep, and withdrawn from the senses."

71. Apollodorus, *Library*, III. 5.2.

72. Virgil, *Georg.*, IV. 454–527.

73. In the twenty-ninth fable, Bacon points out that although those who descended to the infernal regions were not allowed to go back, a single exception was made for those who brought a branch of gold as a present to Proserpina. It was a "unique branch, growing in a vast and dark wood." It did not have a trunk, but "grew like mistletoe on an other tree." Bacon emphasizes that the ancients did not regard the "instauration of natural bodies" as a "desperate thing," but rather as a "thing abstruse and as it were out of the way." This is what he takes to be the

meaning of the fact that the branch had been placed "among infinite other branches in a vast and most dense wood." The branch is represented as golden because gold is the "token of duration," and it is described as having been grafted because the hoped-for effect is the "result of art" (DSV XXIX "Proserpina," VI: 680 ff.). It must have been because the ancients had not yet taken cognizance of the existence of the wood called Eden that they regarded the instauration of corruptible things as a thing out of the way. And it must have been because his having taken cognizance of the existence of Eden compelled him to resort to art that Bacon wrote his twenty-ninth fable.

74. See Robin Hard, *The Routledge Handbook of Greek Mythology* (London: Routledge, 2008), 551 and 683, fn. 3.

75. Nemesis was a goddess who was "venerated by all," and who was moreover "feared by the powerful and fortunate." Bacon points out that her name "openly enough [satis aperte] signifies revenge or retribution," since it was the office and function of this goddess to "preclude constant happiness and perpetual felicity, like a tribune of the plebs who interposes his veto." It was her office and function not only to "castigate insolence," but also to see that "prospering things, however innocent and moderate," were "balanced by the vicissitude of adversary things, as if no one of the human race was to be admitted to the banquets of the gods, except in derision." Bacon explains that the goddess Nemesis is distinguished with a crown on account of the "envious and malignant nature of the vulgar." For "when the fortunate and the powerful are ruined, the people commonly exult at this and put a crown upon the head of Nemesis" (DSV XXII "Nemesis," VI: 662–3).

76. Cf. Ovid, *Metam.*, X. 36–9.

77. Cf. Plato, *Phaedo*, 89c11–90e3.

78. Ovid, *Metam.*, X. 79–82, XI. 1–49. Erathosthenes (c. 24) relates that the reason why Orpheus was torn to pieces was that after having visited the infernal regions he had become unable to recognize Dionysus as the greatest of gods. Cf. Aristophanes, *Frogs*, 628 ff.

79. Cf. Plato, *Phaedo*, 96a6–100d8. It would be a mistake to think that, apart from political philosophy being a different manner of treatment of philosophy, the turn to political philosophy consists in a contraction instead of an extension of the things the philosopher loves. Cf. Plato, *Theaetetus*, 173b9–174b8 with Diogenes Laertius, I. 34. Cf. Xenophon, *Memorabilia*, I. 1.11–6 with IV. 6.1. Cf. also Seth Benardete, *Socrates' Second Sailing* (Chicago: University of Chicago Press, 1989), 173–4.

80. Cf. Aristophanes, *Clouds*, 222–37, and Plato, *Apology*, 18d1–2, 19c2–6. Cf. Leo Strauss, *Socrates and Aristophanes* (Chicago: University of Chicago Press, 1966), 16–49, 173, 193, 311–4.

81. Speaking of his intention, Bacon speaks up for Socratic speech by speaking out against the fate of Socratic speech: "Neither is my meaning as was *spoken* of *Socrates*, to call Philosophy *downe from heauen* to conuerse *vponthe earth*, that is, to *leaue* naturall Philosophie *aside*, and to applye knowledge *onely* to manners, and policie" (AL 1.48, 32, III: 294, emphases mine). Rather, Bacon's meaning

approximates what in almost identical terms was said about Socrates by that familiar "political man," according to whom Socrates "called philosophy away from heaven to establish it in the cities, to introduce it into the households, and to compel it to inquire about men's lives and manners as well as about the good and bad things" (Cicero, *Tusculanae Disputationes*, V. IV (10)). As Socrates was compelled to inquire primarily into manners and policy, without, however, leaving natural philosophy aside when philosophizing (cf. *supra*, fn. 79), the fact that "heavenly" Socratic philosophers applied knowledge only to manners and policy compelled Bacon to put natural philosophy in the forefront, so "that knowledge may not bee . . . as a bond-woman to acquire and gaine to her Masters vse," without, however, leaving the Socratic turn aside when philosophizing. In *Novum Organum*, Bacon himself confirms this interpretation by saying that after Socrates had "brought down [deduxisset]" philosophy from heaven to earth, moral philosophy "became much stronger and averted men's wits from natural philosophy" (NO 1.79, I: 187). Lampert therefore rightly observes that it was "the fate of the second sailing" that Bacon acted on, although by saying that Bacon "reverses the Socratic turn," and undertakes a "new first sailing," he fails in speech to differentiate between speech and deed. Laurence Lampert, *Nietzsche and Modern Times: A Study of Bacon, Descartes, and Nietzsche* (New Haven: Yale UP, 1993), 135, 139, 204.

82. *Supra*, 25–6. Bacon criticizes the reformed school of the Epicureans for proclaiming felicity to be nothing else than the "tranquillity and serenity of the mind, free and exempted from perturbation, as if they had deposed Jupiter again, and restored Saturn and the Golden Age, when there was neither summer, nor winter, spring or autumn, but all after one air and season" (DA 7.1, I: 719). Ovid tells us that in the Golden Age, when there was neither revenge nor law, spring was the only season. But after Saturn had been thrust into Tartarus, and Jupiter had come to rule the world, Jupiter made springtime shorter by adding three seasons to it (*Metam.*, I. 89–125). The possibility of Saturn's power being restored to him thus seems to be dependent on the perturbations caused by the spring of a new era.

83. Cf. Machiavelli, *Principe*, XVIII. In the *Advancement of Learning*, Bacon refers to Machiavelli's treatment of the parable of Cheiron as an example of parabolic poetry being used to "retire and obscure" what is "taught or deliuered," because it involves the "Secrets and Misteries of Religion, Policy, or Philosophy." In "Heathen Poesie, wee see the exposition of Fables doth fall out sometimes with great felicitie," as in "the fable, that *Achilles* was brought vp vnder *Chyron* the *Centaure*, who was part a man, & part a beast, Expounded Ingeniously, but corruptly by Machiauell, that it belongeth to the education and discipline of Princes, to knowe as well how to play the part of the Lyon, in violence, and the Foxe in guile, as of the Man in vertue and Iustice" (AL 2, 74–5, III: 344–5). After having ingeniously expounded the necessity of disgrace or corruption in the central fable of *De Sapientia Veterum*, Bacon prudently decided not to let Machiavelli's parable return in *De Augmentis*. He did elaborate on the lesson of the parable though. For he remarks that like those others who treated the subject of the description of a prince, he "wholly moulded the man who is political for his own fortune according to the perfection

of the art, and not according to vulgar practice." Bacon therefore opposes his own precepts, which he calls "good arts," to the "evil arts" of Machiavelli, referring to examples from *Principe* XVII and XVIII. He adds, however, that "if anyone takes pleasure in this kind of corrupt prudence, this person may advance his fortune more compendiously and more quickly, for in truth it is in life as it is in ways, the shortest way is commonly the foulest and muddiest, and the fairer road is not walked on by many." We conclude that in a way not dissimilar to his predecessor, whom he was compelled to disgrace for prudential reasons, Bacon perfected the princely art by adapting it to vulgar practice, which is foul and muddy. For it is only by means of this conclusion that we can explain Bacon's own conclusion, the conclusion that "if any man should flatter himself for employing his fortune well although he has obtained it by means of evil arts, such compensations of evil are to be approved of after the fact, although such counsels deserve to be condemned" (DA 8.2, I: 789 ff.).

84. Augustine points out that the fact that the length of Noah's ark is six times its breadth and ten times its height (cf. Gen. 6: 15), as is the case with the human body, signifies that Christ appeared in a human body (*Contra Faustum Manichaeum*, XII. 14).

85. In the title and narrative of the fable Bacon uses the name Dionysus, but when interpreting the fable he refers to its protagonist as Bacchus. He thus draws our attention to the traditional understanding of Bacchus as both god and worshipper, before giving it the nontraditional meaning of the god who is his own worshipper (see Walter Burkert, *Griechische Religion der archaischen und klassischen Epoche* (Stuttgart: Verlag W. Kohlhammer, 1977), 253). The change of names as it were etymologically underscores that the subject matter of the twenty-fourth fable is moral practice, seeing that Βακχεία means "rapture" or "exaltation."

86. See Apollodorus, *Library*, III. 4.3, and Ovid, *Metam.*, III. 260–315.

87. Apollodorus, *Library*, III. 5.1.

88. Cf. *supra*, 41 (fn. 20).

89. According to the *Orphic Hymn to Amphietus Bacchus* (52), Bacchus lay asleep for three years. Conti mentions the story that Bacchus was torn to pieces in the war waged by the Titans against Jupiter, and that his "beating [palpitans]" heart was brought to Jupiter by Pallas. After having been buried, Bacchus "revived in his former state [reuxisse integer]." Natalis Comes, *Mythologiae, sive Explicationis Fabularum, Libri decem* (Frankfurt, 1584), V. XIII, 490–1.

90. According to Bacon, the "sacred philosopher" Solomon also noticed this point by saying that the "words of the wise are like goads and fastened nails." The biblical Solomon adds that they are given by one shepherd (Eccl. 12: 11).

91. See Apollodorus, *Library*, III. 5.8, and Hard, *Handbook*, 309 and 644 (fn. 93).

92. Significantly enough, the modern exception to the ancient rule (Apollodorus, *Library*, III. 5.8) occurs in an ancient tragedy (Euripides, *Phoenician Women*, 1504–7). Spenser's Sphinx embodying the Inquisition may serve as an illustration of

the appropriateness of Bacon's distrust of the sextons of Bacchus. Edmund Spenser, *The Faerie Queene*, Book V, Canto 11.

93. Apollodorus, *Library*, III. 5.7. Cf. Friedrich Nietzsche, *Die fröhliche Wissenschaft*, Drittes Buch, Aph. 125 with 153. It seems to be more than a coincidence that the multiheaded Typhon was the father of the multiform Sphinx (*Library*, III. 5.8), because it was Typhon, philosophy's unwitting henchman, who took the father of gods and mortals to the obscure and remote region or cave that was left only by the lover of the father of the father of gods and mortals.

94. The reasonable foundations of the philosopher's empire constitute the subject matter of the fourth chapter.

95. Since Hesiod (*supra*, 57 (265 fn. 56)) relates that the birth of wisdom's son was to be preceded by the birth of wisdom's daughter, we will treat the birth or rebirth of the goddess of wisdom in the context of reconstructing the birth of the philosopher's empire in the fourth chapter, before giving rebirth to wisdom's political son in the fifth chapter.

96. Apollodorus, *Library*, I. 2.1.

97. We do not believe it a coincidence that after having written *De Sapientia Veterum* Bacon decided not to repeat his earlier observation that "for the expressing of affections, passions, corruptions and customes, we are beholding to Poets, more then to the Philosophers workes" (AL 2, 75, III: 346).

98. Bacon emphasizes the ubiquity of perniciousness by subtly changing the Homeric narrative. For whereas Homer's Odysseus ordered his crew to defend him against himself (*Odyss.*, XII. 39–54, 158–200), Bacon's Ulysses defends himself against his crew.

99. Pausanias, *Description of Greece*, IX. 34.3.

100. Bacon elaborates on the nature of the newly born Sirens in the fable in which he discusses the nature of their father, the river god Achelous. Hercules and Achelous disputed on who would marry Deianeira, the daughter of Dionysus. The matter was decided by a fight. Achelous tried a variety of multiple shapes to turn himself into, "which he was at liberty to do." Eventually he appeared in front of Hercules in the form of a "savage and roaring bull" (DSV XXIII "Achelous," VI: 663–4; see Apollodorus, *Library*, I. 8.1, II. 7.5). In the central fable, the fabulist Bacon tells us that, among many other shapes, Jupiter also turned himself into the shape of a bull when pursuing his loves (DSV XVI "Procus Junonis," VI: 654).

101. Although according to Apollodorus (*Library*, Epit. 7.18) the Sirens were daughters of Melpomene, Bacon makes Terpsichore, who is the central Muse mentioned by Hesiod (*Theog.*, 77 ff.), the mother of the Sirens.

102. Declensions of "Sirenes" and "Voluptas" are used twenty-two times altogether in the thirty-first fable. That Bacon had already played with the idea of entering into an alliance with religious methods appears from the following consideration from *Certain Considerations touching the Plantation in Ireland* (1608): "join the harp of David, in casting out the evil spirit of superstition, with the harp of Orpheus, in casting out desolation and barbarism" (*Works* XI, 118). Augustine

contends that Noah's ark is thirty cubits high (Gen. 6: 15) because Christ, "who is our height," gave His sanction to the doctrine of the Gospel in his thirtieth year. The Church Father elsewhere adds that, since it is a "custom of the Holy Scriptures not to reckon what is above certain round numbers," transgressions of round numbers "convey particular meanings." He discusses the example of the number eleven, which denotes "sin" because it transgresses the number ten, which signifies the "perfection of all righteousness and blessedness." See *Contra Faustum Manichaeum*, XII. 14; *Sermones de Scripturis*, LI. 32 ff.

103. Bacon describes the office and function of rhetoric, "if deeply penetrated into," as the "application and commendation of the dictates of reason to the imagination, in order to excite the appetite and the will." He emphasizes that although the imagination performs the office of "messenger or proctor," it can also "receive or usurp no small authority in itself." For "we see [Videmus] that in matters of faith and religion the imagination raises and carries itself above reason, because divine grace uses the motions of the imagination as an instrument of illumination." The end of rhetoric, however, is to "fill the imagination with observations and simulacra, so that it supports reason and does not oppress it." But the possibility of supportive measures becoming oppressive is already indicated by Bacon's own remark that rhetoric treats reason as it is "planted in the opinions of the vulgar" (DA 5.1, I: 615–6; DA 6.3, I: 671 ff.).

104. Bacon gives the example of Petronius who, having been condemned to death, sought enjoyments in the atrium of death. Referring to Tacitus, Bacon points out that Petronius turned to books for consolation, but "only to light verses," and not to books that "teach constancy." According to Tacitus, Petronius listened to his friends, speaking not on the immortality of the soul, but reciting light poems and easy verses (*Ann.*, XVI. 19.2).

105. Apollodorus, *Library*, Epit. 7.19.

Chapter 3

1. Cf. Strauss, *Socrates and Aristophanes*, 168–9, 171, 179, 193.

2. Cf. Xenophon, *Hiero*, 5.1. Bacon points out that princes are best interpreted by their "natures" rather than by their "ends." He explains himself in the following way. "Since princes are at the top of human desires, they hardly [fere] propose to themselves any ends to which they vehemently and constantly aspire, and from the position and distance of which one could deduce the direction and the scale of the rest of their actions," which is "one of the principal causes why their hearts are inscrutable," as "Scripture declares" (DA 8.2, I: 776; Prov. 25: 3).

3. Bacon underscores the central philosophic importance of the question as to the nature of God by mentioning the three objects of philosophy, God, nature, and man three times, changing the sequence only in the center of these three times by mentioning God in the center of the enumeration (DA 3.1, I: 540). In the

Advancement of Learning, Bacon had not yet drawn our attention to this central object of philosophy by means of a subtle change in sequence (AL 2, 76, III: 346).

4. The first book of the *Advancement of Learning* can be thematically divided into thirteen sections: first part: par. 1–3 (Epistle Dedicatory); second part: par. 4–6 (introduction discredits and disgraces; divines); third part: par. 7–15 (politiques); fourth part: par. 16–25 (learned); fifth part: par. 26–37 (studies of the learned); sixth part: par. 38–49 (peccant humours); seventh part: par. 50–65 (attributes and acts of God); eighth part: par. 66–74 (human proofs and emperors); ninth part: par. 75–95 (martial and military virtue); tenth part: par. 96–7 (moral and private virtue); eleventh part: par. 98–101 (power and commandment); twelfth part: par. 102 (immortality and continuance); thirteenth part: par. 103 (custom & habit).

5. See Job 26: 7; 26: 13; 38: 31; 9: 9; 10: 10; 28: 1–2.

6. After mentioning the first two verses of the twenty-eighth chapter of the book of Job, Bacon encourages the attentive reader to read the rest of the chapter by saying: "and so forwards in that Chapter." The final verse of the twenty-eighth chapter states that God said to man: "Behold, the fear of the Lord, that is wisdom" (Job 28: 28).

7. The hundred-and-thirtieth and final aphorism of the first book constitutes the transition to the second book.

8. In a letter to King James (October 12, 1620) on *Novum Organum*, Bacon remarks that the *Advancement of Learning* contains the "same argument" as *Novum Organum*, only "sunk deeper" (*Works* XIV, 120).

9. In an early and unpublished work, Bacon bluntly states that "the diligent mind will perceive that from the fallacious and iniquitous bond between theology and philosophy no less danger threatens than from open hostility" (CV 7, III: 596). But since the danger ensuing from unconcealing hostility could be significantly diminished by prudently concealing it, it was with good reason that Bacon decided not to publish this statement. Instead, he decided to couple encouraging the scientist to separate the realm of the sense from the realm of faith with the use of exoteric–esoteric writing.

10. Cf. the full title of the *Advancement of Learning:* "The Twoo Bookes of Francis Bacon. Of the proficience and *aduancement* of Learning, *diuine* and humane" (emphases mine).

11. In the words of a sagacious theologian: "*Für das Auge des Unglaubens ist . . . Gottes Tun ein geschehenes Weltereignis*" (Rudolf Bultmann, *Zur Frage des Wunders*, in *Glauben und Verstehen*, vol. 1, 9th ed. (Tübingen: Mohr (Siebeck), 1993), 225).

12. In order not to violate the imperative of internal coherence, Bacon earlier in the first book denounced as only "*Platoes* opinion" the idea that "all knowledge is but remembrance, and that the minde of man by nature knoweth all things, and hath but her owne natiue and originall notions (which by the strangenesse and darkenesse of this Tabernacle of the bodie are sequestred) again reuiued and restored" (AL 1.2, 3, III: 262).

13. This is the thirteenth quotation from the Old Testament in the first book of the *Advancement of Learning*. It is also the thirteenth of all biblical quotations or paraphrases in English in the first book, and the paragraph containing this quotation is the thirteenth paragraph containing quotations or paraphrases in English. The word "God" is used in thirteen paragraphs of the central section of the first book.

14. Cf. Johannes Calvin, *Commentaire de M. Iehan Calvin sur l'Epistre aux Romains* (Geneva, 1561), 12.

15. Cf. Calvin, *Commentariorum in Quinque Libros Mosis*, 7–8.

16. That the knowledge of the nature of God also constituted the primary subject matter of the apology to the divines is confirmed by the following signs. The word "God" is used sixty times in the first book of the *Advancement of Learning*, and the sixty-fifth and final paragraph of the section on the attributes and acts of God is the sixtieth paragraph after the paragraph containing the apology to the divines. There are also sixty-five paragraphs between Bacon's use of the word "Antichrist" (AL 1.34), and his use of the word "Sathan" (AL 1.99).

17. In an earlier version, Bacon used the less ambivalent words "leadeth to" (VT, III: 221).

18. That Bacon took the context of this biblical paraphrase into consideration is proved by the fact that in an other work he referred to Jesus' response to the Saducees as the "first canon against heresies delivered upon the case of the resurrection" (FL 7, III: 501).

19. Cf. Luther, *Vorlesungen über 1. Mose*, 9–10; Calvin, *Commentariorum in Quinque Libros Mosis*, 8–9.

20. Cf. AL 2, 189, III: 487: "[T]he Scriptures, which being written to the thoughts of men . . . and particularly of the elect, are not to be interpreted *only* according to the latitude of the proper sense of the place, and respectiuely towards that present occasion, whereupon the wordes were vttered; or in precise congruitie or contexture with the wordes before or after, or in contemplation of the principall scope of the place" (emphasis mine).

21. Since the offensive character of *Novum Organum* calls for a presentation that is in some places more dishonest and in other places more honest than that of some of his more reticent works, Bacon in this work calls natural philosophy religion's "most faithful handmaid," before saying that Jesus' response to the Sadducees "blends and unites into an indissoluble bond information concerning God's will and meditation on His power" (NO 1.89, I: 197).

22. It is for this reason that Bacon remarks that "he who pertinaciously asks how a man can be born when he is old only seems [videri] to be a disciple" of the Pharisee Nicodemus (DA 9.1, I: 833; John 3: 1–13).

23. In the central section of the first book the noun "power" has already been used seven times when the noun "omnipotencie" is used for the first and only time.

24. Bacon implicitly refers to these biblical sources by means of the catchwords "*Scrutamini*" and "*Contemplamini*" (MS 11 "De Haeresibus," VII: 241).

25. Declensions of "voluntas" with regard to God have been used seven times when the noun "omnipotentia" and a declension of "voluntas" with regard to man

are used for the first and only time. Declensions of "potestas" are used thirteen times in the meditation.

26. Cf. Augustine, *Enchiridion*, XI, XCVI.

27. *Tusculanae Disputationes*, IV. 26. That Bacon wanted to refer us to Cicero is proved by the fact that further on in the third book he says that "Sapientia ["sapience"—AL 2, 85, III: 357] was anciently [veteribus] defined as the knowledge of things divine and human" (DA 3.4, I: 568).

28. These observations are in conformity with Bacon's paraphrase of Philo Judaeus (*De Somnis*, I. 83–4) in the apology to the divines, since Bacon's Philo compares the ray of the sun, which "*openeth and revealeth all the terrestrial Globe; but . . . obscureth and concealeth the stars & celestiall Globe*," not to the divine ray but to the "*sence of man.*" Since it is the sense that, by analogy with the sun, not only "*doth . . . discouer naturall thinges*," but also "*darkeneth and shutteth up Diuine*," Bacon concludes that "diuers great learned men haue been hereticall, whilest they haue sought to flye vp to the secrets of the Deitie by the . . . winges of the Sences" (AL 1.6, 8, III: 267).

29. Ovid, *Metam.*, XV. 165. Ovid's Pythagoras adds that the spirit lives "with whatever body it chooses," and that it "will never perish" (166 ff.).

30. Cf. CNR 5, III: 22: "It is sufficiently certain [satis constat] that all things are changed, that nothing really perishes, and that the sum of matter remains entirely the same."

31. Cf. CNR 5, III: 22; *Historia Densi et Rari* (*Works* II, 243); *Sylva Sylvarum*, I. 100 (*Works* II, 383).

32. Cf. Strauss, *Die Religions-Kritik Spinozas*, in Strauss, *Die Religionskritik Spinozas und zugehörige Schriften*, vol. 1 of *Gesammelte Schriften*, ed. Heinrich Meier, 3rd ed. (Stuttgart: Metzler, 2008), 262–3.

33. Bacon himself only gives examples of "much and little," and "similitude and diversity."

34. That Bacon acted with foresight when he decided to postpone his equation of "sapience" with *Philosophia Prima* or "Sophia" is illustrated by the fact that late twentieth century interpreters still characterize Bacon's "sapience" as a "largely biblical definition of wisdom" (John Channing Briggs, "Bacon's Science and Religion," in *Cambridge Companion*, 175).

35. After having introduced *Philosophia Prima* or *Sapientia*, Bacon still "hesitates [haesito]" whether it is to be regarded as a "desideratum," but after having discussed the "axioms" and the "transcendentals," and after having equated *Philosophia Prima* with *Sophia*, he concludes that *Philosophia Prima* is "not undeservedly counted among the desiderata" (DA 3.1, I: 540, 544). Some other reasons for considering *Philosophia Prima* a desideratum can be found in the first book, where in discussing the three central "peccant humors" of learning Bacon observes that "after the distribution of particular Arts and Sciences, men haue abandoned vniuersalitie, or *Philosophia* prima." They therefore "disdain to spell, and so by degrees to read in the volume of Gods works, and contrarywise by continuall meditation and agitation of wit, doe vrge, and as it were inuocate their owne spirits, to diuine,

and giue Oracles vnto them," infecting their "meditations, opinions and doctrines with some conceits which they haue most admired" (AL 1.42–4, 30–1, III: 292–3).

36. To illustrate the power of knowledge of power, Bacon refers to Virgil, who compares the Berecynthian mother to Rome. The Berecynthian mother, "pleased that her sons are gods," embraces "a hundred descendants." Likewise, Rome, "happy with its offspring of men," will make its "empire cover the earth," and make its "spirit equal Olympus" (*Aen.*, VI. 781–7). Virgil refers to Berecynthia once more in the *Aeneid*, when she asks her son Jupiter to use his power "benevolently" (IX. 82–106).

37. Cf. Aristophanes, *Frogs*, 668–71 with Strauss, *Socrates and Aristophanes*, 245. Cf. Calvin, *Institutio*, I. 13.2 and 13.6 with Strauss, *Thoughts on Machiavelli*, 148. Cf. also: Lucretius, *De Rerum Natura*, V. 8–12.

38. Strauss therefore rightly calls natural theology the "forgotten" basis of "modern free thought" (*Progress or Return?* in *The Rebirth of Classical Political Rationalism: An Introduction to the Thought of Leo Strauss*, ed. Thomas L. Pangle (Chicago: University of Chicago Press, 1989), 267).

39. For a conceptual elaboration of the difference between natural theology and natural religion in view of the difference between the philosophic life and the non-philosophic life, see Heinrich Meier, *Über das Glück des philosophischen Lebens: Reflexionen zu Rousseaus Rêveries in zwei Büchern* (München: C. H. Beck, 2011), esp. Erstes Buch, II. Kapitel, and Zweites Buch as a whole.

40. Once again in the words of a sagacious theologian: "In der Frage nach Gott weiß das außergläubige Dasein um Gott." Bultmann draws the boundaries between faith and reason with an accuracy illustrative of the accuracy with which he delineated the boundaries of his own vocation. He points out that faith "wäre damit preisgegeben, daß zugegeben wird: über seine Legitimität könne von einem Standpunkt außerhalb seiner entschieden werden." He therefore concludes that faith does not so much "*kritisiert . . . die außerchristliche Frage nach Gott,*" as "*die Antwort,* die das außerchristliche Fragen sich gibt." Since faith "redet auf Grund der Offenbarung von Gott," and God's revelation "kann nur vernommen werden, im Glauben," that is, in "*Gehorsam,*" it "bestreitet dem Unglauben das Recht, von Gott anders als von der Gottesfrage zu reden." Rudolf Bultmann, *Theologische Enzyklopädie* (Tübingen: Mohr (Siebeck), 1984), 60, 63, 131, and Bultmann, *Die Frage nach der natürlichen Offenbarung*, in *Glauben und Verstehen*, vol. 2, 6th ed. (Tübingen: Mohr (Siebeck), 1993), 79, 86.

41. Augustine points out that "appellamus naturam, cognitum nobis cursum solitumque naturae, contra quem Deus cum aliquid facit, magnalia vel mirabilia nominantur" (*Contra Faustum Manichaeum*, XXVI. 3). According to Aquinas, "illa quae a Deo fiunt praeter causis nobis notas, miracula dicuntur" (*Summa Theologiae*, P. 1, Q. CV. 7). Bultmann is therefore right in saying that the "Gedanke des Mirakels" is not a "Glaubensgedanke." For one must "vorher Gott kennen, um sicher zu gehen"; one must have a "Kriterium" at one's disposal in order to know "ob ein Wunder von Gott stammt oder nicht." And "[w]enn man unter . . . Wunder ein Mirakel versteht, hat man den Gedanken des Tuns Gottes, der zum Wunder gehört, preisgegeben" (*Zur Frage des Wunders*, 216).

42. Spinoza describes a miracle as an event "cujus causa ex principiis rerum naturalium lumine naturali notis explicari nequit" (*Tractatus Theologico-Politicus*, vol. 6 (Hamburg, 1670), 70). Hume defines a miracle as a "violation of the laws of nature" (*Of Miracles*, in *An Enquiry concerning Human Understanding*, ed. E. Steinberg (Indianapolis (IN): Hackett, 1993), 76). According to Rousseau, who uses terms similar to Hume's, "[u]n miracle est, dans un fait particulier, un acte immédiat de la puissance divine, un changement sensible dans l'ordre de la nature, une exception réelle et visible à ses lois" (*Lettres écrites de la Montagne, Lettre Troisième*, in *Œuvres Complètes de Jean-Jacques Rousseau*, vol. 3, Bibliothèque de la Pléiade (Paris: Gallimard, 1964), 736–7).

43. Bacon distinguishes between "natural history," which relates the "deeds and actions of nature," and "civil history," which relates the "deeds and actions of men," but he points out that "divinity undoubtedly shines out in both, although more in the latter" (DA 2.2, I: 495). With regard to the history of learning, which is part of civil history, Bacon emphasizes that he wants events and their causes to be coupled, for this is the "honour and soul of civil history." And with regard to the coupling of events and their causes, the central matter that he wants the historians of learning to record is the "zeal and mixture of religions" (DA 2.4, I: 503).

44. In an early philosophic meditation, Bacon defined a miracle as a "new creation," but "not according to the law of the first creation" (MS 2 "De Miraculis Servatoris," VII: 233). In his *Confession of Faith*, our author explains that "whensoever God doth break the law of Nature by miracles, (which are ever new creations)," they concern "the work of redemption . . . whereto all God's signs and miracles do refer" (CF 9, VII: 221).

45. Bacon confirms this conclusion with regard to the miracles of God's Son by saying that "our Sauiour . . . did first shew his power to subdue ignorance, by his conference with the Priests and Doctors of the lawe; before he shewed his power to subdue nature by his miracles" (AL 1.61, 36, III: 299). That the miracles of Christ presupposed faith was also observed by Rousseau: "tant s'en faut que l'objet réel des miracles de Jésus fut d'etablir la foi, qu'au contraire il commençait par exiger la foi avant que de faire le miracle" (*Lettres écrites de la montagne, Lettre Troisième*, 734–5).

46. In the first book, Bacon pointed out that "delight in deceiuing, and aptnesse to be deceiued, imposture and Credulitie," although they appear to be of a "diuers nature, the one seeming to proceede of cunning, and the other of simplicitie; yet . . . they doe for the most part concurre." Therefore, a "credulous man is a deceiuer" (AL 1.33, 26, III: 287–8). In the central section of the first book, Bacon illustrates this observation by saying that "in the election of those Instruments, which it pleased God to vse for the plantation of the faith, . . . at the first he did employ persons altogether vnlearned, otherwise than by inspiration, more euidently to declare his immediate working, and to abbase all humane wisedome or knowledge" (AL 1.62, 36, III: 299).

47. The increased importance of natural theology for the mature Bacon's thought appears from the fact that the *Advancement of Learning* only mentioned

"the acknowledgement of God," and his "power, prouidence, and goodnesse" (AL 2, 79, III: 350).

48. Although he himself does not confess it until having reached the chapter on physics, the inevitably obscure discussion of angels, spirits, and demons at the end of the chapter on natural theology already displays Bacon's curiosity about the mysteries of faith, a curiosity whose seeds in their turn reach back to the *Advancement of Learning*, where our author called the nature of angels and spirits an "Appendix of Theologie, *both* Diuine and Naturall" (emphasis mine, AL 2, 79, III: 350). Bacon states that the inquiry into the nature of angels and spirits is "neither inscrutable nor interdicted," but that, rather, the "access to it has in great part been opened." For, so he argues, "although Sacred Scripture certainly teaches the faithful to let no man, pressing into what he does not know, deceive them in sublime discourse touching the worship of angels, if one considers this admonition diligently, the sober inquiry into angels, ascending to the knowledge of their nature by the ladder of corporeal things, is not at all prohibited." In fact, this "sober and grounded inquirie" may also "arise out of the passages of holie Scriptures" (AL 2, 79, III: 350). But if we consider "diligently" the Scriptural passage Bacon paraphrased, we indeed find Paul encouraging the faithful not to let themselves be seduced by the worship of "angels not seen"; the reason, however, why the apostle admonishes the faithful to let no man "deceive" them in "subtle discourse" is that he himself is "absent corporally," although he is with them "in spirit, joying and beholding the stedfastness of their faith in Christ, in whom the divine plenitude dwells corporally" (Col. 2: 4–18). In addition to approving of the inquiry into the nature of angels and spirits, Bacon states that he regards the "contemplation and knowledge of the nature, power, and illusions of impure and fallen spirits, not only from passages of Sacred Scripture, but also from reason or experience," as constituting "not the least part of spiritual wisdom." He does not count the knowledge regarding angels and demons among the "desiderata," though, since "many have engaged in it." Rather, he accuses "no small part" of the writers of this kind of being "vain, superstitious, or uselessly subtle" (DA 3.2, I: 546–7). At the beginning of the next chapter, Bacon summarizes his previous discussion of angels, spirits, and demons by saying that the inquiry into spirits is an "appendix of natural theology" (DA 3.3, I: 547), a summary that invites us to "append" the conclusion that, since the light of nature has seen neither angels and spirits, nor demons, it does not know how to distinguish the "sober" from the "sublime," but "subtle," "vain," and "superstitious" discourses of the 'spirited' writers, and that, therefore, in order to achieve "spiritual" wisdom, natural theology can only ascend to the knowledge of the nature of angels, spirits, and demons by climbing the "ladder of corporeal things." But the worship of a corporally absent God, who is with the faithful in spirit until His second corporealization, constitutes one of the greatest mysteries of faith.

49. Homer, *Illiad*, VIII. 19–27. Since Zeus wanted the other gods to submit to his will, he illustrated how much more powerful he was than other gods and mortals by means of the golden cord or chain.

50. Cf. E XVI "Of Atheisme," 51, VI: 413: "while the Minde of Man, looketh upon Second Causes Scattered, it may sometimes rest in them, and goe no further: But when it beholdeth, the Chaine of them, Confederate and Linked together, it must needs flie to *Providence* and *Deitie*."

51. Toward the end of the first book of the *Advancement of Learning*, Bacon describes what it is that the opposite of "a little or superficiall learning" achieves: "It taketh away all leuitie, temeritie, and insolencie, by copious suggestion of all doubts and difficulties, and acquainting the minde to ballance reasons on both sides, and to turne backe the first offers and conceits of the minde, and to accept of nothing but examined and tried. It taketh away vain admiration of any thing, which is the root of all weaknesse" (AL 1.96, 49, III: 314).

52. Cf. René Descartes, *Cogitationes Privatae*, 1619 (*Œuvres de Descartes*, vol. 10, ed. C. Adam and P. Tannery (Paris, 1908), 213): "Vt comoedi, moniti ne in fronte appareat pudor, personam induunt: sic ego, hoc mundi theatrum conscensurus, in quo hactenus spectator exstiti, larvatus prodeo," with *Discours de la Méthode*, 1.5.

53. The God of the Bible "made every thing beautiful in his time" (Eccl. 3: 11, Geneva).

54. The ancients describe Pan or "universal nature" as horned. His horns rise to a point, the tip of which reaches "as far as heaven [usque ad coelum]." Bacon explains that Pan's horns "touch [ferire] heaven," since the "heights of nature or the universal ideas in a certain manner [quodam modo] extend to the divine" (DSV VI "Pan," VI: 635, 637; DA 2.13, I: 522, 525).

55. That philosophy's agon does not become active until philosophy has become active appears from the fact that Bacon's first work of active or political philosophy does not yet call the philosopher theomachic (AL 2, 85, III: 356).

56. Virgil, *Georg.*, I. 278–86.

57. As a distinguishing mark of his power Pan carried a sheep-hook or staff crooked and curved at the top. According to Bacon this sheep-hook, which represents "empire," alludes to the "partly straight and partly oblique ways of nature." The staff is curved chiefly toward the top, because "all works of Divine Providence in the world are mostly wrought by winding and roundabout ways, namely that one seems [videri] to be able [possit] to put in motion [agi], whereas another is in truth moving [agatur]." DSV VI "Pan," VI: 635, 638; DA 2.13, I: 522, 526–7.

58. In a letter (June 1622) to Father Redemptus Baranzano, a professor of philosophy and mathematics at Anneci, Bacon states that there will be "no metaphysics after true physics has been discovered" (*Works* XIV, 375). Since forms are not abstracted from matter, and the inquiry into formal causes is therefore in a way already part of physics, we suspect that by metaphysics Bacon meant the inquiry into final causes.

59. Cf. Diogenes Laertius, X. 38, and Lucretius, *De Rerum Natura*, I. 159 ff. Bacon points out that the fact that Pan did not have any offspring is surprising, since "especially the male gods are very prolific." But he takes it to pertain to the

fact that the "world is sufficient and perfect in itself." For, he argues, "although the world generates through its parts, it cannot generate by means of the whole when no body exists beyond itself" (DSV VI "Pan," VI: 636, 640; DA 2.13, I: 522, 530).

60. Bacon illustrates the possibility of Pan stemming from all the suitors by means of the following quotation from Virgil: "Namque canebat uti magnum per inane coacta—*semina* terrarumque animaeque marisque fuissent—et liquidi simul ignis; ut *his exordia primis*—omnia, et ipse tener mundi concreuerit orbis" (*Eclog.*, VI. 31–4, emphases mine; DSV VI "Pan," VI: 636).

61. Cf. Maimonides, *Guide of the Perplexed*, II. 25.

62. Herodotus, *History*, 2.145; Apollodorus, *Library*, Epit. 7.38; Lucian, *Dialogues of the Gods*, XII. According to the *Homeric Hymn to Pan* (27–47) Hermes begot Pan by the daughter of Dryops, an Arcadian hero.

63. At the beginning of the Praefatio, Bacon remarked that primaeval antiquity, "except what we possess of it in Sacred Scripture," is "covered with oblivion and silence." The silence of antiquity was succeeded by the fables of the poets, which then were succeeded by the writings we possess (DSV Pr., VI: 625). Bacon elsewhere makes it clear that Sacred Scripture is to be counted among these writings, as he remarks that the writings in which the fables are related are the "most ancient of human writings, after the Sacred Scriptures," but that the fables themselves are "far more ancient" (DA 2.13, I: 521).

64. References to God or gods are used seventeen times in the fable.

65. Cf. Rousseau, *Lettre à Beaumont*, 955: ". . . si l'existence éternelle et nécessaire de la matière a *pour nous* ses difficultés, sa création n'en a pas de moindres, *puisque* tant . . . de *philosophes*, qui *dans tous les temps* ont médité sur ce sujet, ont tous *unanimement rejeté la possibilité de la création*, excepté peut-être un très petit nombre qui *paraissent* avoir sincèrement soumis leur raison à l'autorité; sincérité que les motifs de leur intérêt, de leur sûreté, de leur repos, rendent fort suspecte, et dont il sera toujours impossible de s'assurer tant que l'on risquera quelque chose à parler vrai" (emphases mine).

66. Hesiod, *Theog.*, 116–7, 126 ff.

67. Cf. POFCC, III: 86: "It is not written that in the beginning God created matter [hylen], but [that He created] heaven and earth," and E I "Of Truth," 8, VI: 378: "The *first creature* of God, in the workes of the Dayes, was the Light of the Sense; The last, was the Light of Reason; And his Sabbath Worke, ever since, is the Illumination of his Spirit. First he breathed Light, *upon the Face, of the Matter or Chaos*; Then he breathed Light, into the Face of Man; and still he breatheth and inspireth Light, into the Face of his Chosen" (emphases mine). Cf. MS 1 "De Operibus Dei et Hominis," VII: 233: "Vidit Deus omnia quae *fecerant manus* ejus, et erant bona nimis" (emphasis mine) with Gen. 1: 31 (Vulg.): "viditque Deus cuncta quae *fecit* et erant valde bona" (emphasis mine).

68. *Theog.*, 188–200.

69. But cf. Diderot, *Pensées Philosophiques*, XXI: "[S]i quelque chose doit répugner à la raison, c'est la supposition que, la matière s'étant mue de toute

éternité, et qu'ayant peut-être dans la somme infinie des combinaisons possibles un nombre infini d'arrangements admirables, il ne se soit rencontré aucun de ces arrangements admirables dans la multitude infinie de ceux qu'elle a pris successivement. Donc, l'esprit doit être plus étonné de la durée hypothétique du chaos que de la naissance réelle de l'univers" (*Œuvres Complètes de Diderot*, vol. 1 (Paris: Garnier Frères, 1875), 136).

70. Cf. Lucretius, *De Rerum Natura*, I. 1052 ff.

71. The most active political part of the fable on the trials and tribulations of Saturn alludes to the most active part of Bacon's political teaching, which, as we have seen (*supra*, 73 ff.), consisted in the liberation of philosophy from tyrannical politics. The ancient sources relate that Gaia and Coelum had prophesied to Saturn that he would be stripped of his power by his own son (Hesiod, *Theog.*, 459–65; Apollodorus, *Library*, I. 1.5). The representative of the matter of philosophy therefore acted with foresight when he devoured his children, although he could not prevent Jupiter from taking possession of his kingdom. But the tyrant only reigns in relative "security," because the castration of Saturn by Jupiter is nowhere mentioned in the ancient sources, although it is mentioned by Bacon, along with the fact that Saturn has only been "thrust down and overthrown," but not "destroyed and eliminated" (DSV XII "Coelum," VI: 650).

72. "[F]abulam philosophiam continere" is the center of the twenty-five times that the word "philosophia" is used throughout the fables. It is also the center of the three times that the word is used in the twelfth fable. The twelfth fable is the center of seven fables beginning with a conjugation of "trado." Although the concluding sentence of the twelfth fable brings Bacon's cosmogonical transgressions back within the bounds of theological respectability, it must have been a disjunctive reading of sentences like the concluding sentence of the twelfth fable that caused one of the leading scholars of Bacon's natural philosophy to make the typical mistake of concluding that to Bacon natural philosophy was an activity "*bounded* by" revealed theology, and that "[b]y far the most important boundary condition was the belief that the universe was not eternal." Graham Rees, "Bacon's Speculative Philosophy," in *Cambridge Companion*, 132–3, and Rees, in Francis Bacon, *Philosophical Studies. The Oxford Francis Bacon VI*, ed. Rees (Oxford: Clarendon Press, 1996), xlviii–l.

73. Cf. Gen. 1: 24 (Vulg.): "dixit quoque Deus *producat terra* animam viventem in genere suo . . ." (emphasis mine). Declensions of "sacer" and conjugations of "sancio" are used thierteen times in *De Sapientia Veterum*, the center of these thirteen times being "historia sacra" in the thirteenth fable.

74. The only time the word "omnipotentia" occurs in *De Sapientia Veterum* is in the thirteenth fable. The thirteenth fable refers to God three times, the central time of which He is not mentioned by name. Throughout *De Sapientia Veterum*, God is referred to three times without His name being mentioned, the central time of which occurs in the thirteenth fable. References to God or gods occur in seventeen fables, of which the thirteenth fable is the center.

75. Cf. *Descriptio Globi Intellectualis*, *Works* III, 735: "He who knows the catholic passions of matter, and through these passions knows what may be, cannot

not know also what has been, what is, and what will be, according to the sums of things."

76. Hesiod, *Theog.*, 116–23.

77. Cf. Plato, *Symposium*, 178a7–b2.

78. Cf. POFCC, III: 86: "One who philosophizes rightly and orderly must outrightly maintain that primary matter is united with primary form, and also with the first principle of motion, as it is discovered."

79. "Cupido, sive Atomus." Cf. POFCC, III: 111: "The atom is a true being, which has matter, form, dimension, place, resistance, appetite, motion, and emanation. Amid the destruction of all natural bodies it remains unshaken and eternal."

80. In the apology to the divines, Bacon defended learning against the "conceit" that the "ignorance of second causes should make a more deuoute dependance vppon God, which is the first cause," by paraphrasing the question Job asked his friends: "*Will you lye for God, as one man will doe for another, to gratifie him?*" ("Will ye speak wickedly for God's defense, and talk deceitfully for his cause?"—Job 13: 7, Geneva). For, Bacon explains, "certaine it is, that God worketh nothing in Nature, but by second causes," and if one would have it "otherwise beleeued," it is mere "imposture, as it were in fauour towardes God; and nothing else, but to offer to the Author of truth, the vncleane sacrifice of a lye" (AL 1.6, 6 ff., III: 264–7). Wolff is right in suspecting that to Bacon the "tiefste Bedeutung der Leistung der Atomisten . . . lag . . . in ihrem Streben nach einer in sich geschlossenen Welterklärung, allein aus der Natur selbst heraus, in ihrer Konzeption einer absoluten, rein kausalen Gesetzmäßigkeit des Weltgeschehens, in der Auffassung des Kosmos als Mechanismus, in der völligen Ausschaltung des Zweckbegriffes" (Emil Wolff, *Francis Bacon und seine Quellen*, vol. 1 (Berlin: Emil Felber, 1910), 285).

81. For the reasoning underlying this statement the reader is referred to chapter 2.

82. Rees points out that *De Principiis*, which he takes to have been written before 1620, was Bacon's "most extensive examination of various proposals regarding the principles of things." Rees, in *Philosophical Studies*, xxix.

83. Cf. Lucretius, *De Rerum Natura*, I. 151–4 with Pascal, *Pensées*, 198 (ed. Brunschvicg).

84. In the central aphorism on the human intellect in the first book of *Novum Organum,* Bacon observes that the human intellect "swells and cannot stand still or rest, but in vain aspires to go further." It therefore "cannot conceive of any end or limit to the world," but the "notion of there being something beyond always as of necessity springs to mind." Neither can the intellect conceive "how eternity has flowed down to the present day," for the distinction that it has "grown used to," the distinction between infinity "a parte ante" and infinity "a parte post," "cannot be maintained at all." For thence it would follow that "one infinity is greater than another," and that "infinity is being consumed and tends towards the finite." Although the "highest universals in nature" should be "positive, and taken as they are discovered," the human intellect is "unable to stop, but lusts after something still better known." But "straining for what is further off, it falls back on something nearer," namely on "final causes" (NO 1.48, I: 166–7). In other words, Bacon

seems to consider the intellect's own "finiteness" to be the cause of its inability or hampered ability to conceive of an "infinity" that is not also "finite." It is therefore on account of its inability to conceive of infinity within finiteness that the ordinary human intellect brings "eternity" nearer to itself by conceiving a final cause, thereby turning its own finiteness into potential eternity.

85. Simonides (*Frgm.* 575) tells us that he was the merciless son of Aphrodite (Venus) and the war-god Ares.

86. Conti's interpretation of the blindness of Cupid therefore only applies to the younger Cupid (*Mythologiae*, IV. XIV, 413): "[C]aecum Cupidinem ob dixerunt turpitudines, qua ab hominibus dignitatis fuae oblitis committuntur: cum illud potius ad diuinoru consiliorum mirabilitatem spectet, ad que cognoscenda caecum est & infans prorsus humanum genus, cu nulla vis humani ingenij possit ratione diuinae administrationis percipere."

87. Aristophanes, *Birds*, 693–702.

88. And it seems to be because providence also "forces" evil that Night is traditionally described as either an "image of God the Creator," or of the "forces of evil," or so we may conclude from a "solving" reading of a contradiction conveyed by the work of a scholar of Baconian contradictions. Cf. Whitney, *Francis Bacon and Modernity*, 121 with Whitney, "Cupid hatched by Night: The "Mysteries of Faith," and Bacon's Art of Discovery," in *Ineffability. Naming the unnamable: From Dante to Beckett*, ed. Peter S. Hawkins and Anne Howland Schotter (New York: AMS Press, 1984), 54.

89. The words "Cupido" and "Amor" counted together are used seventeen times in the seventeenth fable. Including the title of the book, the word "Sapientia" is used in seventeen places, of which the seventeenth fable constitutes the center.

90. Cf. Friedrich Nietzsche, *Ecce Homo. Wie man wird, was man ist; Warum ich ein Schicksal bin*, Aph. 6, and *Jenseits von Gut und Böse. Erstes Hauptstück: von den Vorurtheilen der Philosophen*, Aph. 23, with *Zur Genealogie der Moral, Erste Abhandlung*, Aph. 1.

91. Diogenes Laertius, X. 134.

92. Ibid.

93. *Pensées Philosophiques*, XVII, 132 (my translation).

94. This is the thirteenth time that the pronoun "I" is used in the *Essays*. This pronoun is absent in twenty-five *Essays*. It is used in thirty-three *Essays*, although, as we have seen (*supra*, 32), it is absent in the seventeenth *Essay*.

95. Paterson is therefore wrong in saying that the principle of the first sentence of the sixteenth *Essay* is that "unworthy or superstitious belief is preferable to disbelief" (Timothy H. Paterson, "On the Role of Christianity in the Political Philosophy of Francis Bacon," *Polity* 19, no. 3 (1987), 424). Despite or because of the fact that he argues *ad hominem*, William Blake was able to make a comment on the sixteenth *Essay* that comes somewhat closer to its principle: "An Atheist pretending to talk against Atheism" (*Complete Writings*, 404).

96. By using the plural form in the second part of the sentence, Bacon seems to allude to the distinction between philosophic men whose minds are brought back to the problem of religion, and non-philosophic men whose minds are brought to

religion by the minds of philosophic men. The word "religion" is used thirty-one times in the *Essays*, the sixteenth and central time of which is to be found in the sixteenth *Essay*, which is also the central *Essay* in which the word "religion" is used.

97. Augustine draws our attention to the significance of the combination of the numbers seven and eight, and of the addition of an extra number to "denote the bond of unity." The number seven signifies the "spirit and the day of rest," and the number eight signifies the "body and the day of resurrection." For, the Church Father explains, "when the saints receive their bodies again after the period of rest of the intermediate state, rest will not cease, but the whole man, body and soul united, will attain to the realization of his hope in the enjoyment of eternal life" (*Contra Faustum Manichaeum*, XII. 15, 17, 19, 21). We mention only the most obvious examples of the numerical signs by means of which Bacon indicates the significance of the sixteenth *Essay*: the word "Almightie" is used only once in the *Essays*, namely in the sixteenth *Essay* in which the word "God" is used (E XLVI "Of Gardens"); declensions of "eye" are used in sixteen *Essays*, and declensions of "lie" are used sixteen times throughout the *Essays*; after the word "God" has been used fifteen times, the word "Idoll" is used one time (E X "Of Love"); the sixteenth *Essay* is the central *Essay* in which declensions of "philosophy" are used; the sixteenth time that the word "wisdom" is used in the *Essays*, which is also the only time the word is used in Latin, is to be found in the sixteenth *Essay*.

98. This is the twenty-second and central time that the word "God" is used in the *Essays*. The word "Truth" is also used twenty-two times in the *Essays*.

99. In his early Sacred Meditation on atheism, Bacon penetrated more deeply into the inward sentiment of the foolish atheist. As it is not said that the fool thought in his heart that there is no God, Bacon argues that the fool not so much "senses inwardly" that there is no God, but that he "wants to believe" that there is no God. Because the fool sees that it would be liberating for him if there were no God, he "strives with all reason to persuade and convince himself of it." He "sets it up as a theme, position, or dogma, which he studies to assert, to maintain, and to confirm." Nevertheless, there "remains in him a spark of the original light whereby we acknowledge divinity," and he "strives in vain to wholly extinguish it, and to tear the sting from his heart." It is therefore not out of his "natural sense and judgment" that he assumes that there is no God. "Rather, as the comic poet [Plautus, *Aulularia*, 383] says: 'Then my mind joined my sentiment,' as if he were different from his mind." Bacon therefore concludes that the foolish atheist said that there is no God, rather than that he "sensed it in his heart" (MS 10 "De Atheismo," VII: 239–40).

100. In his Sacred Meditation, Bacon observes that it was only for "fear of the law and of public opinion" that the fool did not say with his mouth that there is no God. He adds, however, that if these restraints were removed, there would be no heresy "striving more eagerly to spread, sow, and multiply itself," than atheism (MS 10 "De Atheismo," VII: 240).

101. This is the center of the thirteen times that the word "God" is used in the twin *Essays* XVI and XVII. Nowhere in the *Essays* is the word "God" used more often than in the *Essay* on atheism.

102. Diogenes Laertius (X. 123) has "impious."

103. José de Acosta, *Historia Natural y Moral de las Indias* (1590), transl. E. Grimston (London, 1880), vol. 2, ch. V. 3, 301 ff.

104. Without using the word "God," Bacon refers to God or the notion of a god or of gods sixteen times throughout the *Essays*: E III "Dei"; E V "Dei"; E X "Idoll"; E XI "Deus"; E XIII "Deitie"; E XV "Deorum," "Gods"; E XVI "Deitie," "Deos," "Diis," "Gods," "Deus," "Deorum"; E XVII "Deity"; E XIX "Deus," "Dei."

105. In the context of his political argument that atheism is not as great an evil as idolatry, Pierre Bayle refers to Bacon's *Essay* on atheism, in which he finds "un recueil de reflexions très-sensées, avec la detestation d'un aveuglement si énorme" (*Réponse aux Questions d'un Provincial*, vol. 4 (Rotterdam, 1707), ch. 10, 118). Weinberger is therefore right in pointing out that the sixteenth and the seventeenth *Essays* taken together contain an "argument for atheism" (Weinberger, ed., *New Atlantis and the Great Instauration* (Wheeling (IL): Harlan Davidson, 1989), fn. 34, xxiv). The sixteenth and seventeenth *Essays* belonging together is supported by the numerical fact that the words "God" and "superstition" are both used thirteen times altogether in the two *Essays*. In the 1612 edition of the *Essays*, the *Essay* on atheism and the *Essay* on superstition already formed a pair (numbers twenty-seven and twenty-eight).

106. That Bacon argues on the basis of natural theology is confirmed by the fact that in a letter (1608) to the pious Tobie Matthew he states that the gunpowder plot (1605) justifies the "censure of the heathen, that superstition is far worse than atheism." For, he adds, "by how much it is less evil to have no opinion of God at all, than such as is *impious* towards his divine majesty and goodness" (emphasis mine; *Works* XI,10).

107. Plutarch, *De Superstitione*, 170a. The two well-known atheists Pierre Bayle (*Pensées Diverses sur la Comète*, 115) and Denis Diderot (*Pensées Philosophiques*, XII) also referred approvingly to Plutarch's statement on the relative harmlessness of atheism.

108. The precarious character of this rhetorical balancing act was noticed by Shaftesbury: "'Twas good fortune in my Lord BACON's Case, that he shou'd have escap'd being call'd an ATHEIST, or a SKEPTICK, when speaking in a solemn manner of the *religious Passion*, the Ground of SUPERSTITION, or ENTHUSIASM, (which he also terms *a Panick*) he derives it from an Imperfection in the Creation, Make, or Natural Constitution of Man." (Anthony Third Earl of Shaftesbury, *Characteristicks of Men, Manners, Opinions, Times*, vol. 3 (Indianapolis: Liberty Fund, 2001), 45).

109. There are seventeen aphorisms between the first reference to faith (NO 1.28) and the first reference to superstition (NO 1.46) in the first book of *Novum Organum*.

110. References to superstition, faith, and the Christian religion are used seventeen times altogether in the eighty-ninth aphorism of the first book of *Novum Organum*, in which references to divinity occur seventeen times. The center of these seventeen times is to be found in the eighty-ninth aphorism.

111. *A Brief Discourse touching the Happy Union of the Kingdoms of England and Scotland* (1603), *Works* X, 98.

112. The word "prophecie[s]" is used nine times in the thirty-fifth *Essay*.
113. Suetonius, *Vespasian*, 4.
114. Tacitus, *Historiae*, V. 13 (my translation).
115. Suetonius, *Vespasian*, 23.
116. References to religion occur thirty-five times in the *Essays* as a whole.
117. In the thirty-fifth *Essay*, Bacon mentions the prophecy of the daughter of Polycrates, who dreamed that "*Jupiter* bathed her Father, and *Apollo* anointed him." Following Herodotus (III. 124–5), Bacon reports that "it came to passe, that he was crucified in an Open Place, where the Sunne made his Bodie runne with Sweat, and the Raine washed it." After mentioning the crucifixion of Polycrates Bacon is silent on Christ the Saviour for the rest of the *Essays*.
118. Beal (Peter Beal, *Index of Literary Manuscripts*, vol. 1, 1450–1625 (London: Mansell, 1980), 37–8) mentions a copy that was in the possession of one of Bacon's scribes before July 1603. According to Rawley (*Resuscitatio* (London, 1657)), the *Confession* was composed "many years, before [Bacon's] Death."
119. See esp. Vickers, *Critical Edition*, 560–72. Strangely enough, even the fact that his own authority Basil Hall remarks that he knows "of no instance of a private person . . . writing a Confession of faith to clarify his own mind or to organize his thinking" (ibid., 560–1) did not arouse Vickers' suspicion as to Bacon's authenticity.
120. *The Life of the Honourable Author*, *Works* I, 14.
121. Cf. Calvin, *Institutio*, II. 6.1 with II. 12.1.
122. Cf. Calvin, *Institutio*, III. 23.8.
123. Cf. Publilius Syrus, *Sententiae*, 22: "Amare et sapere vix *deo* conceditur" (emphasis mine).
124. Cf. Augustine, *In Joannis Evangelium Tractatus CXXIV*, CX. 6.
125. Cf. Augustine, *Enchiridion*, XXVII.
126. Cf. Calvin, *Institutio*, III. 23.2 with Augustine, *Enchiridion*, CII. Cf. also: Karl Barth, *Der Römerbrief* (Zürich: TVZ, 2005), 73–4.
127. Cf. Calvin, *Institutio*, II. 16.13.
128. Cf. Augustine, *De Civitate Dei*, IX. 15.
129. Cf. Martin Luther, *Vorlesung über den Römerbrief*, in *Der junge Luther*, ed. Erich Vogelsang, vol. 5 of *Luthers Werke in Auswahl*, ed. Otto Clemen, 3rd ed. (Berlin: de Gruyter, 1963), 224, and Calvin, *Romains*, 24.
130. Cf. Calvin, *Institutio*, III. 23.3.
131. Cf. Calvin, *Romains*, 63, with Leibniz, *Théodicée*, Première Partie, 79.
132. In the context of an argument on the causes of longevity, Bacon refers to Simeon, whom he takes to have reached the age of ninety, and whom he refers to as a "religious man full of hope and expectation" (*Historia Vitae et Mortis*, *Works* II, 134–5). The evangelist Luke tells us that the Holy Ghost had revealed to Simeon that he would not see death before he had seen the Lord's Christ. After he had seen the baby Jesus in the temple, he asked the Lord to let him depart in peace, now that his eyes had "seen the Lord's salvation" (Luke 2: 25–35). In other words, either Bacon's Simeon was ninety years old when he visited the temple, and died

after having seen his hope and expectation fulfilled, or Bacon implies that Jesus was not the Christ. Bacon refers to Jesus Christ seventeen times altogether in the *Confession of Faith*. He refers to Christ thirteen times.

133. Tacitus, *Ann.*, V. 10 (my translation).

134. Cf. Friedrich Nietzsche, *Der Antichrist*, Aph. 29–35. That Bacon indeed meant the resurrection of the type of the redeemer by the apotheosis of error in the sixty-fifth aphorism of the first book of *Novum Organum* is further proved by the fact that it is in the sixty-sixth paragraph of the first book of the *Advancement of Learning* that Bacon equates the obtaining of "veneration & adoration as a God" with "that which the Grecians call Apotheosis" (AL 1.66, 38, III: 301).

135. Apollodorus, *Library*, I. 1–2.

136. Ovid, *Metam.*, II. 642–8.

137. Ibid., II. 629–34.

138. Apollodorus, *Library*, III. 10.3–4; Ovid, *Metam.*, XV. 531–40; Virgil, *Aeneid*, VII. 772.

139. Bacon calls attention to the limits of God's compassion by saying that Jupiter pitied the death of Memnon, whereas the ancient sources tell us that it was Memnon's mother Aurora who begged Jupiter for his pity (Ovid, *Metam.*, XIII. 576–622).

140. Bacon modeled his fabulous account of God's course of action on Machiavelli's story of Messer Remirro de Orco (*Principe*, VII). Cesare Borgia, called Duke Valentino by the vulgar, wanted to reduce the province of Romagna to obedience. He employed the cruel Messer Remirro de Orco, to whom he gave the fullest power. After Remirro had reduced the province to peace and unity, Borgia wanted to purge the spirits of the people, and gain them entirely to himself. Because he knew that past rigors had generated some hatred for Remirro, the Duke tried to show that if any cruelty had been committed, it had not come from him, but from the harsh nature of his minister. To this purpose, he had Remirro placed in the piazza at Cesena in two pieces, with a piece of wood and a bloody knife beside him.

141. Apollodorus, *Library*, III. 10.4.

142. The "earth in motion," which signifies the counselor of God in the third in addition to signifying the begetter of the counselor of the philosopher in the second fable (cf. *supra*, 55 ff., and 83 (268 fn. 92)), signifies the "nature of the vulgar" in the ninth fable (DSV IX "Soror Gigantum," VI: 645).

143. Ironically enough, by correcting Bacon's supposed misunderstanding of the passage of Habakkuk Spedding actually approaches Bacon's own corrected version of the passage. For according to the nineteenth-century editor of Bacon's works, the true meaning of the passage is to be paraphrased in the following way: "write so as that the message may be quickly read, in order that the reader may run at once and without loss of time" (*Works* I, 516, fn. 1). Had Kiernan taken Bacon's Latin translation of the passage of Habakkuk into consideration (DA 2.11, I: 516: "*quivis etiam in cursu ea perlegere possit*"), he would not have explained away Bacon's misquotation by saying that it is most likely that Bacon translated the Vulgate. For the Vulgate has: "ut percurrat qui legerit eum" (Kiernan, *Advancement of Learning*, 273). Since Sessions did not take the context of Bacon's misquotation

of Habakkuk into consideration, he did not have an explanation for his observation that Bacon's "personal rendition" of Habakkuk is "almost exactly the same" as the marginal transcription of the verse in the Geneva Bible ("he that runneth may read it"), except for the "key adverb *by*" (Sessions, *Francis Bacon and the Classics*, 245).

144. By saying that poetry "exhibits events and fortunes in accordance with merit and the law of Nemesis," Bacon implies that "true" history is not in accordance with the "law of Nemesis" (DA 2.13, I: 518). The parallel passage in the *Advancement of Learning* accords with this conclusion, since it has that, "because *true Historie* propoundeth the successes and issues of actions, not so agreable to the merits of Vertue and Vice," poetry "faines them more iust in Retribution, and more according to Reuealed Prouidence" (AL 2, 73, III: 343).

145. Hesiod, *Theog.*, 223–4.

146. References to nature occur in twenty-two fables.

147. The importance of Bacon's observation on the magnitude of ancient wisdom is underscored by the fact that it occurs in the exact physical center of Bacon's book on ancient wisdom (see 65 of 129 of the editions of 1609 and 1617, and 100 of 199 of the edition of 1634; see also Studer, *Grapes Ill–Trodden*, 185, fn. 384), as well as by the fact that it is in the context of this observation that the word "sapientia" is used for the ninth and central time in *De Sapientia Veterum*.

148. That Bacon was well aware of the implications of unveiling the true ambition of human wisdom is indicated by the fact that immediately after unveiling this ambition he remarks: "But let us look at human wisdom" (DSV Pr., VI: 626). But that Bacon had great confidence in the art of veiling appears from the fact that he writes to Tobie Matthew (February 17, 1610, *Works* XI, 144–5) that although he "dissenteth in religion," he thinks "the greatest inquisitor in Spain" will allow his book.

149. "Zelotypia" is the thirteenth word in an enumeration of words referring to zeal, gods, religion, and sects in the eighteenth fable. It is also the center of nine references to envy or jealousy in *De Sapientia Veterum* as a whole.

150. This interpretation is supported by the fact that Bacon translates "swan" by "cycnus" only in the central fable and in the appendix to the fable of Diomedes, whereas he uses the noun "olor" to refer to the swans into which the followers of Diomedes had been changed for bewailing the death of their leader.

151. But cf. Martin Luther: "Rechte demut weyß nymmer das sie demutig ist / denn wo sie es wißte / so wurd sie hohmutig von dem ansehen der selben schonen tugent" (*Das Magnificat verdeutschet und ausgelegt*, in *Schriften von 1520–1524*, ed. Otto Clemen, vol. 2 of *Luthers Werke in Auswahl*, ed. Otto Clemen, 5th ed. (Berlin: de Gruyter, 1959), 150).

152. That the *Essay* on praise and the *Essay* on vain glory belong together is underscored by the fact that references to praise and references to vain glory both occur thirteen times altogether in the two *Essays* taken together.

153. "Ego, non Dominus" is the center of seventeen biblical quotations in the ninth book of *De Augmentis*. Biblical quotations occur in seventeen chapters of the partition of the sciences.

154. In the central section of the first book of the *Advancement of Learning*, Bacon remarks that "Saint *Paule*" was the "onely learned amongst the Apostles" (AL 1.62, 36, III: 299).

155. Aristotle, *Politica*, 1253a28–30.

156. That by higher conversation Bacon indeed meant the conversation with God is proved by the fact that he mentions the "Holy Fathers of the Church" as the example of those who were "truly and really" involved in higher conversation (E XXVII "Of Frendship," 81, VI: 437).

157. Bacon illustrates this point by way of discussing the fate of King Henry III of France, who had joined the Holy League of France (1576), a group of radical catholics bent on the "Extirpation of the *Protestants*," before the "same League was turned upon Himselfe," and he was assassinated (1589) by Jacques Clément, a fanatic partisan of the League. Bacon observes that "when the Authority of Princes, is made but an Accessary to a Cause; and . . . there be other Bands, that tie faster, then the Band of Sovereignty, Kings begin to be put almost out of Possession." For, he explains, leagues within the state "raise an Obligation, Paramount to Obligation of Soveraigntie, and make the King, *Tanquam unus ex nobis*," also knowing good and evil (Gen. 3: 22; E XV "Of Seditions and Troubles," 44, VI: 408; E LI "Of Faction," 156, VI: 500).

158. See Calvin, *Institutio*, III. 19.15.

159. This is the center of the thirteen times that the word "religion" is used in the third *Essay*.

160. Cf. Luther, *Vorlesung über den Römerbrief*, 273.

161. Cf. Thomas Aquinas, *Summa Theologiae*, Pt. 2.2, Q. XXV.I.

162. Cf. Calvin, *Institutio*, II. 8.51.

163. Bacon states that the "perfective good" is the highest degree of "passive good" because it is "the greater to raise a thing to a more sublime nature than to preserve it in its existing state." For, he explains, "in all things there are to be found some nobler natures to the dignity and excellence whereof inferior natures aspire as if to their sources and origins." Thus, he goes on to say, "the following was not unaptly said of man: '*Igneus est ollis vigor & coelestis origo.*'" It was Anchises who spoke these words when referring to the seeds of the souls to which fate owed another body, but which had to pass through the river Lethe before being reunited with the body (Virgil, *Aeneid*, VI. 703–51). Bacon remarks that the "depraved and preposterous imitation of the perfective good is the very plague of human life, and a kind of violent whirlwind, dragging off and destroying everything." He takes this imitation to consist in "men hastening with blind ambition to an exaltation that is only local" (DA 7.2, I: 724), as is the case, we observe in passing, when a soul receives another body.

164. C. Thornton Forster and F. H. Blackburne Daniell, eds., *The Life and Letters of Ogier Ghiselin de Busbecq*, vol. 1 (London: C. Kegan Paul & Co, 1881), 224 ff.

165. In *De Augmentis*, Bacon comments on one of Solomon's proverbs: "The righteous man regards the life of his beast, but the tenderness of the impious is

cruel" (Prov. 12:10). He restates the proverb in the following words: "The righteous man pities [miseretur] the life of his beast, but the compassion [misericordia] of the impious is cruel." Bacon observes that in man's nature a "noble and excellent feeling of compassion has been implanted," which "extends even to the brutes that by divine ordinance have been subjected to his command." Therefore, Bacon goes on to argue, this compassion has "some analogy with that of a prince towards his subjects." By way of example, Bacon refers to the Turks, who, although they are "by race and habit a cruel and bloody people," are "wont to give alms to brute creatures, and cannot endure the vexation and torture of animals." However, Bacon emphasizes, "lest every kind of compassion should seem to be countenanced," Solomon "soundly adds that the compassion of the impious is cruel." For as "wicked and criminal men are spared being struck with the sword of justice," this kind of compassion is "more cruel than cruelty itself." And whereas cruelty is practiced on "individual persons," this compassion "arms and unleashes the whole army of criminals upon innocent men, while granting the army of criminals impunity" (DA 8.2, I: 758). This comment amounts to the suggestion that it is because its cruelty is "granted impunity" on account of its piety that the compassion of the "impious" is "more cruel than cruelty itself," and that it is because the pious do not pity the lives of "innocent men" that their cruel compassion extends itself to the brutes or "de-animated" men that have been "subjected to their command" by their divine "prince." It was, however, only by not changing, of all words, the word "impious" that Bacon could draw our attention to this suggestion without appearing to be impious. And it was only Bacon's proficiency in the art of writing that allowed him to underscore this suggestion by declaring that he interprets the proverbs of Solomon "in a natural sense," that is, from the perspective of the impious (DA 8.2, I: 751).

166. That charity is the central subject of the thirteenth *Essay* is indicated by the fact that the words "charity" and "love" occur in thirteen *Essays* altogether, of which the thirteenth *Essay* is the central *Essay*.

167. *Discorsi*, III. 6.14–5.

168. See *Principe*, XIII, and *Discorsi*, II. 20.

169. Cf. *Charge of Owen* (1615, *Works* XII, 164): Religion is "a trumpet that inflameth the heart and powers of a man (above all things) to daring and resolution," for "a votary that hath but an apprehension, though false and erroneous, of the pains of Hell in case he should break his vow when he hath vowed their death [of kings, TvM], doth but despise and scorn the executioner or tormentor here on earth, and thinks only what he shall do and not what he shall suffer." In the second *Essay*, Bacon points out that on the occasion of the suicide of Emperor Otho (69 AD) "*Pitty* (which is the tenderest of Affections) provoked many to die, out of meere compassion to their Soveraigne, and as the truest sort of Followers" (E II "Of Death," 10, VI: 380). We need not be surprised that "compassion" is the Baconian translation of "caritate" (Tacitus, *Hist.*, II. 49). In his *History of the Reign of King Henry VII*, Bacon reports a speech of Dr. Warham, held by order of Henry: "Do you not know that the bloody executioners of tyrants do go to such errands with an halter about their neck, so that if they perform not they are sure to die

for it? And do you think that these men would hazard their own lives for sparing another's?" (*Works* VI, 145). Weinberger notes that no original of this speech is extant, and that the speech is not reported by the historians (Weinberger, ed., *The History of the Reign of King Henry the Seventh* (Ithaca: Cornell UP, 1996), 122, fn. 347). We therefore suspect Bacon of having invented the speech in order to criticize Henry for not having understood the Christian assassin's contempt for death.

170. *De Rerum Natura*, I. 80–3, 101. Lucretius is referred to three times throughout the *Essays*, once in the *Essay* on truth, once in the *Essay* on unity in religion, and once in the *Essay* on atheism.

171. *Discours Préliminaire*, par. 69, in *Encyclopédie*, xvii.

172. There are twenty-two chapters between the chapter on natural theology and the chapter on sacred and inspired theology. Bacon divided knowledge into philosophy and theology in chapter 3.1 of *De Augmentis*, which is the sixteenth part of the book as a whole. The word "theologia" is used in sixteen chapters of the partition of the sciences, and it is used sixteen times in the chapter on sacred theology.

173. The parts of Jesus' commandment that Bacon left out sound even less human: "bless them that curse you," and "pray for them that hurt you" (Matt. 5: 44–5).

174. *Metam.*, X. 330–1.

175. Bacon supports this conclusion in the *Advancement of Learning* by observing that it is the "lawe" of God that "discloseth sinne," and by commending the "diducing of the Lawe of GOD to cases of conscience" (AL 2, 191, III: 489).

176. In the *Advancement of Learning*, Bacon divided the law of God into the "lawe of Nature, the lawe Morall, and the lawe Positiue" (AL 2, 191, III: 489).

177. In a letter to Buckingham (22 Oct. 1623; *Works* XIV, 437), Bacon states that he holds *De Augmentis* "to be very rational."

178. *Discours Préliminaire*, par. 102, in *Encyclopédie*, xxiv–xxv (my translation).

179. At the end of the *Advancement of Learning*, Bacon says that the "good" of his work, "if any bee, is due *Tanquam adeps sacrificij*, to be incensed to the honour . . . of the diuine Maiestie . . ." (AL 2, 193, III: 491). The fat of the sin offering was to be burnt on the altar (Lev. 4: 8–10).

Chapter 4

1. Ovid tells us that Jupiter sent the deluge upon man as a punishment for the impiety that had been displayed by Lycaon (*Metam.*, I. 163–261; see also Apollodorus, *Library*, III. 8.2). According to Apollodorus, the deluge was a local affair (*Library*, I. 7.2).

2. Ovid relates that it was Pyrrha's piety that made her reluctant to throw her mother's bones (*Metam.*, I. 384–8).

3. The subtitle of the fable has "restitution," suggesting that renovation, instauration, and restitution are in a way the same things.

4. Clement of Rome (*First Epistle to the Corinthians*, XXV–VI) and Tertullian (*De Resurrectione Carnis*, XIII) look upon the phoenix as a symbol of the resurrection of Christ.

5. In the first book of *Novum Organum* the seventy-seventh aphorism is the twenty-second and central aphorism in which the word philosophy is used.

6. Cf. Aristotle, *Metaphysica*, 1026a19–22, 1064a28–b7, 1072b28–30.

7. Cf. Aristotle, *Physica*, 195a23–6; *Metaphysica*, 983a30–3.

8. Apophth. 275, *Works* VII, 164. The historical fate of Aristotelianism thus almost gives a touch of irony to one scholar's strong claim that it was actually in order to rule out the possibility of an omnipotent God that Aristotle presented his doctrine of final causes. David Bolotin, *An Approach to Aristotle's Physics: With Particular Attention to the Role of His Manner of Writing* (Albany (NY): SUNY Press, 1998), 31 ff., 152.

9. *Works* XI, 64; Letter to Tobie Matthew (October 10, 1609), *Works* XI, 137. Cf. Aristotle, *Ethica Nicomachea*, 1096a14–8. Although there is one scholar who rightly describes Bacon's scorning of Aristotle as "unavoidable slander" (Jerry Weinberger, *Science, Faith, and Politics: Francis Bacon and the Utopian Roots of the Modern Age: A Commentary on Francis Bacon's Advancement of Learning* (Ithaca: Cornell UP, 1985), 177–8; cf. 168–9, 189–90, 280–1, 292–3, 300), most scholars pay insufficient attention to the fact that the substance of Bacon's disagreement with Aristotle largely concerns the mode of philosophy's political teaching. Of the many comparative treatments of Baconian and Aristotelian doctrines, we mention only the most unpolitical (Fulton H. Anderson, *The Philosophy of Francis Bacon* (Chicago: University of Chicago Press, 1948), 191–216), and the most political (Robert K. Faulkner, *Francis Bacon and the Project of Progress* (Lanham (MD): Rowman & Littlefield, 1992), 34–5, 108–9, 111, 116 ff., 148, 267).

10. It was the opinion of the prudent statesman Pericles that friendship should only extend itself *usque ad aras* (Plutarch, *Praec. Ger. Reip.*, 808b; *De Vit. Pud.*, 531d; Gellius, *Noctes Atticae*, I. 3.20). Bacon illustrates the change of the state of things by means of a quotation from Tacitus (*Ann.*, I. 3–4) on the change of the Roman Republic into the Roman Empire, looked at from the perspective of the latter days of the reign of Augustus. By thus referring to the fundamental change of the state of things that was inaugurated by the first Roman Emperor, our author seems to allude to the more fundamental change of the state of things that was inaugurated by the "emperor" whose ascending years ran parallel to the declining years of the first Roman Emperor (DA 3.4, I: 549).

11. Cf. AL 1.37, 28, III: 290: "[M]any wits and industries haue ben spent about the wit of" Aristotle, "whom many times they haue rather depraued than illustrated."

12. In anticipation of our substantial interpretation of *New Atlantis*, we notice that it was not before the birth, but after the death of Aristotle (322 BCE) that Solamona decided (288 BCE) to forbid strangers entrance to the island of Bensalem (NA 15, III: 144–5).

13. That Bacon was an effective diviner is proved by his philosophic successor Thomas Hobbes, *De Homine*, XI. 6, 15; *Leviathan*, I. 6 (24, 30), XI (in princ.).

14. *Discorsi*, II. 2.2.

15. In one of his early philosophic meditations, Bacon points out that "being well disposed towards one's enemies is among the crown jewels of the Christian law," and is an "imitation of divinity." Although he adds that this virtue of charity "may have something in it of magnanimity rather than of pure charity," he takes the "summit and exaltation of charity" to consist in "being grieved and distressed in the inner chambers of one's heart when evil seizes one's enemy," and in "not rejoicing as if the day of one's revenge had come" (MS 4 "De Exaltatione Charitatis," VII: 235). Since this magnification of charity for the sake of goodness does not answer to the consideration that the magnanimity of charity could also consist in being well disposed toward one's enemies in anticipation of a higher revenge at the end of days, or in anticipating this revenge by ending one's enemies' days, it was with good reason that Bacon decided to republish neither this meditation, nor the meditation in which he equates the duties of charity with the works of compassion (MS 7 "De Hypocritis," VII: 238).

16. *Discorsi*, II. 2.2, III. 1.2.

17. Xenophon, *Memorabilia*, I. 4.5–7, IV. 3.3–12.

18. Bacon himself supports this suspicion by employing examples that only partly support his statement that "all things seem [videantur] to revolve around man and not around themselves." For he says that the revolutions of the stars serve for the "distinction of the seasons and the distribution of the regions of the world," the atmospheres for the "prognostication of weather," and the winds for "navigation, mills, and engines." But human discoveries or inventions are presupposed in order for the stars, the atmospheres, and the winds to start "revolving around man" (DSV XXVI "Prometheus," VI: 670–1).

19. It seems to be in order to make actual history agree with the argument of poetry that Bacon places the "doctrine of the prerogatives and excellencies of the human race" among the "desiderata," and that he commends the "collecting of these miracles of human nature," especially from "credible [fide] history," in what he calls the "record of the triumphs of mankind" (DA 4.1, I: 581 ff.).

20. In the *Advancement of Learning*, Bacon says that if the substance of the human soul was immediately inspired by God, "it is not possible that it should bee (otherwise than by accident) subiect *to the Lawes of Heauen and Earth; which are the subiect of Philosophie*" (AL 2, 103, III: 379). But if the substance of the human soul originates in the dust of the earth, this "accident" can only be philosophy itself.

21. Apollodorus, *Library*, I. 7.1; Ovid, *Metam.*, I. 78–83.

22. Cf. Machiavelli, *Discorsi*, II. 5 and III. 1.1, and Harvey C. Mansfield, *Machiavelli's New Modes and Orders: A Study of the Discourses on Livy* (Chicago: University of Chicago Press, 2001), 300.

23. Duris of Samos, to my knowledge the only ancient source that mentions an affective connection between Prometheus and the goddess of wisdom, only says

that Prometheus was punished because he loved Pallas (*Schol. ad Apollon. Rhod.*, II. 1249).

24. In *A Conference of Pleasure* (1592), fortitude is praised as the noblest of virtues since it leaves "the minde in entire possession." Whereas "other vertues deliuer us from ye rule of vices," fortitude "alone deliuereth us from the servitude of fortune," although it "giueth" the mind "no feeling," but "leaueth it emptye" (ed. James Spedding (London: Longmans et al., 1870), 9–10).

25. Seneca speaks of the "imbecility" and not of the "frailty" of man (*Ad Lucilium*, LIII. 12).

26. This remark constitutes the thirteenth and final section of the twenty-sixth fable.

27. Hesiod tells us that in order to increase his son's reputation Zeus wanted Heracles to set Prometheus free (*Theog.*, 526–33).

28. DSV XX "Erichthonius," VI: 661. The names Pallas and Minerva counted together are referred to seventeen times altogether in the fables of *De Sapientia Veterum*.

29. This sequence of events explains why d'Alembert was able to ignore Bacon's division by placing man immediately after God, whereas Bacon was compelled to let nature intervene. See *Discours Préliminaire*, par. 66, in *Encyclopédie*, xvii.

30. In the expanded version of the fable of Perseus in *De Augmentis*, Bacon strengthens the connection between Medusa, Christ, and Hercules by saying that it was the overcoming of tyrannies, "under which the minds and vigour of the people are succumbed and prostrated, as under the aspect of Medusa," that brought Hercules divinity (DA 2.13, I: 532).

31. The instrumental nature of the virtue of Hercules explains why Bacon says that "extirpers of Tyrants" like Hercules are "iustly" "honoured but with the titles of Worthies or Demy-Gods" (AL 1.66, 38, III: 301).

32. *Supra*, 56 (264 fn. 55), and 59 ff.

33. *Supra*, 85 ff.

34. References to gods and God the Lord occur thirteen times in the fable on the author of man.

35. It seems to be more than a coincidence that Apollodorus, whom Bacon knew well, tells us that Hercules liberated Prometheus so that the two-natured Cheiron could die (*Library*, II. 5.4, 11). Cf. *supra*, 75 (267–8 fn. 83).

36. Aelian (*On Animals*, 6.51) only says that in his anger Zeus bestowed a drug to ward off old age on those who informed him on Prometheus' theft.

37. Cf. Aeschylus, *Prometheus Bound*, 120–3; Machiavelli, *Discorsi*, I. 2.3 (in pr.); Plato, *Statesman*, 274b6–e1.

38. Significantly enough, the only time Bacon refers to Prometheus' invention of fire as theft is in the context of the accusations Jupiter brought against the inventor of fire (DSV XXVI "Prometheus," VI: 670). If the inventor of fire indeed had to anticipate the accusation of theft, the observation of Servius (*Ad Virg. Ecl.*, VI. 42) and Fulgentius (*Myth.*, 2.6) that Pallas assisted Prometheus in obtaining the fire gains extra weight.

39. Having observed that the Renaissance brought the equation of Prometheus and Adam, Blumenberg claims that the projection of Prometheus on Adam could only mean that "der Verlust des Paradieses als *felix culpa* gesehen werden sollte: als die Chance des Menschen, er selbst aus sich selbst zu sein, ganz gleich, wodurch er es geworden war." But what the historian fails to notice is that the possibility for man to be himself by means of himself wholly depends on him not having become himself by having lost his guiltless self. Hans Blumenberg, *Arbeit am Mythos*, 6th ed. (Frankfurt am Main: Suhrkamp, 2006), 393–4, 408.

40. Earlier in the *Essay*, Bacon said that "Scripture exhorteth us: *To possesse our Soules in Patience*," commenting that "Whosoever is out of *Patience*, is out of Possession of his *Soule*" (E LVII "Of Anger," 170, VI: 510). But whereas Scripture "exhorts" believing men to patiently anticipate the "possession of their souls" in the Kingdom of God (Luke 21: 10 ff.), Bacon stimulates angry men to 'possess their souls in patience' in anticipation of the kingdom of man.

41. This is the center of the thirteen times that the word "revenge" is used in the *Essays*.

42. This interpretation is in conformity with Bacon's remark that he considers the "*work [opus]*" of the *Instauratio Magna* to be a "birth of time rather than of wit," if only because he explicitly adds that the "beginning of the thing arose from the mind of one man" (IMEpD, I: 123, emphasis mine).

43. The king of Egypt commanded the midwives of the Hebrew women to kill the child at the time of birth ("partus tempus"—Vulg.), should it be a man ("masculus"—Vulg.). But since the midwives feared God, they let the male children live (Exod. 1: 15 ff.).

44. Abraham Cowley, "To the Royal Society," in Thomas Sprat, *The History of the Royal Society of London. For the Improving of Natural Knowledge* (London, 1667).

45. Cf. *Principe*, VI with *Discorsi*, I. 9.3.

46. *Discorsi*, I. 10.1.

47. That Bacon considered Moses to be, among other things, the inventor of religion is subtly argued in the *Historia Vitae et Mortis*, where the philosopher takes the life and death of Moses as the starting point for his natural history of life and death (cf. *Works* II, 148 with 132–3).

48. That Machiavelli also considered religion to be founded on sovereignty is intimated by the fact that the Florentine philosopher describes not the first but the second Roman king as having been religious (*Discorsi*, I. 19.1).

49. Machiavelli, *Discorsi*, I. 9.2. That the lesser degrees of sovereign honour in a way ensue from the highest degree is illustrated by the fact that Bacon ascribes the second place to "*Legislatores, Lawgivers*; which are also called, *Second Founders*, or *Perpetui Principes*, because they Governe by their Ordinances, after they are gone" (E LV "Of Honour and Reputation," 164, VI: 506).

50. Virgil, *Aen.*, VI. 823–4. It was therefore with good reason that Bacon decided not to let his earlier conclusion that "the case was doubtfull, and had opinion on both sides" (AL 2, 145, III: 431) return in the expanded Latin version of the *Advancement of Learning*.

51. According to Livy, Brutus surrendered to fortune by exacting the punishment from his sons (*Ab Urbe Condita*, II. 5).

52. *Discorsi*, I. 16.3–4, III. 1.6, III. 2, III. 3.

53. Cf. Livy, *Ab Urbe Condita*, II. 1, 6–7 with Machiavelli, *Discorsi*, III. 4.

54. *Discorsi*, III. 2. Cf. Mansfield, *Machiavelli's New Modes and Orders*, 308.

55. See *Principe*, XXV.

56. By casting doubts on the legitimacy of the foundations (1485) of the reign of Henry VII, while emphasizing that Henry's laws were made "out of providence of the future," and by desiring that "those into whose hands" his *History of the Reign of King Henry the Seventh* "shall fall," will "take in good part" his "long insisting" upon Henry's laws, because the subject has "some correspondency with" his own "person," Bacon seems to stimulate his reader to discuss his own case by means of the case of King Henry VII (*Works* VI, 27–32, 36, 92–3, 97, 225–6, 239; cf. Weinberger: *History*, 32, fn. 21, 33, fn. 27, 46, fn. 79, 48, fn. 88, 83, fn. 200, 191–2, fn. 702).

57. Plutarch only says in reply to Jason that one should pass over smaller matters so that in greater matters one may oppose the errors of the people "more stubbornly" (*Praec. Ger. Reip.*, 24; *De San. Praec.*, 24).

58. Although Bacon takes the sting out of this hazardous argument by concluding it with the remark that men should "leave the future to Divine Providence" (DA 7.2, I: 731).

59. Ovid, *Trist.*, I.37.

60. *Gorgias*, 464b2–5d9.

61. The reasonableness of the injustice of rhetoric is underscored by the fact that Bacon emphasizes that Plato's injustice sprang from a "just hatred of the rhetoricians of *his* time" (DA 7.2, I: 672, emphasis mine).

62. *History of the Peloponnesian War*, III. 42.

63. Ibid., III. 49.

64. Considering the fact that a self-sufficient tiller of the ground who practices obedience by bringing an offer to the Lord is necessarily a hypocrite, Bacon may have compared Prometheus the deceiver to a hypocrite in his interpretation of the fable in order to strengthen the similarity between Prometheus and Cain (DSV XXVI "Prometheus," VI: 673).

65. By passing to religion the parable passes to its central section.

66. Cf. Augustine, *De Civitate Dei*, XIV. 28, XV. 1.

67. Hesiod in a way fuses the two parables by saying that Zeus hid the fire from man after Prometheus had attempted to deceive him with his sacrifice, but that it was only after Prometheus had stolen the fire that Zeus sent the evil Pandora (*Theog.*, 535 ff.; *Works and Days*, 45 ff.).

68. *Supra*, 85 ff.

69. Hesiod tells us that it was by the will of Zeus that hope stayed in the vase (*Works and Days*, 96–9).

70. Cf. *supra*, 72, and 89 (270 fn. 104).

71. In his *Historia Vitae et Morti,* Bacon subtly extends the line of Cain to or beyond the Flood. For he says that eleven generations of patriarchs are counted before the Flood, whereas there are only eight generations counted of the sons of Adam by Cain, "which would seem to make Cain's progeny more long-lived." But Scripture does not connect the line of Cain to the Flood, thus suggesting that it had already become extinct before God's punishment flooded the earth (Gen. 4: 16 ff.). This seems to explain why Bacon emphasizes that the longevity of antediluvian men cannot be attributed to "grace or the holy line" (*Works* II, 132–3).

72. *Discorsi,* I. 26, II. 23.

73. The fact that Bacon's philosophic successors were to arrive at different results on the basis of similar premises illustrates that the disagreements between the modern political philosophers largely concern the mode of philosophy's political teaching. Cf. Hobbes, *Elements of Law,* Ep. Ded.; Descartes, *Discours de la Méthode,* 2.2; Spinoza, *Tractatus Politicus,* I. Cf. also: *supra,* 8 (247 fn. 33).

74. The adjective "novum" underscores that the conditions for philosophic method had changed since Andronicus of Rhodes published Aristotle's *Organon* around 40 BCE. For, Bacon argues, "it would be absurd and contradictory to think that things which as yet have never been done can be done except by means which as yet have never been tried" (NO 1.6, I: 157).

75. Unwilling to deprive scientists of the pleasure of disparaging Bacon's method, we will not discuss Bacon's contribution to science, understood in a narrow sense.

76. A comparison of the subtitles of the first and the second book of *Novum Organum* confirms this interpretation. For the first book, which establishes method, carries the subtitle "Aphorismi De Interpretatione Naturae *et* Regno Hominis," whereas the second book, which shows method in action, bears the subtitle "Aphorismorum de Interpretatione Naturae *sive* de Regno Hominis" (emphases mine), which seems to imply that the "Kingdom of Man" had to be established by means of the "Interpretation of Nature" before it could "reign" by interpreting nature.

77. Bacon underscores the importance of natural history for philosophy by introducing it in the 16th and central paragraph of the Distributio Operis. It is hardly a coincidence that the section on idols (NO 1.38–68) consists of thirty-one aphorisms. Cf. *supra,* 268 fn. 84, 269 fn. 102, and 292 fn. 97. In a way natural history seems to be the visualization of sacred history.

78. The maturation of Bacon's argument to obliterate his authorship and therefore his authority can aptly be illustrated by means of a comparison between two early versions of the argument. For whereas in *Temporis Partus Masculus* the narrator told his "son" that because the latter had grown too accustomed to a leader, he should give himself to his "father" in order to be "restored" to himself (TPM, III: 539), in the later *Cogita et Visa,* Bacon states that he did not conduct himself as a leader, so that afterward there would be "no need for a leader" (CV 19, III: 619).

79. Many scholars have been led and led astray by Bacon's rhetoric of novelty. But it was Bacon himself who emphasized that although the things he proposed were

"entirely new in their very kind"—considering that "every *Medicine* is an *Innovation*" (E XXIIII "Of Innovations," 75, VI: 433)—they were "transcribed from a very old model," namely "the world itself and the nature of things and of the mind" (IMEpD, I: 123). Since Bacon elsewhere remarks that *De Augmentis* "exhibits a mixture of new conceits and old; whereas the Instauration gives the new unmixed," we find that some of the philosophic foundations of and of the rhetoric of the *Instauratio Magna* are to be found in *De Augmentis*, especially since Bacon adds that *De Augmentis* "may be some preparative, or key, for the better opening of the Instauration" (AHW EpD, VII: 13).

80. The word idol occurs in twenty-two aphorisms altogether in the first book of *Novum Organum*.

81. Cf. also Ps. 104: 30: "The Lord instaures the face of the earth," which Bacon translates as: "[A]ll things do renew and spring amain; So that the earth, but lately desolate, Doth now return unto the former state." *Translation of Certaine Psalmes into English Verse* (1625), *Works* VII, 283.

82. Augustine argues that the instaured temple signifies the church that was to be built by Christ (*De Civ. Dei*, XVIII. 48).

83. Machiavelli points out that double virtue is shown by those commanders who not only conquer the enemy, but also instruct their army before it "comes hand to hand" with the enemy (*Discorsi*, III. 13.3).

84. Cf. Machiavelli, *Discorsi*, III. 14. Bacon turning the arts of Cain into arts of benevolence is in a way a repetition of Cain himself turning the fruits of the ground into an offering for the Lord. One of the key differences between Bacon and Cain, on the other hand, seems to lie in the fact that Bacon is less of a hypocrite toward the recipients of his arts, seeing that he is honest about their benevolence being only anticipatory.

85. That truth precedes benevolence is indicated by Bacon's remark that fruits are only "as it were sponsors and guarantees of the truth of philosophies" (NO 1.73, I: 182). For, as he explains elsewhere, "unless deeper contemplations of nature are demonstrated and recommended by some evident and excellent utility, they are blown about and extinguished by the wind of vulgar opinion" (CV 9, III: 599). Therefore, they demand to be "delivered with strange advantages of eloquence and power" (FL 8, III: 503).

86. Significantly enough, twenty-two of the twenty-three aphorisms that constitute the section on hope (NO 1.92–114) contain the word hope. The only aphorism in which the word hope is absent is the only aphorism in which God is present (NO 1.93).

87. Bacon emphasizes that he "shall be content" if his "labours bee esteemed, but as the better sorte of wishes: for," he explains, "it requireth some sense, to make a wish not absurd" (AL 2.16, 62, III: 329). But the only reason why Bacon's wishes were not "absurd" was that they were wishes "of the better sort," which explains why Bacon would be "content" if they were esteemed as such.

88. This explains why Bacon says that Aesculapius was a god, although in reality he was only an apotheosized man (cf. DA 4.2, I: 586–7 with Cicero, *De*

Natura Deorum, II. 62). The chapter treating the good of the body is the central chapter of the partition of the sciences in which the word philosophy is used. It is also the sixteenth chapter in which the word nature occurs.

89. *Works* II, 103, 106, 172. The maturation of Bacon's considerations on the rhetorical utility of hope can be aptly illustrated by means of a comparison between the two versions of the fable on war. For whereas in the early version (1609) Bacon had said that the fact that Perseus chose the only Gorgon that was mortal signifies that he "chose a war that could be finished and carried through to the end," and that he "did not pursue vast or infinite hopes" (DSV VII "Perseus," VI: 642), in the later version (1623) he only says that the fact that Perseus chose the mortal Gorgon signifies that he "did not aim at impossibilities" (DA 2.13, I: 533). In the early Sacred Meditation on earthly hope, Bacon still thought that he could counter "etherealized" hope by dismissing earthly hope as such. For there he argued that "the purer the sense of the presence is, without infection or tincture of the imagination, the more prudent and the better is the soul," and that "all hope is to be spent on the future life in heaven" (MS 6 "De Spe Terrestri," VII: 236–7).

90. *Supra*, 65 ff.

91. Cf. Thomas Aquinas, *Summa Theologiae*, Pt. 1.2, Q. LXVI.VI.

92. *Supra*, 151 ff.

93. Bacon emphasizes that knowing "how to put affection in the place of affection and how to use the help of one for the subjugation of the other" is "of great use in morals and civil matters" (DA 7.3, I: 736).

94. This comment is not to be found in Celsus' *De Medicina* (I. 1).

95. Cf. Machiavelli, *Discorsi*, I. 25.

96. Strauss, *Thoughts on Machiavelli*, 176.

97. Cf. *supra*, 153 ff.

98. That the word "reliefe" covers a broad spectrum of meanings is indicated by the fact that, "to speak plainly and clearly," the "true" end of knowledge is the "discovery of all operations and possibilities of operations from immortality (if it were possible) to the meanest mechanical practice," as Bacon speaks somewhat more plainly and clearly in an early work (VT III: 222).

99. Cf. Machiavelli, *Discorsi*, II. 3. In his *Essay* on the extension of the bounds of empire (DA 8.3, I: 792), Bacon underscores the force of love by counseling that now that he cut the tree down, "it is to be procured, That the *Trunck* of *Nebuchanezzars* Tree of *Monarchy*, be great enough, to beare the Branches, and the Boughes; That is, That the *Naturall Subjects* of the Crowne or State, beare a sufficient Proportion, to the *Stranger Subjects*, that they governe" (cf. Dan. 4: 10–37). For "all States, that are liberall of Naturalization towards Strangers, are fit for *Empire*." In the central section of the *Essay*, Bacon goes on to argue that the strangers from another world are to be naturalized by means of the force of their own love. For "*Sedentary and Within-doore Arts*, and delicate Manufactures . . . have, in their Nature, a Contrariety, to a Military disposition." The ancient states "had the use of *Slaves*, which commonly did rid those Manufactures," but since slavery has been abolished, "in greatest part, by the *Christian Law*," that which "commeth nearest

to it, is, to leave" the aforementioned arts and manufactures "chiefly to Strangers" (E XXIX "Of the true Greatnesse of Kingdomes and Estates," 93 ff., VI: 447 ff.). In other words, the Christian law of love liberates Christians from the slavery of militant love by directing them toward peaceful arts.

100. Although more implicitly than in the thirteenth *Essay*, Bacon also distinguishes between goodness and Christian goodness in the context of his treatment of the "culture of the mind." For after having said that the "state of mind" that he proposes was signified by Aristotle as having the "character not of virtue but of divinity" (*Ethica Nicomachea*, 1145a15f.), and that Pliny the Younger presented the virtue of Traian "not as an imitation but as an exemplar of the divine" (*Panegyricus Traiani*, 74), our author concludes that these passages "taste of heathen and profane boasting, grasping at shadows [umbras] greater than the substance," whereas the "true religion and holy Christian faith aims at things themselves." A few lines further down, Bacon says that "by aspiring to a divine similitude of goodness or charity no man was or will ever be endangered." On the contrary, "we are induced to this imitation: 'Love your enemies' etc." (Matt. 5: 44). So, Bacon concludes, "in the very archetype of the divine nature the heathen religion gives these words: 'Optimus Maximus,' whereas Sacred Scripture declares: 'His compassion is over all His works'" (Ps. 145: 9). Having reminded his reader that he does not strive after "beautiful things," but after "utility and truth," Bacon goes on to recall Virgil's parable of the 'two gates of sleep' (*Aen.*, VI: 893–6): "through the gate of horn an easy exit is given to the true shadows [umbris], but through the gate of ivory the Manes send false dreams to heaven." Bacon comments that "the magnificence of the gate of ivory is remarkable, but that the true dreams pass through the gate of horn" (DA 7.3, I: 741 ff.). We, on the other hand, comment that although Bacon's "heathen religion" of goodness 'tastes somewhat of boasting,' since it presents an "archetype of the divine nature" that is unattainable by "virtue," its "shadows" and "dreams" aim at the "substance" of things, because they serve "truth and utility." The Christian dreams of charity, by contrast, not only bring man in "danger," since they induce him to "imitate" divine compassion, but are also "false," since they aim at "shadows" of the substance of things.

101. The serpent's venom is in its proverbial tail, since the Geneva Bible (1 Cor. 8: 1) has "edifieth" instead of "buildeth up."

102. *Ad Att.*, XIII. 42.

103. *Nemean Odes*, V. 31–2.

104. The God whom the causes and beginnings of the greatest things are referred to constitutes the seventh and central reference to God in the first book of *Novum Organum*.

105. In his early (1608) fragment *Of the True Greatness of the Kingdom of Britain*, Bacon states that scarcely any of the "great monarchies of the world" did not have "their foundations in poverty and contemptible beginnings," in which point they "conform to the heavenly kingdom, of which it is pronounced, *Regnum Dei non venit cum observatione*" (*Works* VII, 56).

106. In his *Historia Vitae et Mortis,* Bacon argues that it is on account of their "fervency and inexperience of evil" that the young are "inclined to religion and devotion" (*Works* II, 212).

107. Augustine, *Contra Faustum Manichaeum,* XII. 8; *De Trin.,* IV. 4.7.

108. Cf. *supra,* 33–4. Considering the fact that in his *History of the Reign of King Henry the Seventh* Bacon repeatedly deviates from the historians by either praising Henry's foresight or criticizing his lack of foresight, one of the main goals of this "mirror for princes" seems to be to teach future princes the art of foresight, which Henry lacked to the extent that he could "see clear . . . through superstition" only for his "times," which were "full of mutations and rare accidents," especially "in matters of religion and the state ecclesiastical" (*Works* VI, 25, 31, 48, 59, 69, 73–4, 83, 120, 131, 156, 159–60, 178–9, 237–8, 244; cf. Weinberger: *History,* 31, fn. 19,47, fn. 84, 57, fn. 109, 64, fn. 136, 67, fn. 144, 68, fn. 151,75, fn. 167, 100, fn. 264, 132, fn. 389, 136, fn. 403, 202, fn. 743, 203, fn. 747, 210, fn. 797; *The History of the reigns of K. Henry VIII, K. Edward, Q. Mary, and part of Q. Elizabeth, Works* VI, 19).

109. Cf. Suetonius, *Divus Augustus,* 28.

110. *Imago Civilis Julii Caesaris* and *Imago Civilis Augusti Caesaris, Works* VI, 335 ff.

111. Augustine argues that the number six signifies both the building of the second temple and the building of the body of Jesus Christ (*De Trin.,* IV. 5.9).

112. *Historia Naturalis et Experimentalis Ad Condendam Philosophiam: sive Phaenomena Universi* (1622), *Works* II, 9.

113. In his letter to Father Fulgentio (1625), Bacon says that "the ages and posterity" will "*perhaps*" make the sixth part of the *Instauratio* flourish (*Works* XIV, 532; emphasis mine).

114. *Imago Civilis Julii Caesaris, Works* VI, 335.

115. In *De Sapientia Veterum,* seventeen fables occur between the first and the sixth and final time that the word "instauratio" is used.

116. Cf. Machiavelli, *Discorsi,* III. 35.2. Bacon warns that the "prudence of having foresight with regard to one's own affairs," if it is "professed and openly advertised," has always been regarded not only as "unpolitical," but also as "unfortunate and inauspicious." The central example by which he illustrates this point consists in an uncommented quotation from the prophet Habakkuk, who said "they exult and sacrifice in their nets" about the impious, who remain as yet unpunished for their impiety (Hab. 1: 13–7; DA 8.2, I: 770–1). Since this is the only one of Bacon's examples that has an "inauspicious" but "open" and not "unfortunate" end, we suspect that Bacon counsels the impious to have "foresight with regard to their own affairs," without, however, "professing and advertising" their "prudence" openly. This suspicion is proved correct when we consider that Bacon elsewhere uses the same quotation from Habakkuk as an example of that which "carries no burden with it" (DA 6.3, I: 684–5).

117. Cf. Rousseau, *Du contrat social,* II. VII. 9–10.

118. In his letter to Fulgentio, Bacon says that he hopes that his *Instauratio* "appears [videntur]" to proceed from the "providence and immense goodness of God" (*Works* XIV, 532).

119. The "divine goodness" is referred to three times altogether in the Praefatio and the Distributio Operis of the *Instauratio Magna*, with in each case sixteen paragraphs in between the references. Significantly enough, the center of these three paragraphs, which is the fifteenth paragraph of the Distributio Operis, is the only paragraph in which the "miseries of men" are referred to (IMDO 15, I: 139–40). That this is not a coincidence is proved by the fact that an early version of one of the paragraphs of the Praefatio that does *not* refer to the divine goodness, contained a reference to the divine goodness (cf. TPM, III: 527 with IMPr. 4, I: 131). References to philosophy occur sixteen times altogether in the Praefatio and the Distributio Operis. The twin center of these references is constituted by the fifteenth and sixteenth paragraphs of the Distributio Operis.

120. Cf. Machiavelli, *Discorsi*, II. 13 with II. 14.

121. Blake comments on the final sentence of the first *Essay*: "Bacon put an End to Faith" (*Complete Writings*, 398).

122. Cf. Augustine, *De Doctrina Christiana*, I. XXXVIII.42.

123. Cf. *supra*, 27 ff. In his letter to Father Fulgentio, Bacon calls his *Essays* "Faithful [fideles] discourses, or the interiors of things [Interiora rerum]" (*Works* XIV, 531). Cf. also DA 7.2, I: 729–30, where Bacon reserves the treatment of "frauds, cautions, impostures, and vices" for a treatise on the "interiors of things [*Interioribus Rerum*]."

124. Significantly enough, all four references (*Essays* XIII, XV, XXXIX, and LVIII) to Machiavelli in the *Essays* concern the matter of sects.

125. Machiavelli, *Discorsi*, II. 5.1.

126. God is not mentioned in the *Essay* on the vicissitude of sects and religions.

127. On the Baconian presupposition that the Christian religion is a form of superstition, we find Bacon subtly arguing that the human foundation of the Christian religion is built on the sand (cf. Matt. 7: 24–7; Luke 6: 48–9). For it is by means of an implicit reference to Plutarch's condemnation of superstition that he illustrates his observation that "human foundations are sometimes built upon the sand" (DA 8.2, I: 791; Plutarch: *De Superstitione*, 165A).

128. Cf. *supra*, 85 ff.

129. *Supra*, 69 and 90.

130. Cf. the movement from the first toward the last *Essay* with the movement from book I toward book VI of Lucretius' *De Rerum Natura*. Cf. Leo Strauss, *Notes on Lucretius*, in Strauss, *Liberalism Ancient and Modern* (New York: Basic Books, 1968), 82 ff.

131. We call to mind that it was the most profound critic of those wanting to "rendre la philosophie populaire" (Denis Diderot, *Pensées sur l'Interpretation de la Nature*, XL, in *Œuvres Complètes de Diderot*, vol. 2 (Paris: Garnier Frères, 1875)

38) who confessed that the Prometheus on the frontispiece of his *Discours sur les sciences et les arts* is "le Citoyen de Génève" (Jean-Jacques Rousseau, *Lettre à Lecat*, in *Œuvres Complètes de Jean-Jacques Rousseau*, vol. 3, Bibliothèque de la Pléiade (Paris: Gallimard, 1964), 102; cf. Heinrich Meier, *Les rêveries du Promeneur Solitaire: Rousseau über das philosophische Leben*, 2nd ed. (München: Carl Friedrich von Siemens Stiftung, 2010), 13–20).

Chapter 5

1. *Contra Faustum Manichaeum*, XII. 14, 18.

2. If understood in this sense, Weinberger is right in calling *New Atlantis* "at once the core and the prolegomenon to Bacon's teaching" ("Science and Rule in Bacon's Utopia: An Introduction to the Reading of the New Atlantis," *American Political Science Review* 70 (1976), 885).

3. Bacon adds that such a circle of tales is "not fit" for a "Writing" like the *Essays*, which we take to be an implicit reference to the tale of *New Atlantis*, especially since the fifty-eighth paragraph of *New Atlantis* discusses the subject matter of the fifty-eighth *Essay*. For the Father of Salomon's House says that the fellows "declare natural divinations of diseases, plagues, swarms of hurtful creatures, scarcity, tempests, earthquakes, great inundations, comets . . . and divers other things," and that they "give counsel . . . what the people shall do for the prevention and remedy of them" (NA 58, III: 166).

4. That *New Atlantis* as it were presupposes a forestallment of the Parousia is also indicated by the fact that whereas thirty is the number of Jesus Christ, the section describing the "true state of Salomon's House" contains thirty-one paragraphs beginning with the words "We have" (NA 21–58, III: 156–66). Cf. *supra*, 295 fn. 77.

5. Paradoxically enough, if Rawley is right in saying that Bacon "designed" *New Atlantis* "for this place," that is, as an appendix to *Sylva Sylvarum*, the significance of the work only increases. For had Bacon only "devised" the "fable" to the "end that he might exhibit therein a model or description of a college instituted for the interpretation of nature," as the same Rawley claims (NA Pr., III: 127), he would have prefixed it to the work in which he interprets nature.

6. *Works* II, 337.

7. The Latin translation of Plato's *Critias* that was probably used by Bacon bears the title: "Critias, Siue, Atlanticvs" (Plato, *Platonis Avgvstiss Philosophi, omnium quae exstant operum*, vol. 3, ed. Joannes Serranus [Jean de Serres] and Henricus Stephanus [Henri Estienne], (Geneva, 1578), 105). Bacon himself refers to Plato's *Critias* as *Atlanticus* in his *Essay* on prophecies (E XXXV "Of Prophecies," 114, VI: 465).

8. *The Life of the Honourable Author*, *Works* I, 9–10. The scholars who claim that *New Atlantis* was written in 1624 or 1625 provide no or no convincing

evidence for their claim. See Spedding, *Works* III, 121; Weinberger, *New Atlantis and the Great Instauration*, xxxvii; Bronwen Price, *Francis Bacon's New Atlantis: New Interdisciplinary Essays* (Manchester: Manchester UP, 2002), 1, 23 (fn. 2).

9. In the *Advancement of Learning*, Bacon made the following comment on the subject of government: "it is a part of knowledge, secret and retyred in both these respects, in which things are deemed secret: for some things are secret, because they are hard to know, and some because they are not fit to vtter" (AL 2, 179, III: 473–4). Since by uttering himself in a way fitting the "secret" nature of government Bacon drew the art of government out of "retirement," it was with good reason that he decided not to let this remark return in *De Augmentis*.

10. The Latin edition that Bacon probably used has "opus habent" (Plutarch, *Lysander*, 19, in *Vitae Parallellae seu Comparatae*, vol. 2, transl. W. Xylander (Basilea, 1579), 324). The original Greek has: "δέονται."

11. Cf. Plato, *Timaeus*, 19b–c.

12. In the preface to the *Historia Ventorum* (1622), Bacon predicts that "if human diligence does not fail, the winds can be employed for many other purposes than navigation and the turning of mills" (*Works* II, 19).

13. To the extent that Bacon did not intend this symbolic meaning in *New Atlantis*, Paterson is right in saying that the wings hanging downward are a "symbol of knowledge which is directed towards earth rather than heaven." T. H. Paterson, *The Politics of Baconian Science: An Analysis of Bacon's "New Atlantis,"* unpubl. diss. (Yale University, 1982), 100.

14. In the next paragraph, this suspicion is implicitly confirmed when the narrator tells his crewmembers that since they "are come . . . amongst a Christian people, full of piety and humanity," they should not show their "vices or unworthiness" before their hosts (NA 3, III: 134).

15. Our attention is drawn to this point by the fact that after one of the representatives of the island had already gone aboard the ship, another representative did not go on board because he had been "warned by the Conservator of Health of the city that he should keep a distance." A few moments later, however, a third representative did go aboard, holding a scarlet orange in his hand, of which the crew initially assumed that it was a "preservative against infection," but of which they learned the next day that it was actually used as a "remedy for sickness taken at sea" (NA 1–3, III: 130–4). The thirty-third paragraph is the final paragraph of the first uninterrupted series of paragraphs beginning with the words "We have," which words are used seventeen times altogether in this first series of paragraphs. The central paragraph of this series is moreover the paragraph describing the "Chambers of Health" (NA 27, III: 158). References to Christianity and Jesus Christ the Savior and Messiah occur seventeen times altogether in *New Atlantis*, and there are seventeen paragraphs between the last use of the word "sickness" and the first use of the word "diseases" (NA 3, 21, III: 134, 157).

16. In the previous paragraph the narrator had compared the situation of the crew in the wilderness of the waters with the situation of Jonah, who was between the jaws of a whale before the Lord made the whale vomit him out upon the land

in response to his remark that "salvation is of the Lord" (NA 3, III: 134; Jonah 2: 10). The evangelist Matthew compares the three days and nights Jonah spent in the whale's belly to the three days and nights the Son of Man was to spend in the heart of the earth (Matt. 12: 40), which makes the crew's "picture" of their "salvation in heaven" even more earthly.

17. Significantly enough, in *New Atlantis* as a whole the adjective "happy" is used twice as often as the adjective "holy."

18. The crew's aptness to "think there was somewhat supernatural in this island, but yet rather as angelical than magical" (NA 11, III: 140) was therefore justified to the extent that the supernatural is identical to the poetical.

19. This interpretation is supported by the fact that if the events related in the separate paragraphs had occurred chronologically, nine days would already have been spent by the time the narrator said that "[b]y that time six or seven days were spent" (NA 17, III: 151). The reason why the narrator seems to doubt whether it was on the seventh or on the eighth day that he went abroad will be discussed below.

20. In the context of his discussion of dramatic poetry, Bacon emphasizes that it is "one of the great secrets of nature" that "men's minds are more open to affections and impressions when they are congregated than when they are alone" (DA 2.13, I: 519).

21. Pseudo-Dionysius Areopagita, *De Caelesti Hierarchia*, 7.

22. Thomas Hobbes was a careful reader of Bacon's thought, as appears from a careful reading of chapter III. XXXVII.13 of *Leviathan*.

23. Considering that the governor says that navigation "increased . . . within these six-score years" (NA 12, III: 140), it seems reasonable to take 1612 (1492 + 120) as the year of the voyage.

24. Further on in the fourteenth paragraph, the governor substitutes "main accident of time" for "Divine Revenge," thereby indicating that it was not in order to explain what had caused the inundation that he mentioned the divine revenge (NA 14, III: 143). In the *Historia Vitae et Mortis*, Bacon counts the Flood itself among the great accidents (*Works* II, 148). Inundations, floods, deluges, and earthquakes, understood as natural accidents, occur 14 times in the fifty-eighth *Essay*.

25. Plato, *Timaeus*, 25c; Acosta, *Historia Natural y Moral de las Indias*, vol. 1, ch.III. XXVI, 178. Further on in the paragraph, the governor contradicts Acosta once more. For he says that the inhabitants of America started going naked after having gone apparelled, whereas according to Acosta it was exactly the other way around (vol. 2, ch. VII. II–III, 449–52).

26. That the seemingly obvious does not invite questioning is illustrated by the fact that only one scholar questioned the title of Bacon's fable in writing. Tobin L. Craig, "On the Significance of the Literary Character of Francis Bacon's New Atlantis for an Understanding of His Political Thought," *The Review of Politics* 72 (2010), esp. 228–39.

27. *Timaeus*, 22a–23d.

28. Herodotus, *History*, I. 29–30; Plutarch, *Life of Solon*, 25–6.

29. After saying that the Coyans made an expedition "upon this *our* island," the governor says that the voyage of the Coyans "upon *us*" would not have had "better fortune, *if they had not* met with enemies of greater clemency. For the king of *this* island . . . dismissed them all in safety" (NA 14, III: 142, emphases mine). In other words, it is the Bensalemites that would have been the enemies 'of lesser clemency' had the Coyans not met with the clement king of the island of Atlantis.

30. The name of the first king of Atlantis was Atlas (*Critias*, 114a).

31. This accords with Socrates' remark that only he who is both a philosopher and a statesman is capable of setting words in motion (*Timaeus*, 19b–20c).

32. Cf. *Critias*, 120e ff. Significantly enough, God is not mentioned in the fourteenth paragraph of *New Atlantis*, although the fourteenth paragraph is the center of the thirteen paragraphs in which God and the divine occur.

33. *Discorsi*, II. 5.2.

34. Plato, *Nomoi*, 676a ff. Cf. Leo Strauss, *The Argument and the Action of Plato's Laws* (Chicago: University of Chicago Press, 1975), 41–2; Seth Benardete, *Plato's Laws: The Discovery of Being* (Chicago: University of Chicago Press, 2000), 92.

35. This sequence of events is in accordance with the Athenian stranger's narrative (*Nomoi*, 678b7 ff.).

36. Cf. Rousseau, *Du contrat social*, II. VII. 1.

37. This makes it almost ironical that throughout *New Atlantis* the word "humanity" is used where one would expect the word "charity," which is used not once in the work describing Bacon's charitable programme in motion. The ambivalence of Bensalemite goodness is indicated by the fact that the words "good" and "goodness" occur in thirteen paragraphs altogether, which reminds us of the twofold goodness treated in the thirteenth *Essay*.

38. Cf. Machiavelli, *Discorsi*, III. 17 (ad fin.). Further on in the fifteenth paragraph, the governor stimulates us to compare Solamona with Solomon by saying that *he* thinks Solamona found "himself to symbolize *in many things* with that king of the Hebrews" (NA 15, III: 145, emphases mine).

39. The word "nature" is used in nine paragraphs, of which the seventeenth paragraph constitutes the center. The seventeenth paragraph is also the central paragraph of the paragraphs in which the pronoun "I" occurs.

40. The fact that what transpires in the following paragraphs is not meant for the crew as a whole is indicated by the fact that on the day of the coming of one of the Fathers of Salomon's House fifteen days had gone by since the crew had been told to leave the coast within sixteen days, and that at the end of the fifteenth paragraph the narrator emphasizes that there was still reason for having certain crew members "look to" their "ship" (NA 15, III: 147). After all, when the governor told the crew that they had been given "licence" to stay on land for the space of six weeks, he explicitly added that "in this point" the "law" was "not precise" (NA 4, III: 135). The exclusivity of the second half of *New Atlantis* is also indicated by the fact that the governor describes his relation of Salomon's House in the fifteenth paragraph as a digression (NA 15, III: 145), which implies that knowledge of science was not a precondition for the crew's conversion to the island

devoted to modern science, although some of the ten auditors of the governor-priest may have become 'initiates' after their conversion. This implication is underscored by the fact that when the crew had come to take themselves for "free men," they were still physically restrained, which makes us suspect that Salomon's House was located more than a mile and a half from the walls of the city of Renfusa, the new home of the majority of the crew (cf. NA 4 with NA 16, III: 135, 147).

41. In the fifteenth paragraph the governor says that every twelve years two ships with three "Fellows or Brethren" of Salomon's House in each are sent into the world, whereas according to the Father of Salomon's House it is twelve fellows that sail into foreign countries (NA 15, 46, III: 146, 164).

42. Plato, *Nomoi*, 951a4–c6, 953c4–d2. As the narrator distinguishes himself ("I") from the crew ("we"), Joabin distinguishes himself ("I") from the Bensalemites ("we" and "they").

43. That it is Joabin who took the initiative of the three conversations he had with the narrator is indicated by the fact that with regard to the second and third conversations the narrator says that Joabin came to him "again" (NA 18, III: 154–5).

44. Plato, *Nomoi*, 771e1–2a8; Thomas More, *Utopia*, II (*De Coniugiis*).

45. Cf. Machiavelli, *Discorsi*, I. 19.2.

46. The hazardousness of this comparison compelled the narrator to obscure it, which he was able to do successfully, as appears from the fact that the few scholars who actually paid attention to the "Jewish dreams" of Joabin unanimously concluded that the Jew was comparing Eliah to Christ. But after having said that the Bensalemite Jews "give unto our Saviour many high attributes," the narrator actually says that "surely *this man* [Joabin, TvM] would ever acknowledge that Christ was born of a Virgin," before saying that "*he* [Joabin, TvM] would tell how God made *him* [Joabin, TvM] ruler of the Seraphims," and that "*they* [Bensalemite Jews, TvM] call *him*" [Joabin, TvM] the "Eliah of the *Messiah*" (NA 17, III: 151, emphases mine).

47. Bacon directs us to these verses by spelling "Elias" and "Sarepta." For the King James Bible has "Elias" and Sarepta" in its translation of the verses of Luke, whereas it spells "Elijah" and "Zarephath" in its translation of the original story of Elijah's visit to the widow of Zarephath (1 Kings 17).

48. Bensalem's lawlessness is subtly alluded to by the description of the "Feast of the Family" in the sixteenth paragraph, a "most natural, pious and reverend custom," which shows the nation of Bensalem "to be compounded of all goodness." It is granted to "any man that shall live to see thirty persons descended of his body" (cf. 2 Sam. 5: 4, 6: 1). This man, the "Father of the Family," is called "*Tirsan*," spelled backwards "nasrit," which is an allusion to the "Nazarite," who during the time that he 'separates himself unto the Lord' is not allowed to drink wine (Num. 6: 1–21). The son of the Tirsan, by contrast, is called "Son of the Vine." In light of the foregoing we need not wonder that the goodness that is shown in this "pious" and "natural" custom constitutes the seventeenth reference to goodness or the good in *New Atlantis*, that during the preparatory consultations "direction is given touching marriages," and that it is in the sixteenth paragraph that the unborn

"*Son of Bensalem*," that is, the son and therefore the prince of peace, receives the "*blessing of the everlasting Father, the Prince of Peace, and the Holy Dove,*" which implies that there shall be no end of the increase of his government and of peace, on the throne of David (Isa. 9: 5–6; NA 16, III: 147–51).

49. Cf. *supra*, 175 ff.

50. This frontispiece is to be found in the first edition of *Sylva Sylvarum* (London, 1627). For reasons remaining obscure to me, Spedding as well as all other modern editors of *New Atlantis* decided not to reprint the frontispiece, despite the fact that in Bacon's time a frontispiece was "always closely related to the contents of the book for which it was made" (M. Corbett and R. Lightbown, *The Comely Frontispiece: The Emblematic Title-Page in England 1550–1660* (London: Routledge et al., 1979), 35, 45–6).

51. If understood in this narrow sense, White is right in saying that *New Atlantis* is the "most important testament to Bacon's political philosophy" (*Peace among the Willows*, 108).

52. That the Father indeed "carried the true state of Salomon's House" is indicated by the fact that the word "Bensalem" is used for the last time in the seventeenth paragraph, and that biblical quotations and paraphrases occur only in the first seventeen paragraphs.

53. That Bacon does not lay down any rules for the consultations of the thirty-six members of Salomon's House can only be explained by the fact that the thirty-six angels that represent future scientists could still be expected to submit to the verdict of the philosopher who created them, in case they should disagree among themselves. The eccentricity of the wise reminds us of the reasons for the obscurity of the Athenian stranger's account of the Nocturnal Council in Plato's *Nomoi* (*Nomoi*, 951d8–952b6, 962d2–5, 968a7–b2, 968c3–e8; cf. Strauss, *The Argument and the Action of Plato's Laws*, 181, 184–5).

54. Considering the abundance of Cherubim in *New Atlantis*, this would explain Spedding's remark that there is "no single work of" Bacon "which has so much of himself in it" (*Works* III, 122).

55. The increasing attention that is being given to this still relatively unknown work is largely the merit of Laurence Lampert, who published a creditable edition of the dialogue, with helpful notes, and a thoughtful interpretive essay. Francis Bacon, *An Advertisement touching a Holy War*, ed. Laurence Lampert (Prosp. Heights (IL): Waveland Press, 2000).

56. *Works* XIV, 352.

57. Cf. *supra*, 175 ff.

58. That the work dedicated to Bishop Lancelot Andrewes of Winchester does not merely "contemplate the temple" is indicated by the fact that the Latin version of the Epistle Dedicatory adds: "Exceptis paucis alicubi inspersis, quae ad Religionem spectant" (*Works* VII, 15, fn. 3).

59. Virgil, *Aeneid*, IV. 625.

60. That the dialogue took place in this year appears from the fact that Martius mentions three actions that had taken place "within the space of fifty years" (AHW 7, VII: 18–9), of which the earliest was the Battle of Lepanto (1571).

61. Aristotle, *De Caelo*, 268b11–9b17, 270a13–b31.

62. Baldassare Castiglione, *Libro del Cortegiano*, 1.26. That Bacon had Castiglione's courtier in mind when he designed the character of Pollio also appears from the chapters 1.28, 1.34, 2.7, 2.11, 2.22, 2.38, 2.41, and 4.5, among others.

63. Virgil, *Eclogues*, IV. 13–4; Horace, *Carmina*, II. 1.7–8. Cf. Jean-Jacques Rousseau, *Émile*, Livre Quatrième, in *Œuvres Complètes de Jean-Jacques Rousseau*, vol. 4, Bibliothèque de la Pléiade (Paris: Gallimard, 1969), 558.

64. Bacon subtly draws our attention to the common root of the different appearances of zealotry by mentioning Zebedaeus before Gamaliel in the description of the setting of the dialogue, after having mentioned Gamaliel before Zebedaeus in the characterization of the persons.

65. Lucretius, *De Rerum Natura*, I. 29–37.

66. Martius is mentioned twenty-five times in *An Advertisement touching a Holy War*, of which the thirteenth and central time authorizes his response to what Pollio had said in the thirteenth paragraph. That Martius indeed became Pollio's instrument during his central appearance in the dialogue is also indicated by the fact that the soldier went on to emphasize the "humanity and civility" of the "government of the Incas," whereas his former spiritual commander was to argue the next day that "any nation that had only policy and moral virtue" might by the "law of nature" have "subdued" the Incas, "though the propagation of the faith . . . were set by, and not made part of the case." Zebedaeus supported this statement by explicitly referring to Garcilaso de la Vega, whom he perceived to have been read by Martius. But whether or not Martius had actually read the *Comentarios Reales de los Incas* (1609), his description of the Incas was in accordance with that of the man whose commentaries had definitely been read by Zebedaeus, before the zealot decided to abuse them in order to make them fit his own conclusion, a self-justifying but ultimately self-incriminating procedure which he applied to his reading of more than one author, as we shall see. For whereas the "Romish Catholic Zelant" justified his conclusion that the subjection of the Incas by the Roman Catholic Spaniards was lawful by referring to the Incas' "nakedness" and their "eating of men," we find in Vega that the humane and civilized Incas actually detested and prohibited nakedness and the eating of human flesh (AHW 14, 25, VII:21–2, 34). See Garcilaso de la Vega, *Wahrhaftige Kommentare zum Reich der Inka*, transl. W. Planckmeyer (Berlin: Rütten & Loening, 1983), 66–7, 85, 271–2, 289–90, 377, 381, 384.

67. *Works* XIII,158; *Considerations touching a War with Spain* (1624), *Works* XIV, 470, 475–6.

68. This doctrine deserves "some holy war amongst all Christian Princes of either religion . . . for the extirping and razing of" its authors "from the face of the earth, as the common enemies of mankind." *Charge against Owen, Works* XII, 157, 165.

69. *Observations on a Libel* (1592), *Works* VIII, 186. Francesco Guicciardini: *Storia d'Italia*, XII. XIX.

70. Cf. *supra*, 159 ff.

71. Niketas Choniates, *O City of Byzantium: Annals of Niketas Choniates*, transl. H. J. Magoulias (Detroit (MI): Wayne State UP, 1984), 121 ff. (nos. 213–9).

72. Although the most significant part of the preceding discussion had eluded Martius, the simple-minded soldier remained mouldable, as appears from his confession that it was his "opinion that a war upon the Turk is more worthy than upon any *other* gentiles, *infidels, or savages*, . . . though facility and hope of success mought (perhaps) invite some *other* choice" (AHW 17, VII: 23, emphases mine).

73. The fact that it was only "the heat of the day" (AHW 5, VII: 18) at the beginning of the short dialogue underscores the gravity of Pollio's last say at the end of the day.

74. Hippocrates, *17th Letter (to Damagetus)*, ad fin.

75. Spedding points out that the last part of Pollio's speech is not to be found in the original manuscript, which can be explained by the fact that Bacon was not a prophet (*Works* VII, 25, fn. 2).

76. The next morning, Pollio himself made a necessary contribution to making his last intervention seem merely ludicrous. For the author of the dialogue tells us that after having said that "he had dreamt of nothing but Janizaries and Tartars and Sultans all the night long," the courtier delivered "some sporting speeches" on "how the war was already begun" (VII: 25).

77. It is in the twenty-second of the paragraphs describing the "preparations and instruments" of Salomon's House that the instruments of war are discussed (NA 42, III: 163–4). Pollio is mentioned twenty-two times in *An Advertisement touching a Holy War*.

78. Zebedaeus calls Martius' "invective" against the Turks a "true charge." That the zealot's speech indeed contains the true charge is underscored by the fact that the adjective "true" constitutes the center of the seventeen times that the word "truth" and its derivatives occur in *An Advertisement touching a Holy War*. The true object of Zebedaeus' charge is alluded to by the fact that the center of the seventeen references to religion, faith, and piety is occupied by the word "superstition."

79. Aristotle, *Politica*, 1254a23–4.

80. Ibid., 1254b16–21.

81. Francisco de Vitoria, *De Indis*, in *Political Writings*, eds. A. Pagden and J. Lawrence (Cambridge: Cambridge UP, 1991), esp. 239–46, 250–1, 253, 263–77.

82. The God that is equated with natural reason constitutes the ninth and central reference to God or gods in *An Advertisement touching a Holy War*.

83. The closest Zebedaeus came to distinguishing the "law divine" from the "laws of nature and nations" of which it was the "perfection" was by saying that there are "nations that are outlawed and proscribed by the law of nature and nations, *or* by the immediate commandment of God" (AHW 25, VII: 31, emphasis mine).

84. Aratus, *Phaenomena*, 5.

85. Terence, *Heauton Timorumenos*, 77.

86. Eusebius, *De Vita Constantini*, I. XXVIII.

87. The fact that this anti-ecclesiastical history remained as yet to be written provides us with an explanation for the fact that we find the words "The rest was not perfected" at the end of the dialogue; words which for the rest originate from Rawley, as the Doctor of Divinity himself in a way confirms by telling us that during

his lifetime Bacon worked on a translation of the *"Dialogue of the Holy War"* into Latin (*Life of the Honourable Author*, *Works* I, 10).

Epilogue

1. Cf. *supra*, 89.
2. Cf. *supra*, 35 f. (256 fn. 9), 86 ff., and 199.
3. Cf. Thomas Hobbes, *Leviathan*, ch. XLVI. 14 with XLVI. 1, V. 17, and IX. 1.
4. Cf. Plato, *Theaetetus*, 172c1–6a2.
5. Cf. Aristotle, *Ethica Nicomachea*, 1123a34–1124a20.
6. After the Vicar had delivered his teaching on natural religion, the proselyte remarked that he believed he had heard "the divine Orpheus singing the first hymns and teaching men the cult of the gods" (*La Profession de foi du Vicaire Savoyard*, par. 106 (Rousseau, *Émile*, 606)). Rousseau himself draws our attention to the importance of the proselyte's remark by making Orpheus *"enseignant aux hommes le culte des Dieux"* the subject of the frontispiece belonging to the third volume of the first edition of *Émile*, and by adding an explicit reference to the page on which the proselyte's remark was to be found.
7. Cf. *supra*, 266–7 fn. 81.
8. Cf. Friedrich Nietzsche, *Morgenröthe*, Erstes Buch, Aph. 95.
9. Friedrich Nietzsche, *Götzen-Dämmerung oder Wie man mit dem Hammer philosophirt*; *Das Problem des Sokrates*, Aph. 10.
10. Friedrich Nietzsche, *Ecce homo. Wie man wird, was man ist*; *Warum ich so klug bin*, Aph. 4 (my translation).
11. Cf. *supra*, 146 ff., 180, 189 ff.
12. Cf. Rousseau, *Les rêveries du Promeneur Solitaire*, Cinquième Promenade, par. 13–6.
13. Aubrey, *Francis Bacon*, 16. Cf. Letter (1626) to the Earl of Sarundel and Surry, *Works* XIV, 550.
14. Jardine and Stewart, *Hostage to Fortune*, 502–11.
15. Cf. Nietzsche, *Die fröhliche Wissenschaft*, Viertes Buch, Aph. 324.
16. Pliny the Younger, *Epistularum Libri Decem*, VI. 16.
17. Plato, *Phaedo*, 118a7–8; cf. *Euthyphro*, 5c9–d5.

Bibliography

Primary Sources

Acosta, José de. *Historia Natural y Moral de las Indias* (1590). Translated by E. Grimston. London, 1880.

Aesop. *The Complete Fables*. Translated by Olivia and Robert Temple. London: Penguin, 1998.

Alembert, Jean le Rond d'. "Discours Préliminaire." In *Encyclopédie ou Dictionnaire Raisonnée des Sciences, des Arts et des Métiers*, vol. 1. Paris, 1751.

Alfarabi. *The Philosophy of Plato*. In Alfarabi. *Philosophy of Plato and Aristotle*. Translated by Mushin Mahdi. Revised edition. Ithaca: Cornell UP, 2001.

Aubrey, John. *Francis Bacon*. In *Aubrey's Brief Lives*, edited by Oliver Lawson Dick. Boston: David R. Godine, 1999.

Bacon, Francis: *A Conference of Pleasure*, edited by James Spedding. London: Longmans et al., 1870.

———. *An Advertisement touching a Holy War*, edited by Laurence Lampert. Prosp. Heights (IL): Waveland Pr., 2000.

———. *De Sapientia Veterum Liber*. London, 1609 and 1617.

———. *Philosophical Studies. The Oxford Francis Bacon VI*, edited by Graham Rees. Oxford: Clarendon Press, 1996.

———. *The Advancement of Learning*, edited by William Aldis Wright. Oxford: Clarendon Press, 1869.

———. *The Advancement of Learning. The Oxford Francis Bacon IV*, edited by Michael Kiernan. Oxford: Clarendon Press, 2000.

———. *The Essayes or Counsels, Civill and Morall. The Oxford Francis Bacon XV*, edited by Michael Kiernan. Oxford: Clarendon Press, 2006.

———. *The Works of Francis Bacon*. 14 vols., edited by James Spedding, Robert Leslie Ellis, and Douglas Denon Heath. London: Longmans & Co, 1870.

Bancroft, Richard. *A Sermon preached at Paules Crosse, 9 February 1589*. London, 1589.

Barth, Karl. *Der Römerbrief*. Zürich: TVZ, 2005.

Bayle, Pierre. *Réponse aux Questions d'un Provincial*, vol. 4. Rotterdam, 1707.

Blake, William. *Annotations to Bacon.* In Blake. *Complete Writings,* edited by Geoffrey Keynes. London: Oxford UP, 1969.

Bultmann, Rudolf. *Die Frage nach der natürlichen Offenbarung.* In Bultmann. *Glauben und Verstehen,* vol. 2. 6th edition. Tübingen: Mohr (Siebeck), 1993.

———. *Theologische Enzyklopädie.* Tübingen: Mohr (Siebeck), 1984.

———. *Zur Frage des Wunders.* In Bultmann. *Glauben und Verstehen,* vol. 1. 9th edition. Tübingen: Mohr (Siebeck), 1993.

Calvin, Johannes. *Commentaire de M. Iehan Calvin sur l'Epistre aux Romains.* Geneva, 1561.

———. *Commentariorum in Quinque Libros Mosis.* In Calvin. *Opera quae Supersunt Omnia,* edited by G. Baum et al., vol. 23. Brunswick, 1882.

———. *Harmonia ex Evangelistis Tribus.* Geneva, 1582.

Choniates, Niketas. *O City of Byzantium: Annals of Niketas Choniates.* Translated by H. J. Magoulias. Detroit (MI): Wayne State UP, 1984.

Comes, Natalis. *Mythologiae, sive Explicationis Fabularum, Libri decem.* Frankfurt, 1584.

Cowley, Abraham. "To the Royal Society." In Sprat, Thomas. *The History of the Royal Society of London. For the Improving of Natural Knowledge.* London, 1667.

Descartes, René. *Cogitationes Privatae* (1619). In *Œuvres de Descartes,* vol. 10, edited by C. Adam and P. Tannery. Paris, 1908.

Diderot, Denis. *Pensées Philosophiques.* In *Œuvres Complètes de Diderot,* vol. 1. Paris: Garnier Frères, 1875.

———. *Pensées sur l'Interpretation de la Nature.* In *Œuvres Complètes de Diderot,* vol. 2. Paris: Garnier Frères, 1875.

———. *Prospectus.* In *Œuvres Complètes de Diderot,* vol. 13. Paris: Garnier Frères, 1876.

Hume, David. *Of Miracles,* in *An Enquiry concerning Human Understanding,* edited E. Steinberg. Indianapolis (IN): Hackett, 1993.

———. *The History of England,* vol. 4. Boston: Little, Brown, and Company, 1854.

Jewel, John. *An Apologie or Answere in Defence of the Churche of Englande.* London, 1564.

Leibniz, Gottfried Wilhelm. *Nouveaux Essais sur l'Entendement Humain.* In *Œuvres Philosophiques de Leibniz,* vol. 1. 2nd edition. Paris, 1900.

———. "Epistolae mutuae G. G. Leibnitii et. Fr. Guil. Bierlingii." In Leibniz. *Opera Philologica.* Vol. 5 of *Opera Omnia,* edited by Ludovicus Dutens. Geneva, 1768.

Lessing, Gotthold Ephraim. *Leibniz von den ewigen Strafen.* In Lessing. *Werke 1770–1773.* Vol. 7 of *Werke und Briefe in zwölf Bänden,* edited by W. Barner et al. Frankfurt am Main: Deutscher Klassiker Verlag, 2000.

Lipsius, Justus. *In Cornelii Tacit. Hist. Lib. Notae.* In Cornelius Tacitus. *Annalium et Historiarum Libri qui extant,* edited by Justus Lipsius. Lyon, 1576.

Luther, Martin. *Das Magnificat verdeutschet und ausgelegt.* In *Schriften von 1520–1524,* edited by Otto Clemen. Vol. 2 of *Luthers Werke in Auswahl,* edited by Otto Clemen. 5th edition. Berlin: de Gruyter, 1959.

———. *Vorlesung über den Römerbrief.* In *Der junge Luther*, edited by Erich Vogelsang. Vol. 5 of *Luthers Werke in Auswahl*, edited by Otto Clemen. 3rd edition. Berlin: de Gruyter, 1963.

———. *Vorlesungen über 1. Mose von 1535–45.* In *Martin Luthers Werke. Kritische Gesamtausgabe*, vol. 42. Weimar: H. Böhlaus Nachfolger, 1911.

———. *Predigt über Johannes 17: 1–3 (15 August 1528).* In *Predigten*, edited by Emanuel Hirsch. Vol. 7 of *Luthers Werke in Auswahl*, edited by Otto Clemen. 3rd edition. Berlin: de Gruyter, 1962.

Machiavelli, Niccolò. *Discorsi Sopra la Prima Deca di Tito Livio*, vol. 1. Rome: Salerno Ed., 2001.

———. *Discourses on Livy*, edited by Nathan Tarcov and Harvey C. Mansfield. Chicago: University of Chicago Press, 1996.

Maimonides, Moses: *The Guide of the Perplexed.* Translated by Shlomo Pines. Chicago: University of Chicago Press, 1963.

Maistre, Joseph de. *Examen de la Philosophie de Bacon; où l'on traite différentes questions de Philosophie Rationelle*, vol. 2. Paris/Lyon: Poussielgue-Rusand, 1836.

Montaigne, Michel de. *Essais*, edited by Pierre Villey, vol. 2. New edition. Paris: PUF, 1992.

Plato. *Platonis Avgvstiss Philosophi, omnium quae exstant operum*, edited by Joannes Serranus [Jean de Serres] and Henricus Stephanus [Henri Estienne], vol. 3. Geneva, 1578.

Plutarch. *Lysander.* In Plutarch. *Vitae Parallellae seu Comparatae*, vol. 2. Translated by W. Xylander. Basle, 1579.

Rawley, William. *Resuscitatio.* London, 1657.

Rousseau, Jean-Jacques: *Discours sur les sciences et les arts.* In *Œuvres Complètes de Jean-Jacques Rousseau*, vol. 3, Bibliothèque de la Pléiade. Paris: Gallimard, 1964.

———. *Discours sur l'inégalité*, edited by Heinrich Meier. 5th edition. Paderborn: Ferdinand Schöningh, 2001.

———. *Émile.* In *Œuvres Complètes de Jean-Jacques Rousseau*, vol. 4, Bibliothèque de la Pléiade. Paris: Gallimard, 1969.

———. *Les rêveries du Promeneur Solitaire.* In *Œuvres Complètes de Jean-Jacques Rousseau*, vol. 1, Bibliothèque de la Pléiade. Paris: Gallimard, 1959.

———. *Lettre à Lecat.* In *Œuvres Complètes de Jean-Jacques Rousseau*, vol. 3, Bibliothèque de la Pléiade. Paris: Gallimard, 1964.

———. *Lettre à Monseigneur de Beaumont.* In *Œuvres Complètes de Jean-Jacques Rousseau*, vol. 4, Bibliothèque de la Pléiade. Paris: Gallimard, 1969.

———. *Lettres écrites de la Montagne.* In *Œuvres Complètes de Jean-Jacques Rousseau*, vol. 3, Bibliothèque de la Pléiade. Paris: Gallimard, 1964.

Shaftesbury, Anthony Third Earl of. *Characteristicks of Men, Manners, Opinions, Times*, vol. 3. Indianapolis: Liberty Fund, 2001.

Spinoza, Baruch de. *Tractatus Theologico-Politicus*, vol. 6. Hamburg, 1670.

Swift, Jonathan. *A Tritical Essay upon the Faculties of the Mind.* In *The Writings of Jonathan Swift*, edited by R. A. Greenberg and W. B. Piper. New York: Norton, 1973.

———. *The Battel of the Books*. In Swift. *A Tale of a Tub and Other Works*, edited by Marcus Walsh. Cambridge: Cambridge UP, 2010.

Tacitus, C. Cornelius. *Opera quae Exstant*, edited by Justus Lipsius. Antwerp, 1585.

Toland, John. *Clidophorus: or of the Exoteric and Esoteric Philosophy*. In Toland. *Tetradymus*. London, 1720.

———. *Pantheisticon*. London, 1751.

Vega, Garcilaso de la. *Wahrhaftige Kommentare zum Reich der Inka*. Translated by W. Planckmeyer. Berlin: Rütten & Loening, 1983.

Vitoria, Francisco de: *De Indis*. In Vitoria. *Political Writings*, edited by A. Pagden and J. Lawrence. Cambridge: Cambridge UP, 1991.

Secondary Sources

Anderson, Fulton H. *The Philosophy of Francis Bacon*. Chicago: University of Chicago Press, 1948.

Beal, Peter. *Index of Literary Manuscripts. 1450–1625*, vol. 1. London: Mansell, 1980.

Benardete, Seth. *Plato's Laws: The Discovery of Being*. Chicago: University of Chicago Press, 2000.

———. *Socrates' Second Sailing*. Chicago: University of Chicago Press, 1989.

Blumenberg, Hans. *Arbeit am Mythos*. 6th edition. Frankfurt am Main: Suhrkamp, 2006.

———. *Die Legitimität der Neuzeit*. Frankfurt am Main: Suhrkamp, 1996.

Bolotin, David. *An Approach to Aristotle's Physics: With Particular Attention to the Role of His Manner of Writing*. Albany (NY): SUNY Press, 1998.

Briggs, John Channing. "Bacon's Science and Religion." In *The Cambridge Companion to Bacon*, edited by Markku Peltonen. Cambridge: Cambridge UP, 1996.

Burger, Ronna. *Aristotle's Dialogue with Socrates: On the Nicomachean Ethics*. Chicago: University of Chicago Press, 2008.

Burkert, Walter. *Griechische Religion der archaischen und klassischen Epoche*. Stuttgart: Verlag W. Kohlhammer, 1977.

Corbett, Margery, and R. W. Lightbown. *The Comely Frontispiece: The Emblematic Title-Page in England 1550–1660*. London: Routledge, 1979.

Craig, Tobin L. "On the Significance of the Literary Character of Francis Bacon's New Atlantis for an Understanding of His Political Thought." *The Review of Politics* 72, no. 2 (2010): 213–39.

Deleyre, Alexandre. *Analyse de la Philosophie du Chancelier François Bacon*, vol. 2. Amsterdam: Artskée & Merkus, 1755.

Farrington, Benjamin. *The Philosophy of Francis Bacon: An Essay on its Development from 1603 to 1609 with New Translations of Fundamental Texts*. Chicago: University of Chicago Press, 1966.

Fattori, Marta. "Sir Francis Bacon and the Holy Office." *British Journal for the History of Philosophy* 13, no. 1 (2005): 21–49.

Faulkner, Robert K. *Francis Bacon and the Project of Progress.* Lanham (MD): Rowman & Littlefield, 1992.

Fish, Stanley E. "Georgics of the Mind: The Experience of Bacon's Essays." In Fish. *Self-consuming Artifacts: the Experience of Seventeenth-Century Literature.* Berkeley: University of California Press, 1972.

Forster, C. Thornton, and F. H. Blackburne Daniell, eds. *The Life and Letters of Ogier Ghiselin de Busbecq*, vol. 1. London: C. Kegan Paul & Co, 1881.

Gadamer, Hans-Georg. *Wahrheit und Methode: Grundzüge einer philosophischen Hermeneutik.* In *Gesammelte Werke*, vol. 1. Tübingen: J. C. B. Mohr (Paul Siebeck), 1990.

Gaukroger, Stephen. *Francis Bacon and the Transformation of Early-Modern Philosophy.* Cambridge/New York: Cambridge UP, 2001.

———. *The Emergence of a Scientific Culture: Science and the Shaping of Modernity 1210–1685.* Oxford: Clarendon Press, 2006.

Gillespie, Michael Allen. *The Theological Origins of Modernity.* Chicago: University of Chicago Press, 2008.

Hard, Robin. *The Routledge Handbook of Greek Mythology.* London: Routledge, 2008.

Jardine, Lisa. *Francis Bacon: Discovery and the Art of Discourse.* London/New York: Cambridge UP, 1974.

Jardine, Lisa, and Alan Stewart. *Hostage to Fortune: The Troubled Life of Francis Bacon 1561–1626.* London: Phoenix Giant, 1999.

Kennington, Richard. "Blumenberg and the Legitimacy of the Modern Age." In *The Ambiguous Legacy of the Enlightenment*, edited by W. A. Rusher. Lanham (MD): UP of America, 1995.

Lampert, Laurence. *Nietzsche and Modern Times: A Study of Bacon, Descartes, and Nietzsche.* New Haven: Yale UP, 1993.

Lemmi, Charles W. *The Classic Deities in Bacon: A study in Mythological Symbolism.* New York: Octagon Books, 1971.

Löwith, Karl. *Meaning in History.* Chicago: University of Chicago Press, 1949.

Macaulay, Thomas Babington: "Lord Bacon." In *Critical and Historical Essays, contributed to the Edinburgh Review*, vol. 2. 5th edition. London: Longman et al., 1848.

Mansfield, Harvey C. *Machiavelli's New Modes and Orders: A Study of the Discourses on Livy.* Chicago: University of Chicago Press, 2001.

Meier, Heinrich. *Das theologisch-politische Problem: Zum Thema von Leo Strauss.* Stuttgart: Metzler, 2003.

———. *Die Lehre Carl Schmitts: Vier Kapitel zur Unterscheidung Politischer Theologie und Politischer Philosophie.* 3rd edition. Stuttgart: Metzler, 2009.

———. *Les rêveries du Promeneur Solitaire: Rousseau über das philosophische Leben.* 2nd edition. München: Carl Friedrich von Siemens Stiftung, 2010.

———. *Über das Glück des philosophischen Lebens: Reflexionen zu Rousseaus Rêveries in zwei Büchern.* München: C. H. Beck, 2011.

———. *Warum Politische Philosophie?* Stuttgart: Metzler, 2000.

Paterson, Timothy H. "On the Role of Christianity in the Political Philosophy of Francis Bacon." *Polity* 19, no. 3 (1987): 419–42.

———. *The Politics of Baconian Science: An Analysis of Bacon's "New Atlantis."* Unpublished dissertation. Yale University, 1982.
Peltonen, Markku. *Classical Humanism and Republicanism in English Political Thought 1570–1640.* Cambridge/New York: Cambridge UP, 1995.
———. "Introduction." In *The Cambridge Companion to Bacon*, edited by Markku Peltonen. Cambridge: Cambridge UP, 1996.
Pérez-Ramos, Antonio. *Francis Bacon's Idea of Science and the Maker's Knowledge Tradition.* Oxford: Clarendon Press, 1988.
Pesic, Peter. *Labyrinth: A Search for the Hidden Meaning of Science.* Cambridge: MIT Press, 2001.
Pippin, Robert. "Blumenberg and the Modernity Problem." *Review of Metaphysics* 40, no. 3 (1987): 535–57.
Pott, Henry. *The Promus of Formularies and Elegancies by Francis Bacon.* London: Longmans, 1883.
Price, Bronwen. *Francis Bacon's New Atlantis: New Interdisciplinary Essays.* Manchester: Manchester UP, 2002.
Rees, Graham. "Bacon's Speculative Philosophy." In *The Cambridge Companion to Bacon*, edited by Markku Peltonen. Cambridge: Cambridge UP, 1996.
Rossi, Paolo. "Baconianism." In *The Dictionary of the History of Ideas: Studies of Selected Pivotal Ideas*, edited by Philip P. Wiener, vol. 1. New York: Charles Scribner's Sons, 1973–74.
———. "Bacon's Idea of Science." In *The Cambridge Companion to Bacon*, edited by Markku Peltonen. Cambridge: Cambridge UP, 1996.
———. *From Magic to Science.* Translated by S. Rabinovitch. Chicago: University of Chicago Press, 1968.
———. *The Birth of Modern Science.* Translated by C. de Nardi Ipsen. Oxford: Blackwell Publishers, 2000.
Sessions, William A. "Francis Bacon and the Classics: the Discovery of Discovery." In *Francis Bacon's Legacy of Texts: "The Art of Discovery grows with Discovery,"* edited by William A. Sessions. New York: AMS Press, 1990.
Strauss, Leo. "An Untitled Lecture on Plato's Euthyphron." *Interpretation* 24, no. 1 (1996): 3–24.
———. *Die geistige Lage der Gegenwart.* In Strauss. *Philosophie und Gesetz – Frühe Schriften.* Vol. 2 of *Gesammelte Schriften*, edited by Heinrich Meier. Stuttgart: Metzler, 1997.
———. *Die Religions-Kritik Spinozas.* In Strauss. *Die Religionskritik Spinozas und zugehörige Schriften.* Vol. 1 of *Gesammelte Schriften*, edited by Heinrich Meier. 3rd edition. Stuttgart: Metzler, 2008.
———. *Natural Right and History.* Chicago: University of Chicago Press, 1971.
———. *Notes on Lucretius.* In Strauss. *Liberalism Ancient and Modern.* New York: Basic Books, 1968.
———. "On the Interpretation of Genesis." *L'Homme. Revue française d'anthropologie* 21, no. 1 (1981): 5–20.

———. *On Tyranny: Revised and Expanded Edition*, edited by Victor Gourevitch and Michael S. Roth. New York: Free Press, 1991.

———. *Persecution and the Art of Writing*. Glencoe (IL): Free Press, 1952.

———. *Philosophie und Gesetz: Beiträge zum Verständnis Maimunis und seiner Vorläufer*. In Strauss. *Philosophie und Gesetz – Frühe Schriften*, vol. 2 of *Gesammelte Schriften*, edited by Heinrich Meier. Stuttgart: Metzler, 1997.

———. *Progress or Return?* In *The Rebirth of Classical Political Rationalism: An Introduction to the Thought of Leo Strauss*, edited by Thomas L. Pangle. Chicago: University of Chicago Press, 1989.

———. *Socrates and Aristophanes*. Chicago: University of Chicago Press, 1966.

———. *The Argument and the Action of Plato's Laws*. Chicago: University of Chicago Press, 1975.

———. *The City and Man*. Chicago: Rand McNally, 1964.

———. *Thoughts on Machiavelli*. Chicago: University of Chicago Press, 1958.

Strauss, Leo, and Hans-Georg Gadamer. "Correspondence concerning Wahrheit und Methode." *Independent Journal of Philosophy* 2 (1978): 5–12.

Studer, Heidi. *Grapes Ill-Trodden: Francis Bacon and the Wisdom of the Ancients*. Unpublished dissertation. Canada, University of Toronto, 1992.

Vickers, Brian. *Francis Bacon*. Harlow: Longman Gr. LTD, 1978.

———. *Francis Bacon and Renaissance Prose*. Cambridge: Cambridge UP, 1968.

———, ed. *Francis Bacon: A Critical Edition of the Major Works*. Oxford: Oxford UP, 1996.

Webster, Charles. *The Great Instauration: Science, Medicine and Reform 1626–1660*. London: Duckworth, 1975.

Weinberger, Jerry. "Science and Rule in Bacon's Utopia: An Introduction to the Reading of the New Atlantis." *American Political Science Review* 70 (1976): 865–85.

———. *Science, Faith, and Politics: Francis Bacon and the Utopian Roots of the Modern Age: A Commentary on Francis Bacon's Advancement of Learning*. Ithaca: Cornell UP, 1985.

———, ed. *New Atlantis and the Great Instauration*. Wheeling (IL): Harlan Davidson, 1989.

———, ed. *The History of the Reign of King Henry the Seventh*. Ithaca: Cornell UP, 1996.

White, Howard B. *Peace among the Willows: The Political Philosophy of Francis Bacon*. The Hague: Martinus Nijhoff, 1968.

Whitehead, Alfred North: *Science and the Modern World*. New York: Free Press, 1967.

Whitney, Charles. "Cupid hatched by Night: The 'Mysteries of Faith,' and Bacon's Art of Discovery." In *Ineffability. Naming the unnamable: From Dante to Beckett*, edited by Peter S. Hawkins and Anne Howland Schotter. New York: AMS Press, 1984.

———. *Francis Bacon and Modernity*. New Haven: Yale UP, 1986.

Wolff, Emil. *Francis Bacon und seine Quellen*, vol. 1. Berlin: Emil Felber, 1910.
Zagorin, Perez. *Francis Bacon*. Princeton: Princeton UP, 1998.
———. *Ways of Lying: Dissimulation, Persecution, and Conformity in Early Modern Europe*. Cambridge: Harvard UP, 1990.

Index

Achelous, 86, 269n100
Acosta, 134, 211–2
Actaeon, 63–5, 69
adversity, 2, 138, 174–7, 237, 240, 262n25
Aesculapius, 146–8, 241–2, 296n88
Antichrist, 22, 222, 240, 272n16
antichristianism, 105, 107, 182
Apollo, 52–4, 56–7, 60–1, 68, 134, 146–8
Aristophanes, 73, 93, 95, 129
Aristotelian (political) philosophy, 132, 168–9
Aristotelianism, 239, 290n8
Aristotle, 35, 155, 168–9, 224, 233, 290n8, 290n9, 290n11, 290n12, 295n74
atheism, 105–6, 110–2, 114, 132–4, 135, 139, 162, 175, 199, 240, 281n95, 282n99, 282n100, 282n101, 283n105, 283n106, 283n107, 283n108
St. Augustine, 31–2, 195–6, 201, 268n84, 269–70n102, 282n97, 296n82, 299n111
Bacchus, 63, 66, 76, 78, 79–80, 83, 268n85
St. Bartholomew, 14, 162, 209–10

Bayle, Pierre, 283n105

Caesar, Augustus, 30, 38, 196, 257n14

Caesar, Julius, 30, 183, 196–7, 259n15, 260n19
Cassandra, 52–4, 56–7, 69
Cato, 39, 53, 264n50
charity, 1, 9, 158–61, 170–1, 186, 191, 193–4, 218, 235, 288n166, 291n15, 298n100, 304n37
Cheiron, 75, 147, 267n83, 292n35
Cicero, Marcus Tullius, 30, 53, 108, 194, 223, 246n47, 267n81
Coelum, 120–2, 124, 127, 146
compassion, 147–8, 159–160, 285n139, 288n165, 288n169
Cupid, 125–9, 150–1
Cyclopes, 146–8

d'Alembert, Jean le Rond, 4–5, 163, 165, 222, 240
Daedalus, 1
David, 31, 207, 216–8, 269n102, 306n48
Democritus, 117, 120, 122, 132, 230, 255n3
Deucalion, 167, 176
Diderot, Denis, 4, 131, 222, 278–9n69, 283n107, 300n131
Diomedes, 150–2
Dionysus, 64, 76–80, 82, 87–9, 268n85
dissimulation, 29–30, 38, 45, 47, 62–3, 75, 254–5n54

echo, 58, 61, 69

Encyclopédie/Encyclopédistes, 4–5, 163, 222
Endymion, 61–3, 64, 69, 74
envy, 30–1, 148, 164, 197–8, 255n56, 262–3n33
Epicurus, 131–3, 162
Epimetheus, 185–6
Eros, 7, 86, 182
Eusebius of Caesaria, 236
exoteric-esoteric writing, 18–23, 26, 30, 37, 251n30, 251n31, 251–2n33, 271n9

Fall, 40–1, 43, 119, 120, 130, 135, 140, 142, 172, 178
foresight, 3, 4, 53, 138, 185, 196–7, 209, 214–5, 220, 273n34, 279n71, 299n108, 299n116
fortitude, 2, 174–6, 237, 292n24

good, communicative, 36, 158, 181–2, 223, 237, 256n9; Private, 182, 237, 256n9
grace, 31–2, 68, 74–5, 98, 142, 144, 146, 148, 153–4, 156, 158–9

Hercules, 173–7, 219, 269n100, 292n30
Hermes, 119–20
Hobbes, Thomas, 219, 291n13, 303n22
humanity, 161, 214–5, 241, 304n37

idols, 20, 186–88, 239, 245–6n22, 263n40, 282n97, 283n104, 295n77
immortality, 12, 67, 70, 72, 186, 189, 190, 270n104, 297n98
Instauration, 66–7, 167, 187–9

jealousy, 30, 54, 71, 84, 110, 139–41, 147, 152, 156, 198, 286n149
Juno, 25, 55, 57, 73–5, 76, 78–9, 82, 83–5, 88–90, 240
Jupiter, 25–6, 34, 55–8, 68, 73–4, 76, 78, 80, 82, 83–5, 114, 118, 120–4, 129, 134, 146–8, 173–4, 178–9, 184–5, 190, 197
justice, 9, 43, 53–4, 120, 142–4, 146, 148–9, 152–6, 158–9, 171, 178–9, 183–4, 184–5, 223, 254n50, 258n15, 294n61

Kingdom of Man, 9, 187, 194–5, 199, 237, 239, 293n40, 295n76

law, 31–2, 37, 40, 42, 44–6, 53–4, 64, 68, 72, 116, 128–9, 150–2, 158, 163–5, 179, 180, 184, 205, 213–5, 216–8, 228–9, 232–6, 289n175, 289n176, 308n83
Leibniz, Gottfried Wilhelm, 252n33
Lessing, Gotthold Ephraim, 252n33
love, 52, 58–9, 61–3, 64–5, 74–5, 128–9, 140–1, 152–3, 158, 263n40, 264n52, 269fn100, 297–8n99

Machiavelli, Niccolò, 45, 161–2, 171, 180, 181, 183, 186, 198–9, 213, 223, 236, 252n37, 257n15, 259n15, 267–8n83, 293n48, 296n83, 300n124
Maimonides, Moses, 23
Memnon, 147–8
Mercury, 55–6, 59–61, 68, 118, 177
mercy, 70, 104, 152–3, 158
Metis, 57, 84
Minerva, 176–7, 219
miracles, 109, 111–2, 130, 132, 177, 208–9, 275n42, 275n44, 275n45
moderation, 14, 34, 37, 98, 199, 236
Montaigne, Michel de, 28, 254n51, 260n18
morality, 26, 41–3, 64–5, 76–80, 97, 128–9, 131, 163–4, 185, 189–90, 191, 193, 195, 222–3, 244n14, 250n25
Moses, 91, 180–1, 206, 293n47
Muses, 66, 78, 80–2, 85–9

Narcissus, 58, 61, 68

INDEX

natural theology, 96, 108, 110–1, 113–4, 151, 155, 239, 274n38, 274n39, 275n47, 276n48, 283n106, 289n172
nemesis, 149–50, 266n75, 286n144
Nietzsche, Friedrich, 240–1, 252n33

Oedipus, 80–1, 83
omnipotence, 5, 97–8, 102–9, 111–3, 122, 124, 127, 129–30, 139, 272n23, 272n25, 279n74
Orpheus, 65–73, 76, 78, 82, 86–9, 91, 177, 186, 240

Pallas (Athena), 25, 55, 57, 84–5, 173–4, 176
Pan, 59–61, 68, 118–20, 122, 130–1
Pandora, 85, 90, 184–5
St. Paul, 27, 154, 157–9, 162, 176, 188, 192, 209, 226, 235, 259n15, 276n48, 287n154
Pentheus, 63–5, 69, 76, 78, 82
philanthropy, 1, 7–9, 34, 94, 182, 191, 192, 223, 236, 237, 239, 241, 247n30
Philosophia Prima, 9, 108–9, 115, 171, 238, 273n34, 273n35
Philosophia Secunda, 220, 238
Plato, 53, 184, 211, 213, 215, 217
Platonic political philosophy, 73
pride, 41, 59, 74, 75, 77, 135
Prometheus, 85, 171–8, 184–6, 200, 239, 241
prophecy, 9, 52–4, 90, 101, 107, 112, 123–5, 135, 137–9, 149, 188, 205, 217–8, 231, 258–9n15, 264n52, 308n75
Proserpina, 265–6n73
prosperity, 31, 36, 199, 202, 220, 266n75
Proteus, 123–5
providence, 5–6, 60, 117–8, 128–9, 149–50, 171, 173, 197, 239, 241, 260n15, 265n59, 277n50, 277n57, 281n88, 300n118
prudence, 18, 20–1, 29, 39, 47, 48, 56, 84, 98, 117, 176, 250n27, 267–8n83, 271n9, 290n10, 299n116

Pyrrha, 167, 185

Rousseau, Jean-Jacques, 5, 240, 253–4n50, 261n20, 261n21, 275n45, 278n65, 300–1n131

Saturn, 34–5, 74–5, 84, 120–2, 124, 127, 128, 147, 267n82, 279n71
serpent, 28, 55–6, 91, 184–5, 193
serpentine prudence, 28, 46, 56
Shaftesbury, Anthony Third Earl of, 283n108
sin, 26, 40–2, 104, 106, 119–20, 140–1, 143, 147, 148, 152–3, 158–9, 183, 220, 270n102, 289n175
Sirens, 86–90
Socrates, 39, 53, 73, 184, 240–1, 242, 266–7n81, 304n31
Socratic turn, 73, 267n81
Solomon, 88, 116, 136, 188, 207, 208, 215, 217, 221, 239, 256n13, 268n90, 287–8n165, 304n38
Sphinx, 80–3, 88
Styx, 69–70, 85
superstition, 32, 69, 111, 112, 134–5, 145, 155–6, 161, 205–6, 269n102, 283n110, 300n127, 308n78
Swift, Jonathan, 8, 27

Thracian women, 66, 71
Tithonus, 190
Toland, John, 21, 251n33
truth, 6, 8, 9, 21, 27–9, 30, 36, 45, 52, 54, 68, 99–100, 138, 145, 155, 169, 182, 189, 194, 198–9, 220, 222, 236, 238–9, 251n33, 254n53, 268n83, 277n57, 282n98, 296n85, 298n100, 308n78
Typhon, 55–8, 88
tyranny, 30, 40, 55–6, 74–5, 76–7, 85, 122, 156, 162, 171, 183, 227, 234, 240–1, 262n26, 279n71, 292n31

Ulysses, 86–8, 118, 120

Venus, 120–4, 128–9, 150–1